AF699364

Applications of Hyperstructure Theory

Advances in Mathematics

VOLUME 5

Series Editor:

J. Szep, *Budapest University of Economics, Hungary*

Advisory Board:

S-N. Chow, *Georgia Institute of Technology, U.S.A.*

G. Erjaee, *Shiraz University, Iran*

W. Fouche, *University of South Africa, South Africa*

P. Grillet, *Tulane University, U.S.A.*

H.J. Hoehnke, *Germany*

F. Szidarovszky, *University of Arizona, U.S.A.*

P.G. Trotter, *University of Tasmania, Australia*

P. Zecca, *Università di Firenze, Italy*

The titles published in this series are listed at the end of this volume.

Applications of Hyperstructure Theory

by

PIERGIULIO CORSINI

Department of Mathematics and Computer Sciences, University of Udine, Udine, Italy

and

VIOLETA LEOREANU

Faculty of Mathematics, "Al.I Cuza" University, Iaşi, Romania

KLUWER ACADEMIC PUBLISHERS

DORDRECHT / BOSTON / LONDON

A C.I.P. Catalogue record for this book is available from the Library of Congress.

ISBN 1-4020-1222-5

Published by Kluwer Academic Publishers,
P.O. Box 17, 3300 AA Dordrecht, The Netherlands.

Sold and distributed in North, Central and South America
by Kluwer Academic Publishers,
101 Philip Drive, Norwell, MA 02061, U.S.A.

In all other countries, sold and distributed
by Kluwer Academic Publishers,
P.O. Box 322, 3300 AH Dordrecht, The Netherlands.

Printed on acid-free paper

All Rights Reserved
© 2003 Kluwer Academic Publishers
No part of this work may be reproduced, stored in a retrieval system, or transmitted
in any form or by any means, electronic, mechanical, photocopying, microfilming, recording
or otherwise, without written permission from the Publisher, with the exception
of any material supplied specifically for the purpose of being entered
and executed on a computer system, for exclusive use by the purchaser of the work.

Printed in the Netherlands.

Contents

Introduction

Some mathematical disciplines can be presented and developed in the context of other disciplines, for instance Boolean algebras, that Stone has converted in a branch of ring theory, projective geometries, characterized by Birkhoff as lattices of a special type, projective, descriptive and spherical geometries, represented by Prenowitz, as multigroups, linear geometries and convex sets presented by Jantosciak and Prenowitz as join spaces. As Prenowitz and Jantosciak did for geometries, in this book we present and study several mathematical disciplines that use the Hyperstructure Theory.

Since the beginning, the Hyperstructure Theory and particularly the Hypergroup Theory, had applications to several domains. Marty, who introduced hypergroups in 1934, applied them to groups, algebraic functions and rational fractions. New applications to groups were also found among others by Eaton, Ore, Krasner, Utumi, Drbohlav, Harrison, Roth, Mockor, Sureau and Haddad. Connections with other subjects of classical pure Mathematics have been determined and studied:

- *Fields* by Krasner, Stratigopoulos and Massouros Ch.
- *Lattices* by Mittas, Comer, Konstantinidou, Serafimidis, Leoreanu and Călugăreanu
- *Rings* by Nakano, Kemprasit, Yuwaree
- *Quasigroups* and *Groupoids* by Koskas, Corsini, Kepka, Drbohlav, Nemec
- *Semigroups* by Kepka, Drbohlav, Nemec, Yuwaree, Kemprasit, Punkla, Leoreanu
- *Ordered Structures* by Prenowitz, Corsini, Chvalina

- *Combinatorics* by Comer, Tallini, Migliorato, De Salvo, Scafati, Gionfriddo, Scorzoni
- *Vector Spaces* by Mittas
- *Topology* by Mittas , Konstantinidou
- *Ternary Algebras* by Bandelt and Hedlikova.

In the 1940's, Prenowitz represented several kinds of Geometries (Projective, Descriptive, Spherical) as hypergroups, and later, with Jantosciak, founded Geometries on *Join Spaces*, a special hypergroups, which in the last decades were shown to be an useful instrument in the study of several matters: graphs, hypergraphs, binary relations, fuzzy sets and rough sets.

In 1978 Tallini established another link between geometries and a type of hypergroups he called Steiner hypergroups.

Connections between Hyperstructures and Binary Relations in the most general meaning, were considered for the first time in 1996, by Rosenberg. Afterwards they were studied also by Corsini, and by Corsini and Leoreanu (2000), but in special cases Hyperstructures had been already associated with binary relations, by Chvalina in 1994 with order relations, by Corsini (2000) and by Leoreanu (2000) with hypergraphs (a setting more general than symmetric relations), and by Nieminen, Corsini, Rosenberg, with graphs.

In 1996 Corsini introduced join spaces associated with Fuzzy Sets. These structures have been studied again by Corsini, Leoreanu, Tofan. The ideas of associating a hyperstructure with a fuzzy set and of considering algebraic structures endowed with a fuzzy structure, have been brought forward also by several Iranian scientists as Zahedi, Ameri, Borzooei, Hasankhani, Bolurian.

It is known that Fuzzy Sets, introduced by Zadeh in [429]), are a powerful tool in several applied sciences (see for instance Dubois and Prade [137]) and so, in view of the above correspondence, hyperstructures could as well be. The same is true for Hyperstructures associated with Rough Sets (see Corsini [76], Leoreanu [232]). Rough Sets introduced by Shafer, were analyzed by Pawlak and used by him and others as a mathematical tool in studying the Artificial Intelligence.

There existed till now two books on general theory of Hyperstructures (one by Corsini [437] on the basic theory of Hypergroups, the else by Vougiouklis [440], mostly on representations of hypergroups and on Hv-structures, that are hyperstructures satisfying conditions weaker than the classic ones) and others on particular sectors and applications.

Another important book for the applications in Geometry and also for the clearness of the exposition is that one by Prenowitz–Jantosciak [439].

Finally, we mention certain Doctoral theses, whose reading can be useful to deeper the knowledge both for the basics and the applications.

Konguetsof, L.	1964	Paris University, France
Koskas, M.	1967	Paris University, France
Stratigopoulos, D.	1969	Louvain University, Belgium
Mittas, J.	1969	Athens University, Greece
Konstantinidou, M.	1977	University of Thessaloniki, Greece
Sureau, Y.	1980	Université de Clermont II, France
Vougiouklis, T.	1980	Democritus University, Xanthi, Greece
Ioulidis, S.	1981	University of Thessaloniki, Greece
Serafimidis, Ch.	1983	University of Thessaloniki, Greece
Freni, D.	1985	Université de Clermont II, France
Massouros, Ch.	1988	Technical University of Athens, Greece
Spartalis, St.	1990	Democritus University, Xanthi, Greece
Yuwaree, Punkla	1991	University of Chulalongkorn, Bangkok, Thailand
Massouros, G.	1993	Technical University of Athens, Greece
Guţan, C.	1994	Université de Clermont II, France

Dramalidis, A.	1996	Democritus University, Greece
Yatras, C.	1996	Democritus University, Greece
Hasankhani, A.	1997	Shahid Bahonar Univ. of Kerman, Iran
Ameri, R.	1997	Shahid Bahonar Univ. of Kerman, Iran
Mouèka, J.	1997	Military University of the Ground Forces Vykov/ Masaryk University, Brno
Leoreanu, V.	1998	"Babeş Bolyai" University, Cluj–Napoca, Romania
Hort, D.	1999	Faculty of Education, Masaryk University, Brno
Borzooei, R.A.	2000	Shahid Bahonar Univ. of Kerman, Iran

By this book we present some of the numerous applications of hyperstructures, especially those from the last fifteen years, to the subjects:

1. Some topics of Geometry
2. Hypergraphs and Graphs
3. Binary Relations
4. Lattices
5. Fuzzy Sets and Rough Sets
6. Automata
7. Cryptography
8. Codes
9. Median Algebras, Relation Algebras, C-Algebras
10. Artificial Intelligence
11. Probabilities

This work, a survey of the most recent applications of Hyperstructure Theory, is based on many papers, some of which contain more detailed presentation. We hope this book will get a progress of science through a study in depth of these applications.

ACKNOWLEDGEMENTS. We warmly thank Professor Jeno Szep for encouraging us to write this book. Our thanks to Professors James Jantosciak, Ivo Rosenberg and Thomas Vougiouklis for careful reading of the manuscript and for their precious suggestions. We thank to lady Elena Mocanu for her patience and disponibility in the Latex processing of the manuscript.

Basic notions and results on Hyperstructure Theory

The most important notions and results, obtained on Hyperstructure Theory, are presented here. For more details, see [437].

Let H be a non–empty set and denoted by $\mathcal{P}^*(H)$ the set of all non–empty subsets of H.

1. Definition. A *n–hyperoperation* on H is a map $f : H^n \to$ $\to \mathcal{P}^*(H)$. The number n is called the *arity* of f.

2. Definition. A set H, endowed with a family Γ of hyperoperations, is called a *hyperstructure* (or a *multivalued algebra*).

3. Definition. If Γ is a singleton, that is $\Gamma = \{f\}$ where the arity of f is 2, then the hyperstructure is called a *hypergroupoid.*

Usually, the hyperoperation is denoted by "$\circ$" and the image of the pair (a, b) of H^2 is denoted by $a{\circ}b$ and called the *hyperproduct* of a and b.

If A and B are non–empty subsets of H, then $A \circ B = \bigcup\limits_{\substack{a\in A\\ b\in B}} a{\circ}b.$

4. Definition.

(i) A *semihypergroup* is a hypergroupoid $(H, \circ)$ such that $\forall\,(a, b, c) \in H^3,\ \ (a{\circ}b){\circ}c = a{\circ}(b{\circ}c).$

(ii) A *quasihypergroup* is a hypergroupoid $(H, \circ)$ which satisfies the *reproductive law*:

$$(*)\qquad \forall\, a \in H,\ H{\circ}a = a{\circ}H = H.$$

(iii) A *hypergroup* is a semihypergroup which is also a quasihypergroup.

5. Definition. Let $(H, \circ)$ be a hypergroupoid. An element $e \in H$ is called an *identity* or *unit* if

$$\forall\, a \in H,\quad a \in a{\circ}e \cap e{\circ}a.$$

6. Definition. Let $(H, \circ)$ be a hypergroup, endowed with at least an identity. An element $a' \in H$ is called an *inverse* of $a \in H$ if there is an identity $e \in H$, such that

$$e \in a{\circ}a' \cap a'{\circ}a.$$

7. Remark. Sometimes, more general structures are considered, for instance the *Wall–hypergroup* (see [423]) *of dimension n*, which is a non–empty set H, endowed with a hyperoperation "$\circ$", such that for any $(a, b) \in H^2$, the hyperproduct $a{\circ}b$ is a set of n elements of H, not *necessarily distinct elements.* Moreover, the associativity law is valid, there is at least one identity and any element has an inverse in a Wall hypergroup.

8. Definition. We say that two binary hyperoperations $< \circ_1 >$, $< \circ_2 >$ on the same set H are *mutually associative* (m.a.) if $\forall\,(x, y, z) \in H^3$, we have

$$(x \circ_1 y) \circ_2 z = x \circ_1 (y \circ_2 z) \text{ and}$$
$$(x \circ_2 y) \circ_1 z = x \circ_2 (y \circ_1 z).$$

We also say that the pair $((H, \circ_1), (H, \circ_2))$ is m.a.

The mutual associativity of two hyperoperations has been introduced by P. Corsini. In [73], he has started to investigate the problem of determining pairs of finite quasihypergroups which are mutually associative (m.a.).

9. Definition. A semihypergroup $(H, \circ)$ is called *simplifiable on the left* if: $\forall (x, a, b) \in H^3,\ x \circ a \cap x \circ b \neq \emptyset \Longrightarrow a = b$.

Similarly, we can define the simplifiability on the right.

F. Marty [248] proved that any hypergroup simplifiable on the left (or on the right) is a group. Later, M. Koskas [213] gave a simplier proof for the same result.

In [227] it is proved the following:

10. Theorem. *Let $(H, \circ)$ be a semihypergroup such that $\forall t \in H$, $t \circ H = H$ and $\exists s_0 \in H$, $H \circ s_0 = H$.*

(i) *If H is simplifiable on the left, then H is a group;*

(ii) *If H is simplifiable on the right, then H is a group.*

11. Definition

H is said to be of *type C on the right* (see [383]) if $\exists e \in H$, called a *scalar identity on the right*, such that:

1) $\forall x \in H,\ x \circ e = x$

2) $\forall (x, y, z) \in H^3,\ x \circ y \cap x \circ z \neq \emptyset \Longrightarrow e \circ y = e \circ z$.

Relation β and quotient hypergroupoids

Let $(H, \circ)$ be a hypergroupoid and let ρ be an equivalence relation on H.

12. Definition. We say that ρ is *regular on the right* if the following implication holds:

$$a \rho b \Longrightarrow \forall u \in H,\ \forall x \in a \circ u,\ \exists y \in b \circ u : x \rho y \text{ and } \forall \bar{y} \in b \circ u,\ \exists \bar{x} \in a \circ u : \bar{x} \rho \bar{y}$$

Similarly, the *regularity on the left* can be defined .

We say that ρ is *regular* if it is regular on the right and on the left.

13. Definition. We say that ρ is *strongly regular on the right* if the following implication holds:

$$a \rho b \Longrightarrow \forall u \in H,\ \forall x \in a \circ u,\ \forall y \in b \circ u : x \rho y.$$

Similarly, the *strong regularity on the left* can be defined.

We say that ρ is *strongly regular* if it is strongly regular on the right and the left.

14. Definition. Let $(H, \circ)$ be a hypergroupoid. We define the relation β on H, as follows:

$$a\beta b \Longleftrightarrow \exists n \in \mathbb{N}^*,\ \exists(x_1, x_2, ..., x_n) \in H^n : a \in \prod_{i=1}^{n} x_i \ni b.$$

Notice that β is a reflexive and a symmetric relation on H, but generally, not a transitive one.

Let us denote by β^* the transitive closure of β.

The following results hold:

15. Theorem. *If $(H, \circ)$ is a hypergroupoid, then β^* is the smallest equivalence strongly regular on H, with respect to the inclusion.*

16. Theorem. *If H is a hypergroup, then $\beta^* = \beta$.*

17. Notation. $\forall\,(a, b) \in H^2$, $a/b = \{x \mid a \in x{\circ}b\}$ and $b\backslash a = \{y \mid a \in b{\circ}y\}$.

18. Definition. Let $(H, \circ)$ and $(K, *)$ be hypergroupoids and $f : H \longrightarrow K$. We say that:

(i) f is a *homomorphism* if $\forall\,(a, b) \in H^2$, $f(a \circ b) \subset f(a) * f(b)$;

(ii) f is a *good homomorphism* if $\forall\,(a, b) \in H^2$, $f(a{\circ}b) = f(a) * f(b)$;

(iii) the homomorphism f is *strong on the left* if $f(c) \in f(a) * f(b) \Longrightarrow \exists a' \in H : f(a) = f(a')$ and $c \in a'{\circ}b$.

Similarly, we can define a homomorphism, which is *strong on the right.*

If a homomorphism f is strong on the right and on the left, we say that f is a *strong homomorphism.*

(iv) f is a *very good homomorphism* if f is a good homomorphism and moreover, $\forall\,(x, y) \in H^2$, $f(x/y) = f(x)/f(y)$ and $f(x\backslash y) = f(x)\backslash f(y)$.

Now, some basic results about quotient hypergroupoids are presented.

19. Theorem. *Let $(H,\circ)$ be a semihypergroup and ρ an equivalence relation on H.*

(i) *If ρ is regular, then H/ρ is a semihypergroup, with respect to the following hyperoperation:*

$$\forall\,(\bar{x},\bar{y}) \in (H/\rho)^2,\ \ \bar{x} \otimes \bar{y} = \{\bar{z} \mid z \in x{\circ}y\}.$$

(ii) *Conversely, if the hyperoperation "$\otimes$" is well-defined on H/ρ, then ρ is regular.*

(iii) *In the above–mentioned hypothesis, the canonical projection $\pi{:}H{\rightarrow}H/\rho$ is a good epimorphism and if $(H,\circ)$ is a hypergroup, then $(H/\rho,\otimes)$ is also a hypergroup, denoted by H/ρ.*

20. Theorem. *Let $(H,\circ)$ be a semihypergroup and ρ a strongly regular equivalence relation on H. Then:*

(i) *H/ρ is a semigroup;*

(ii) *if H is a hypergroup, then H/ρ is a group;*

(iii) *if S is a semigroup and $f: H \longrightarrow S$ is a homomorphism, then the equivalence relation R associated with f, as follows $aRb \Longleftrightarrow f(a)=f(b)$, is strongly regular.*

21. Corollary. *If $(H,\circ)$ is a hypergroup, then H/β is a group.*

Moreover, β is the smallest equivalence relation ρ on H, such that H/ρ is a group.

Complete parts, subhypergroups and the heart of hypergroup

22. Definition. Let $(H,\circ)$ be a semihypergroup and A a non–empty subset of H. We say that A is a *complete part* of H if the following implication holds:

$$\forall n \in \mathbb{N}^*,\ \forall\,(x_1,...,x_n) \in H^n,\ \prod_{i=1}^{n} x_i \cap A \neq \emptyset \Longrightarrow \prod_{i=1}^{n} x_i \subset A.$$

23. Definition. If $(H, \circ)$ is a semihypergroup and $A \subset H$, $A \neq \emptyset$, then the *complete closure of* A *in* H is the intersection of all the complete parts of H, which contain A. It will be denoted by $\mathcal{C}(A)$.

Some basic results concerning $\mathcal{C}(A)$:

24. Theorem. *Let* $(H, \circ)$ *be a semihypergroup and* $A \subset H$, $A \neq \emptyset$. *We consider* $K_0(A) = A$ *and* $\forall n \in \mathbb{N}$,

$$K_{n+1}(A) = \left\{ a \in H \mid \exists p \in \mathbb{N}^*,\ \exists (x_1, ..., x_p) \in H^p : a \in \prod_{i=1}^{p} x_i \text{ and } \prod_{i=1}^{p} x_i \cap K_n(A) \neq \emptyset \right\}$$

Let $K(A) = \bigcup_{n \in \mathbb{N}} K^n(A)$. *Then* $\mathcal{C}(A) = K(A)$.

Let $(H, \circ)$ be a semihypergroup.

25. Theorem.

(i) *The relation* K *defined as follows*

$$aKb \Longleftrightarrow x \in \mathcal{C}(\{y\})$$

is an equivalence relation on H.

(ii) $\forall (a, b) \in H^2$, *we have* $aKb \Longleftrightarrow a\beta^* b$.

26. Theorem. *If* A *is a non-empty subset of a semihypergroup* $(H, \circ)$, *then* $\mathcal{C}(A) = \bigcup_{a \in A} \mathcal{C}(a)$.

The following theorem characterizes the semihypergroups for which the relation β is transitive.

27. Theorem. ([152] and Th.47, Ch.3) *Let* H *be a semihypergroup. The relation* β *is transitive in* H *if and only if* $\forall x \in H$, $\mathcal{C}(x) = K_1(x)$.

Let $\varphi_H : H \longrightarrow H/\beta$ be the canonical projection.

28. Definition. The *heart* of a hypergroup H is $\omega_H = \{x \in H \mid \varphi_H(x) = 1\}$, where 1 is the identity of the group H/β.

29. Theorem. *If A is a non–empty subset of a hypergroup H, then $\mathcal{C}(A) = A \circ \omega_H = \omega_H \circ A$.*

30. Corollary. *If A and B are non–empty subsets of a hypergroup $(H, \circ)$, such that one of A and B is complete, then $A\circ B$ and $B\circ A$ are complete parts.*

31. Definition. Let $(H, \circ)$ be a hypergroupoid and A a non–empty subset of H. We say that

(i) A is *reflexive* in H if $\forall\,(x, y) \in H^2$, from $x\circ y \cap A \neq \emptyset$ it follows $y\circ x \cap A \neq \emptyset$;

(ii) A is *invariant* (or *normal*) in H if $\forall\, x \in H$, we have $x\circ A = A\circ x$.

(iii) A is *invertible on the left* in H if $\forall\,(x, y) \in H^2$, the following implication holds: $y \in A\circ x \Longrightarrow x \in A\circ y$.

Similarly, we define the *invertibility on the right*. We say that A is *invertible* if it is invertible on the right and on the left.

32. Definition. Let $(H, \circ)$ be a hypergroupoid and K a non-empty subset of H.

K is called a *subhypergroupoid* of H if $K\circ K \subset K$. A subhypergroupoid

K of H is called a *subhypergroup* of H, if $(K, \circ)$ is a hypergroup.

Now, we define some important **types of subhypergroups:**

33. Definition. Let $(H, \circ)$ be a hypergroup and K a subhypergroup of it. We say that:

(i) K is *closed on the left* in H if $\forall\, a \in H$, $\forall\,(x,y) \in K^2$, from $x \in a{\circ}y$ follows $a \in K$.

Similarly, we can define the notion *closed on the right.*

K is closed in H if it is closed on the right and on the left.

(ii) K is *ultraclosed on the left* in H if $\forall\, x \in H$, $K{\circ}x \cap (H-K){\circ}x \neq \emptyset$.

Similarly, we can define the notion *ultraclosed on the right.*

K is *ultraclosed* if it is ultraclosed on the right and on the left.

We characterize ultraclosed subhypergroups:

34. Theorem. *Let $(H,\circ)$ be a hypergroup, I_p the set of partial identities, that is $I_p = \{e \in H \mid \exists x \in H : x \in e{\circ}x \cup x{\circ}e\}$.*

Let K be a subhypergroup of H.

K is ultraclosed if and only if it is closed and contains I_p.

35. Definition. Let $(H,\circ)$ be a hypergroup and K_1, K_2 subhypergroups of H. We say that K_2 is K_1*–conjugable* if the following conditions hold:

1) $K = K_1 \cap K_2 \neq \emptyset$;
2) K_2 is closed in H;
3) $\forall\, x \in K_1,\ \exists x' \in K_1$ such that $x'{\circ}x \subset K$.

The following characterization holds:

36. Theorem. *A subhypergroup K of a hypergroup H is a complete part of H if and only if K is H–conjugable.*

We state some connections between complete parts, invertible, closed, ultraclosed subhypergroups:

37. Theorem. *Let $(H,\circ)$ be a hypergroup and K a subhypergroup of H. The following statements hold:*

(i) *if K is a complete part of H, then K is ultraclosed in H;*

(ii) *if K is ultraclosed in H, then K is invertible in H;*

(iii) *if K is invertible on the right (on the left)in H, then K is closed on the left (on the right)in H.*

38. Theorem. *If K is a subhypergroup and a complete part of a hypergroup H, then K is invariant in H if and only if it is reflexive in H.*

39. Remark. In [437, pp.52–53] examples are given of non–closed subhypergroups, ultraclosed but not complete parts subhypergroups, invertible but not ultraclosed subhypergroups, closed but not invertible subhypergroups.

40. Theorem. *The heart of a hypergroup H is the intersection of all subhypergroups of H, which are complete parts.*

41. Definition. The intersection of all ultraclosed subhypergroups of a hypergroup H is called *nucleus* of H.

By *C.U.* it is denoted the class of hypergroups, whose ultraclosed subhypergroups are all complete parts.

Several important classes of hypergroups

I. Regular hypergroups, complete hypergroups and canonical hypergroups.

42. Definition. A hypergroup H is *regular* if it has at least one identity and each element has at least one inverse.

A regular hypergroup $(H, \circ)$ is called *reversible* if for any $(x, y, z) \in H^3$, it satisfies the following conditions:

1) if $y \in a \circ x$, then there exists an inverse a' of a, such that $x \in a' \circ y$;

2) if $y \in x \circ a$, then there exists an inverse a'' of a, such that $x \in y \circ a''$.

If H is regular, we denote by E the set of identities of H and for any $a \in H$, by $i(a)$ the set of inverses of a.

43. Theorem. *If H is a regular reversible hypergroup and $\{A_i\}_{i\in I}$ is a family of its invertible subhypergroups, then $A = \bigcap_{i\in I} A_i$ is an invertible subhypergroup.*

In [437, p. 63] it is presented an example of regular hypergroup, which is not reversible.

44. Definition. A semihypergroup$(H, \circ)$ is called *complete* if

$$\forall\,(x, y) \in H^2, \quad \mathcal{C}(x \circ y) = x \circ y,$$

where C was defined in 23.

Some results about the complete hypergroups:

45. Theorem. *A semihypergroup H is complete if $H = \bigcup_{s\in S} A_s$, where S and A_s satisfy the conditions:*

1) *$(S, \circ)$ is a semigroup;*
2) *$\forall\,(s, t) \in S^2$, $s \neq t$, we have $A_s \cap A_t = \emptyset$;*
3) *if $(a, b) \in A_s \times A_t$, then $a \circ b = A_{st}$.*

46. Theorem. *If H is a complete hypergroup, then*

1) *ω_H is the set of identities of H and*
2) *H is a regular reversible hypergroup.*

47. Definition. A hypergroup H is *flat* if for any subhypergroup K of H, the following equality holds: $\omega_K = \omega_H \cap K$.

48. Theorem. *Every complete hypergroup is flat.*

49. Theorem. *Let H be a regular reversible hypergroup. If A is a closed subhypergroup, then A is invertible.*

50. Theorem. *Let $f : H \rightarrow H'$ be a very good epimorphism of hypergroups and let K be a subhypergroup and a complete part of H. Then $f(K)$ is a complete part and a subhypergroup of H'.*

51. Theorem. *If $f : H \to H'$ is a very good epimorphism between hypergroups, then $f(\omega_H) = \omega_{H'}$.*

52. Theorem. *If H and H' are complete hypergroups and $f : H \to H'$ is a good homomorphism, then f is very good.*

53. Definition. We say that a hypergroup H is *canonical* if

1) it is commutative
2) it has a scalar identity
3) every element has a unique inverse
4) it is reversible.

54. Remark. Not all subhypergroups of a canonical hypergroup are canonical (see Th. 200 [437]).

Let $(H, +)$ be a canonical hypergroup and $x \in H$. For any $n \in \mathbb{Z}$, we define

$$nx = \begin{cases} \underbrace{x + x + \cdots + x}_{n \text{ times}} & , \text{ if } n > 0 \\ 0 & , \text{ if } n = 0 \\ \underbrace{(-x) + \cdots + (-x)}_{(-n) \text{ times}} & , \text{ if } n < 0, \end{cases}$$

where $\forall\, x \in H$, we denote by "$-x$" the inverse of x.

We can verify that

$$mx + nx = \begin{cases} (m+n)x, & \text{if } mn \geq 0 \\ (m+n)x + \min\{|m|, |n|\} \cdot (x - x), & \text{if } mn < 0. \end{cases}$$

55. Definition. Let $(H, +)$ be a canonical hypergroup and $x \in H$. We say that the *order of x is infinite* ($o(x) = \infty$) if $\forall\, (h, k) \in \mathbb{Z}^2$, where $h \neq 0$, we have $0 \notin hx + k(x - x)$.

56. Theorem. *Let $(H, +)$ be a canonical hypergroup and $x \in H$. Then $o(x) = \infty$ if and only if $\forall\, (m, n) \in \mathbb{Z}^2$, $m \neq n$ we have $mx \cap nx = \emptyset$.*

57. Definition. Let $(H,+)$ be a canonical hypergroup. Let us suppose that there exists $(m,n) \in \mathbb{Z}\times\mathbb{N}$, $m \neq 0$, such that $0 \in mx + n(x-x)$.

Let $h = \min\{r \in \mathbb{N}^* \mid \exists n \in \mathbb{N} : o \in rx + n(x-x)\}$. The number h is called *the principal order of* x.

58. Theorem. *Let* $(H,+)$ *be a canonical hypergroup and* $x \in H$. *We have* $0 \in mx + n'(x-x)$ *if and only if ord* x *divides* m.

59. Definition. Let h divide m and $q = \min\{s \in \mathbb{N}^* \mid mx + +s(x-x) \ni 0\}$. The couple (h,q) is called the *order of* x.

60. Definition. A canonical hypergroup $(H,+)$ is called *strongly canonical* if it satisfies the following conditions:

1) $\forall\,(x,a) \in H^2,\ x \in x+a \Longrightarrow x = x+a$;
2) $(x+y)\cap(z+w) \neq \emptyset \Longrightarrow x+y \subset z+w$ or $z+w \subset x+y$.

II. Join spaces.

61. Definition. A commutative hypergroup $(H,\circ)$ is called a *join space* if $\forall\,(a,b,c,d) \in H^4$, the following implication holds:

$$a/b \cap c/d \neq \emptyset \Longrightarrow a{\circ}d \cap b{\circ}c \neq \emptyset.$$

If A and B are subsets of a hypergroup H, we denote by A/B the set $\bigcup\limits_{\substack{a\in A\\ b\in B}} a/b$.

62. Theorem. *A commutative hypergroup is canonical if and only if it is a join space with a scalar identity.*

63. Theorem. *A is a closed subhypergroup of a join space H if and only if* $A/A = A$.

64. Theorem. *Let A, B, C, D be non–empty subsets of a join space* $(H,\circ)$. *We have:*

1) *if* $A \subset B$ *and* $C \subset D$, *then* $A/C \subset B/D$;

2) $A \cap B/C \neq \emptyset$ *if and only if* $A \circ C \cap B \neq \emptyset$;

3) $A/(B \circ C) = (A/B)/C$;

4) $A/(B/C) \subset (A \circ C)/B$;

5) $A \circ (B/C) \subset (A \circ B)/C$;

6) $B \subset A/(A/B)$.

65. Definition. A join space H is called *geometric* if $\forall\, x \in H$, we have $x \circ x = \{x\} = x/x$.

66. Definition. For a closed subhypergroup N of join space H and $(x, y) \in H^2$, we write xJ_Ny if $x \circ N \cap y \circ N \neq \emptyset$.

67. Theorem. *The relation* J_N *is an equivalence relation. The equivalence class of* $a \in H$ *is* $(a)_N = (a \circ N)/N = N/(N/a)$. *In particular,* $\forall\, x \in N$, $(x)_N = N$.

68. Theorem. *If* H *is a join space and* N *is a closed subhypergroup of* H, *then the equivalence relation* J_N *is regular and the quotient* H/J_N *is canonical.*

69. Theorem. ([312]) *The following statements (concerning the canonical hypergroup* $(H/J_N, \otimes)$*) hold:*

1) *the identity element is* N *and* $\forall\, n \in N$, *we have* $(n)_N = N$;

2) $(a')_N$ *is the inverse of* $(a)_N$ *if and only if* $N \cap a \circ a' \neq \emptyset$;

3) *if* $(a')_N$ *is the inverse of* $(a)_N$, *then* $(a')_N = N/a$.

If $A_1, A_2, ..., A_n$ are subsets of a hypergroup, we denote by $< A_1, ..., A_n >$ the closed subhypergroup generated by $\bigcup_{i=1}^{n} A_i$.

70. Theorem. *If H is a join space and A is a subhypergroup of H, then A is ultraclosed if and only if it is a complete part of H.*

Let $(H, \circ)$ be a hypergroup. Let us denote

$$I_p = \{e \in H \mid \exists x \in H : x \in e \circ x \cup x \circ e\}.$$

For $n \in \mathbb{N}^*$, set

$$I_p^n = \underbrace{I_p \circ \cdots \circ I_p}_{n \text{ times}}.$$

We obtain:

71. Theorem. *Let $(H, \circ)$ be a join space. Then*

$$\omega_H = \bigcup_{n \in \mathbb{N}^*} (I_p^n / I_p^n).$$

72. Definition. Let H be a join space. If H has a scalar identity e, we set $E = \{e\}$, otherwise $E = \emptyset$.

Furthermore, we define

$$\lhd \emptyset \rhd = E \text{ and if } A \in \mathcal{P}^*(H),\ \lhd A \rhd = < A > .$$

73. Definition. A join space H is called an *exchange space* if it satisfies the following conditions:

(I) if $a \in \lhd b \rhd$, $a \notin E$, then $\lhd a \rhd = \lhd b \rhd$;

(II) if $c \in \lhd a, b \rhd$ and $c \notin \lhd b \rhd$, then $\lhd c, b \rhd = \lhd a, b \rhd$.

For an exchange space, $< >$ will mean $\lhd\ \rhd$.

74. Theorem. *If A and B are non–empty subsets of a join space, such that $< A > \cap < B > \neq \emptyset$, then $< A, B > = < A > / < B >$.*

75. Theorem. *Let H be a join space with a scalar identity e. Then H satisfies* (I) *if and only if it satisfies* (II).

76. Definition. Let A be a subset of a join space H. A is called *independent* if $\forall\, a \in A$, we have $a \notin < A - \{a\} >$.

77. Definition. A subset A of a closed subhypergroup S of a join space H is called a *basis* of S if it is independent and furthermore $< A > = S$.

78. Theorem. *Let A be a subset of an exchange space H, and let $(x,y) \in H^2$. If $y \in < A, x >$ and $y \notin < A >$, then $< A, x > =$ $= < A, y >$.*

79. Theorem. *All complete commutative hypergroups are join spaces, but there are commutative regular reversible hypergroups, which are not join spaces.*

III. Quasi–canonical Hypergroups. Cogroups.

Let $(H, \circ)$ be a hypergroup and $x \in H$.

We denote by $i_\ell(x)$ the set of $x' \in H$ such that $e \in x' \circ x\}$ for a left identity e and by $i_r(x)$ the set $x'' \in H$ such that $e \in x \circ x''\}$ for a right identity$e$.

We also denote by $i(x)$ the set of all inverses of x.

80. Definition. A hypergroup H is called *feebly quasi–canonical* if it is regular, reversible and satisfies the condition:

$$\forall\,(x,a)\in H^2, \forall\,\{u,v\} \subset i_\ell(x), \forall\,\{w,z\} \subset i_r(x),\ u\circ a = v\circ a,\, a\circ w = a\circ z.$$

If H is also commutative, we say that H is *feebly canonical.*

We denote by $F.Q.C.$ and by $F.C.$ the classes of feebly quasi–canonical, respectively feebly canonical hypergroups.

81. Theorem. *Let $H \in F.Q.C.$ and K be a subhypergroup of H. Then K is ultraclosed if and only if it is a complete part of H.*

82. Theorem. *Let $H \in F.Q.C.$ Then the following conditions are equivalent:*

a) $\forall\, x \in H,\ \operatorname{card} i(x) = 1;$

b) *H has exactly one identity, which is a scalar.*

83. Definition. A hypergroup in $F.Q.C.$, satisfying the equivalent conditions a) or b) of the above theorem, is called *quasi–canonical* (or a *polygroup*).

We denote by $Q.C.$ the class of quasi–canonical hypergroups.

Clearly, the canonical hypergroups are the commutative quasi–canonical hypergroups.

Let $(H, \circ)$ be a feebly quasi–canonical hypergroup and let R be the following relation on H: $xRy \Longleftrightarrow \exists z \in H : \{x, y\} \subset i(z)$.

84. Theorem.

(i) *The relation R is a regular equivalence relation.*

(ii) *The quotient H/R is a hypergroup, with respect to the hyperoperation*
$$\bar{x} \otimes \bar{y} = \{\bar{z} \mid z \in x{\circ}y\}.$$

Moreover, the canonical projection $p : H \longrightarrow H/R$ is a good epimorphism.

85. Theorem. *If $H \in F.Q.C.$, then H/R is quasi–canonical.*

86. Theorem. *The following conditions are equivalent for $H \in F.Q.C.$*

(i) *H is complete*

(ii) *ω_H is the set of identities of H*

(iii) *H/R is a group.*

87. Definition. A *weak left cogroup* is a regular reversible hypergroup $(H, \circ)$, endowed with a left scalar identity "e" and satisfying $x{\circ}y \cap z{\circ}y \neq \emptyset \Longrightarrow x \in z{\circ}e$.

A weak left cogroup is called a *left cogroup* if it also satisfies $\forall\,(x, y, z) \in H^3,\ \text{card}(x{\circ}y) = \text{card}(x{\circ}z)$.

Let $(H, \circ)$ be a weak left cogroup with a left scalar identity e. The following relation R defined on H: $xRy \iff x \in y{\circ}e$ is an equivalence relation.

88. Theorem. *The quotient H/R endowed with the structure $a{\circ}e \otimes b{\circ}e = \{v{\circ}e \mid v \in a{\circ}b\}$ is a quasi–canonical hypergroup.*

89. Theorem. *Let H be a subhypergroup of a cogroup C. Then*
(i) *H is an invertible part of H and a subcogroup of C;*
(ii) *if* $\operatorname{card} C < \aleph_0$, *then the order of H divides the order of C.*

90. Definition. A partial hyperalgebraic structure $< H, \circ, I, ^{-1} >$ is called a *quasi–canonical hypergroupoid* if "$\circ$" is a partial binary hyperoperation on H, i.e. a map from H^2 into $\mathcal{P}(H)$, $I \subseteq H$ and $^{-1}$ is a unary operation on H, such that the following conditions hold for any $(x, y, z) \in H^3$:

1) $(x{\circ}y){\circ}z = x{\circ}(y{\circ}z)$, which should be interpreted as follows: if either side is non–empty, then both sides are non–empty and the sets are equal.
2) $x{\circ}I = I{\circ}x = x$;
3) $x \in y{\circ}z \iff y \in x{\circ}z^{-1} \iff z \in y^{-1}{\circ}x$.

Quasi–canonical hypergroupoids are also called *polygroupoids*. They were introduced by S. Comer and correspond to the atom structures of systems of relations. Comer generalized polygroupoids to partial multi–valued loops.

IV. Cyclic hypergroups.

91. Definition. A hypergroup H is called *cyclic with a generator* x if $\varphi_H(H)$ is a cyclic group generated from $\varphi_H(x)$.

92. Definition. An element x of a hypergroup H is called *periodic of period* $p(x) = n$ if $x^n \subset \omega_H$ and $n = \min\{k \in \mathbb{N} \mid x^k \subset \omega_H\}$.

93. Definition. A semihypergroup H is called *s-cyclic* with *s-generator* $h \in H$ if for all $x \in H$ we have $x \in h^n for some\ n \in \mathbb{N}$.

94. Theorem. *If H is a cyclic and complete hypergroup, then it is commutative.*

95. Definition. If H is an s-cyclic semihypergroup with s-generator h, we call the *cyclicity* of $a \in H$ the integer

$$m = \min\{q \in \mathbb{N}^* - \{1\} \mid a \in h^q\}.$$

We write $\text{cycl}(a) = m$.

96. Theorem. *A cyclic and complete semihypergroupis a join space.*

97. Theorem. *If H is a cyclic and complete hypergroup and h is its s-generator, such that* $\text{cycl}(h) = r$, *then* $H = \bigoplus_{t=2}^{r} h^t$.

98. Theorem. *Every hypergroup $< H, \circ >$ is embeddable in a cyclic hypergroup $< K, \otimes >$, with $\omega_K = H$.*

V. K_H–hypergroups.

99. Definition. Let $<H, \circ>$ be a hypergroupoid and let $\{A(x)\}_{x \in H}$ be a family of pairwise disjoint non–empty sets.Let $K_H = \bigcup_{x \in H} A(x)$ and let us define

$$\forall\, a \in K_H, \quad g(a) = x \Longleftrightarrow a \in A(x).$$

We define in K_H the hyperoperation:

$$\forall\, (a, b) \in K_H^2, \quad a \Box b = \bigcup_{z \in g(a) \circ g(b)} A(z).$$

100. Theorem.

1) *$(H, \circ)$ is a semihypergroup if and only if $< K_H, \Box >$ is a semihypergroup;*

2) $(H, \circ)$ *is a hypergroup if and only if* $< K_H, \Box >$ *is a hypergroup.*

101. Notation. For any $P \in \mathcal{P}^*(H)$, set $K(P) = \bigcup_{x \in P} A(x)$.

102. Theorem.
1) $E(K_H) = K(E_H)$;
2) $\forall a \in K_H,\ i(a) = K(i(g(a))) = g^{-1}(i(g(a)))$.

103. Theorem.
1) *If* P *is a complete part of* $< H, \circ >$, *then* $K(P)$ *is a complete part of* $< K_H, \Box >$.
2) *If* P *is a non-empty part of a semi-hypergroup* H, *then* P *is a subhypergroup of* H *if and only if* $K(P)$ *is a subhypergroup of* K_H.

104. Theorem. $\forall (x, y) \in H^2$, $\forall (u, v) \in A(x) \times A(y)$, *if* $u \beta_{K_H} v$ *then* $x \beta_H y$.

105. Theorem. *If* H *is a hypergroup, then* $\omega_{K_H} = K(\omega_H)$.

106. Theorem. *If* H *is a hypergroup, then:*
1) $< K_H, \Box >$ *is regular if and only if* $(H, \circ)$ *is regular;*
2) $< K_H, \Box >$ *is reversible if and only if* $(H, \circ)$ *is reversible;*
3) $< K_H, \Box >$ *is feebly quasi–canonical if and only if* $(H, \circ)$ *is feebly quasi–canonical.*

Hyperrings, hypermodules and vector hyperspaces

107. Definition. A *(Krasner) hyperring* is a hyperstructure $< A, +, \cdot, 0 >$ where:
1) $(A, +)$ is a canonical hypergroup;
2) $(A, \cdot)$ is a semigroup endowed with a two–sided absorbing element 0;
3) the product distributes from both sides over the sum.

108. Definition. A *hyperfield* is a Krasner hyperring $(K, +, \cdot, 0)$, such that $(K - \{0\}, \cdot)$ is a group.

109. Definition. Let x be an element of a hyperring A. If $o(x) = \infty$, we say that the *characteristic of* x is zero and we set: $\chi(x) = 0$. If $o(x) \neq \infty$, we set $\chi(x) = h$, where h is the principal order of x in the canonical hypergroup $< A, + >$.

110. Definition. We call the *characteristic* $\chi(A)$ *of* A the least common multiple (ℓ.c.m.) of $\chi(x)$ for $x \in A$ if it exists and is $\neq 0$, otherwise we set $\chi(A) = 0$.

111. Remark. If $y = ax$, then $\chi(y)$ divides $\chi(x)$.

112. Definition. If $< A, +, \cdot >$ is a hyperring and B is a non–empty subset of A, we say that $< B, +, \cdot >$ is a *subhyperring* of A if:

> $(B, +)$ is a canonical subhypergroup of $< A, + >$ and $(B, \cdot)$ is a subsemigroup of $(A, \cdot)$.

We say that B is a *left hyperideal* of A if $(B, +)$ is a canonical subhypergroup of A and $A \cdot B \subset B$.

Similarly, we can define the notion of *right hyperideal* and of the two–sided hyperideal of A.

113. Proposition. *The heart* ω_A *of* $< A, + >$ *is a hyperideal of* $< A, +, \cdot >$.

114. Proposition. *Let* A *and* B *be respectively a hyperring and a two–sided hyperideal of* A. *If in the quotient* $A/B = (A, +)/(B, +)$ *we set* $(x + B)(y + B) = xy + B$, *then the structure* $(A/B, +, \cdot)$ *is a hyperring.*

115. Definition. Let A be a hyperring. We say that $< M, +, \circ >$ is a *right* A*–hypermodule* if

1) $(M, +)$ is a canonical hypergroup;

2) $\circ$ is a scalar single–valued operation, that is a function which associates with any pair $(x, a) \in M \times A$ an element $x \circ a \in M$, such that $\forall (x, y) \in M^2$, $\forall (a, b) \in A^2$, the following conditions hold:
   1°. $(x + y) \circ a = x \circ a + y \circ a$;
   2°. $x \circ (a + b) = x \circ a + x \circ b$;
   3°. $x \circ (a \cdot b) = (x \circ a) \circ b$;
   4°. $x \circ 0 = 0$.

If A is endowed with a unit 1, M is called *unitary* if $\forall x \in M$, $x \circ 1 = x$.

116. Definition. If K is a hyperskewfield, then a right unitary hypermodule $< V, +, \circ >$ is called *K–vectorial hyperspace.*

117. Definition. If M and M' are right A–hypermodules (where A is a hyperring) and $f : M \longrightarrow M'$ is a map, we say that f is a *homomorphism* if: $\forall (x, y) \in M^2$, $f(x + y) = f(x) + f(y)$ and $\forall (x, a) \in M \times A$, $f(x \circ a) = f(x) \circ a$.

118. Proposition. *Let M be a right A hypermodule and N a subhypermodule of M (that is a canonical subhypergroup such that $\forall a \in A$, $N \circ a \subset N$). If we set $\forall (x, y) \in M^2$, $xRy \iff x + N =$ $= y + N$ and we define on the quotient M/R, $\forall (x, y) \in M^2$, $(x+M)+(y+N)=\{v+N \mid v \in x+y\}$, $\forall a \in A$, $(x+N) \circ a = x \circ a+N$, then we obtain on M/R (denoted by M/N) a structure of right hypermodule.*

H_v-structures

One of the topics of great interest, in the last years, is the study of *weak hyperstructures*, so-called *H_v–structures.* The class of H_v–structures is the largest class of algebraic hyperstructures.

These structures satisfy weak axioms, where the non–empty intersection replaces the equality.

This topic was introduced in 1990 by Vougiouklis ([413]) and studied by himself and then by R. Migliorato and their students. R. Ameri has introduced the categories of H_v–groups and H_v-modules.

Vougiouklis abbreviated the weak associativity by WASS and the weak commutativity by COW.

119. Definition. A hypergroupoid $(H, \cdot)$ is called an *H_v–group* if the weak associativity is satisfied, that is:

$$(\alpha) \qquad \forall (x, y, z) \in H^3, \ x \cdot (y \cdot z) \cap (x \cdot y) \cdot z \neq \emptyset$$

and also the reproductive axiom holds:

$$\forall x \in H, \ x \cdot H = H \cdot x = H.$$

A hypergroupoid which satisfies only (α) is called *H_v-semigroup.*

120. Definition. Let $(H_1, \cdot)$ and $H_2, *)$ be two H_v-groups. A map $f : H_1 \to H_2$ is called a *weak homomorphism* if:

$$\forall (x, y) \in H_1^2, \ f(x \cdot y) \cap f(x) * f(y) \neq \emptyset.$$

Let $(H, \cdot)$ be an H_v-group. The relation β^* is the smallest equivalence relation on H, such that the quotient H/β^* is a group.

It is called the *fundamental group* and β^* is called the *fundamental equivalence* relation on H.

The relation β is defined on an H_v-group in the same way as in a hypergroup.

Finally β^* is the transitive closure of β.

121. Definition. An H_v-group $(H, \circ)$ is called an *H_b-group* if there exists a group operation "$\cdot$" on H, such that $\forall (x, y) \in H^2$, we have $x \cdot y \in x \circ y$.

122. Definition. An *H_v-ring* is a hyperstructure $(R, +, \cdot)$, where both hyperoperations "$+$" and "$\cdot$" are weakly associative, "$\cdot$" weakly distributed over "$+$" from both sides and "$+$" is reproductive.

Let $\mathcal{U}$ be the set of all finite polynomials of elements of R over $\mathbb{N}$.

Let us define the relation γ on R, as follows:

$$x\gamma y \Longleftrightarrow \exists u \in \mathcal{U}, \text{ such that } \{x,y\} \subseteq u.$$

Let γ^* denote the transitive closure of γ.

Note that γ^* is the smallest equivalence relation on R such that the quotient R/γ^* is a ring.

The relation γ^* is called the *fundamental relation* of R and is the main tool for the study of H_v-rings.

123. Definition. The H_v-ring $(R,+,\cdot)$ is called an *H_v–field* if the ring R/γ^* is a field.

Let us denote by ω^* the kernel of the canonical map

$$\pi : R \rightarrow R/\gamma^*.$$

124. Definition. An H_v-ring $(R,+,\cdot)$ is called a *reproductive H_v–field* if the following condition holds:

$$\forall x \in R-\omega^*,\ x\cdot(R-\omega^*) = (R-\omega^*)\cdot x = R-\omega^*.$$

The importance of the reproductivity with respect to the hyperoperation "$\cdot$" consists in the representations in the diagonal form.

125. Definition. A matrix whose entries are elements of an H_v-ring is called *H_v-matrix*.

H_v-matrices have been especially studied by Vougiouklis.

126. Definition. A cow group $(M,+)$ is called a *left H_v-module* over an H_v-ring R, if for every $a \in R$ there is a map $(a,x) \longmapsto ax$ from $R\times M$ into $\mathcal{P}^*(M)$ such that $\forall\,(a,b) \in R^2$, $\forall\,(x,y) \in M^2$, we have

$$a(x+y) \cap (ax+ay) \neq \emptyset;$$
$$(a+b)x \cap (ax+bx) \neq \emptyset$$
$$(ab)x \cap a(bx) \neq \emptyset.$$

The *fundamental relation* ε^* in M over R is the smallest equivalence relation on M, such that M/ε^* is a module over the ring R/γ^*.

ε^* is constructed as follows:

Let $(M, +)$ be an H_v-module over an H_v-ring R. Let $\mathcal{U}$ be the set of all expressions consisting of finite hyperoperations either on R and M or the external hyperoperation applied on finite sets of elements of R and M.

Define a binary relation ε on M by:

$$x\varepsilon y \Longleftrightarrow \exists u \in \mathcal{U}, \text{ such that } \{x, y\} \subset u$$

and denote by ε^* the transitive closure of the relation ε.

In the *fundamental module* $(M/\varepsilon^*, \oplus, \odot)$ over R/γ^*, the hyperoperations $\oplus$ and $\odot$ are defined as follows:

$$\forall\,(x, y)\in M^2,\ \varepsilon^*(x) \oplus \varepsilon^*(y) = \varepsilon^*(z) \text{ for any } z\in\varepsilon^*(x) + \varepsilon^*(y)$$
$$\forall\, a \in R,\ \forall\, x\in M,\ \gamma^*(a) \odot \varepsilon^*(x) = \varepsilon^*(z), \text{ for any } z\in\gamma^*(a) \cdot \varepsilon^*(x).$$

Definition 127. An H_v-semigroup $(H, \cdot)$ is called *h/v-group* if the quotient H/β^* is a group.

Remark 128. In a similar way as above, the h/v–rings, h/v–fields, h/v–modulus, h/v–vector spaces are defined. These structures has been studied by T. Vougiouklis.

Chapter 1

Some topics of Geometry

• *Several branches of geometry can be treated as certain kinds of hypergroups, known as join spaces. Introduced by W. Prenowitz and studied afterwards by him together with J. Jantosciak, the concept of a join space is "sufficiently general to cover the theories of ordered and partially ordered linear, spherical and projective geometries, as well as abelian groups".*

• *If we consider a spherical geometry and identify antipodal points, we obtain a projective geometry. This construction can be described in the context of join spaces as follows:*

Let J be the set of points of a spherical join space and for any $a \in J$, let $\bar{a} = \{a, a^{-1}\}$. Let $\overline{J} = \{\bar{a} \mid a \in J\}$. We define on $\overline{J}$ the following hyperoperation:

$$\bar{a} \circ \bar{b} = \{\bar{x} \mid x \in \bar{a} \cdot \bar{b}\},$$

where "$\cdot$" is the hyperoperation of the spherical join space.

Theorem. (see [168]) *$(\overline{J}, \circ)$ is a projective join space, such that $\forall\, \bar{a} \in \overline{J}$, $\bar{a} \circ \bar{a} = \bar{a}/\bar{a} = \{\bar{e}, \bar{a}\}$, where $\bar{e}$ is the identity.*

The results of §1, §2, §3 of this chapter are due to W. Prenowitz and W. Prenowitz–J. Jantosciak. Using the notion of join space, they have rebuilt several branches of geometry.

We start by presenting some important and interesting examples of join spaces, suggested by three types of geometris:

1) Affine join spaces over ordered fields

Let L be a vector space over an ordered field K. We define the following hyperoperation on L:

$$\forall (x,y) \in L^2,\ x \circ y = \{\alpha x + \beta y \mid \alpha > 0,\ \beta > 0,\ \alpha + \beta = 1\}$$

Then $(L, \circ)$ is a join space, called *the affine join space over* K.

2) Ray spaces over ordered fields

Let L be a vector space over an ordered field K. Given $x \in L$, the *ray* $\vec{x}$ is the set $\{\lambda x \mid \lambda > 0\}$. Let R be the family of rays of L. Let us define on R the following hyperoperation$\otimes$:

> $\forall (\vec{x}, \vec{y}) \in R^2$, $\vec{x} \otimes \vec{y}$ is the set of rays determined by the elements of $x \circ y$, where "$\circ$" is the hyperoperation defined in 1).

Then $(R, \otimes)$ is again a join space, called *the ray space of* L.

We can obtain the following interesting isomorphism:

Let L be a real vector space, with an inner product, let S be a hypersphere of L centered at 0, the zero of L and the bijection function $x \longrightarrow \vec{x}$ from S onto $R - \{\vec{0}\}$. The (open) minor arc $\widehat{xy}$ of a great circle with endpoints x and y is mapped onto $\vec{x} \otimes \vec{y}$. Let e be an "ideal element", introduced to correspond to $\vec{0}$ and let $S^* = S \cup \{e\}$. Then S^* can be converted into a join space isomorphic to $(R, \otimes)$, where the hyperproduct of two distinct nonopposite points x and y of S is $\widehat{xy}$.

3) Projective join spaces over a division ring

Let M be a left module over a division ring R. For $x \in M$ we denote by x^* the linear manifold of M determined by $x \in M$, that is $x^* = \{\lambda x \mid \lambda \in R,\ \lambda \neq 0\}$.

We define the following hyperoperation $\square$ on the family L of all linear manifolds of M.:

> $\forall (x^*, y^*) \in L^2$, $x^* \square\, y^*$ is the set of linear manifolds determined by the elements of $x^* + y^*$,

where "+" is the addition in M applied to subsets of M. Then $(L, \square\ )$ is a join space, called the *linear manifold space of* M (or a *projective join space over* R). We have: $\forall\, a^* \in L$, $a^* \square\, 0^* = a^*$ and $(0^* \in a^* \square\, x^* \Longleftrightarrow x^* = a^*)$.

Notice that if we define a *point* to be any element of $L - \{0^*\}$ and a *line* as any set of the following type $x^* \square\, y^* \cup \{x^*, y^*\}$ (where $x^* \neq y^*$), then the sets of points and lines form an analytic projective geometry over R. Moreover, all analytic projective geometries can be obtained by this construction.

Now, we present some important connections between *classical geometries* and *join spaces*, established by W. Prenowitz and then, by him and by J. Jantosciak.

§1. Descriptive geometries and join spaces

Essentially, a descriptive geometry is the linear geometry of a convex region.

The Euclidean, hyperbolic and other classic geometries are examples of descriptive geometry.

Descriptive geometries were studied by Coxeter, Pasch, Peano, Hilbert, Moore, Russell and their work culminated in the definitive treatment by Veblen.

1. Definition. A *descriptive geometry* is a pair (S, R), where S is a set of elements, called *points* and R is a ternary relation on S, called *betweenness*, satisfying the following conditions: For $(a, b) \in S^2$, $a \neq b$, the *line* ab is the set $\{x \in S \mid x = a$ or $x = b$ or $(x, a, b) \in R$ or $(a, x, b) \in R$ or $(a, b, x) \in R\}$.

P1) if $(a, b, c) \in R$, then a, b, c are distinct;

P2) if $(a, b, c) \in R$, then $(c, b, a) \in R$ and $(b, c, a) \notin R$;

P3) if $(a,b,c,d) \in S^4$, $a \neq b$, $c \neq d$ and $\{c,d\} \subset ab$, then $a \in cd$.

P4) if $(a,b) \in S^2$, $a \neq b$, there is $c \in S$, such that $(a,b,c) \in R$;

P5) there exist three points not in the same line;

P6) (*the Transversal Postulate*) if $(a,b,c) \in S^3$, $a \neq b \neq c \neq a$, $a \notin bc$ and if $(d,e) \in S^2$, such that $(b,c,d) \in R$ and $(c,e,a) \in R$, then there is $f \in de$, such that $(a,f,b) \in R$.

2. Definition. If $(a,b) \in S^2$, $a \neq b$, then the set $[a,b] = \{x \in S \mid (a,x,b) \in R\}$ is called by *segment* $[a,b]$.

The set $\{x \in S \mid (x,a,b) \in R\}$ is called a *ray* and it is said to *emanate* from a.

We characterize descriptive geometries in terms of join spaces.

We define on S the following hyperoperation $\forall (x,y) \in S^2$, $x \neq y$, we have $x \circ y = \{t \mid (x,t,y) \in R\}$ and $x \circ x = \{x\}$.

We obtain that $(S, \circ)$ is a join space, called the *descriptive join space* or the *associated join space* of the descriptive geometry (S,R).

Indeed, the associativity of "$\circ$" is essentially an algebraic restatement of the Transversal Postulate P6); however, it has greater deductive power, since no restriction on a,b,c is assumed.

From P4), it results that $\forall (a,b) \in S^2$, $a \neq b$, we have $a/b \neq \emptyset$. We call a/b the *extension of a from b*.

Notice that $a/a = \{a\}$ and $a \circ b = b \circ a$ for any $(a,b) \in S^2$.

The implication $a/b \cap c/d \neq \emptyset \Longrightarrow a \circ d \cap b \circ c \neq \emptyset$ is in essence a reformulation of the Transversal Postulate P6), of Peano, which may be stated in its conventional form: "Segments which join two vertices of a triangle to respective points of their opposite sides intersect".

Notice that the line ab is the set $a \circ b \cup a/b \cup b/a \cup \{a,b\}$.

Now, let us consider a join space$(J, \circ)$, for which

$$(\tau) \qquad \forall a \in J,\ a \circ a = a/a = \{a\}.$$

Define on J the following ternary relation:

$$(a, b, c) \in R \Longleftrightarrow a \neq c \text{ and } b \in a \circ c.$$

3. Theorem. *If* $(a, b, c) \in R$, *then* $a \circ b \cap b \circ c = \emptyset$.

Proof. Supose to the contrary that there is $(a, b, c) \in R$, such that $a \circ b \cap b \circ c \neq \emptyset$. Since $(a, b, c) \in R$, we get $\beta \in a \circ c$. We obtain $a \in (b \circ c)/b$ and $a \in b/c$. So $(b \circ c)/b \cap b/c \neq \emptyset$, whence $b \circ c = \{b\}$. Hence $c \in b/b = \{b\}$, that means $b = c$. Since $a \in (b \circ c)/b$, it results $a = b = c$, contrary to hypothesis. Therefore $a \circ b \cap b \circ c = \emptyset$. ■

4. Corollary. *If* $(a, b, c) \in R$, *then* a, b, c *are distinct.*

5. Theorem. *If* $(a, b, c) \in R$ *and* $(b, c, d) \in R$, *then* $(a, b, d) \in R$ *and* $(a, c, d) \in R$.

Proof. We have $b \in a \circ c$ and $c \in b \circ d$, whence $b/a \cap b \circ d \neq \emptyset$, so $b \in a \circ b \circ d$. Thus $\{b\} = b/b \cap a \circ d \neq \emptyset$, that is $b \in a \circ d$. If $a = d$, then $b = a = d$, a contradiction to $(b, c, d) \in R$. Then $a \neq d$ and since $b \in a \circ d$, we get $(a, b, d) \in R$.

Similarly, we obtain $(a, c, d) \in R$. ■

In a similar way, we can prove the following results:

6. Theorem. *If* $(a, b, c) \in R$ *and* $(a, c, d) \in R$, *then* $(a, b, d) \in R$ *and* $(b, c, d) \in R$.

7. Theorem. *If* $(a, b, x) \in R$ *and* $(a, b, y) \in R$, *then* $(x, a, y) \notin R$ *and* $(x, b, y) \notin R$.

Similarly, if $(a, x, b) \in R$ *and* $(a, y, b) \in R$, *then* $(x, a, y) \notin R$ *and* $(x, b, y) \notin R$.

The following theorem establishes a first connection between the conditions of a join space and the postulates of a descriptive geometry.

8. Theorem. *The ternary relation* R *on* J *satisfies postulates* P1), P2), P4), P6).

Proof. P1) is a consequence of Theorem 3.

By Theorem 5, $(a,b,c)\in R$ and $(b,c,a)\in R$ imply $(a,b,a)\in R$, a contradiction with corollary 4. Therefore, we have P2).

P4) is essentially a restatement of the fact that $\forall (a,b) \in J^2$, there is $x \in J$, such that $a \in b \circ x$.

Now, let us verify P6). Suppose a, b, c are distinct and such that a does not belong to the line bc and $(b,c,d) \in R$, $(c,e,a) \in R$, that is $c \in b \circ d$ and $e \in c \circ a$. From here, we obtain $a \circ b \cap e/d \neq \emptyset$.

Let $f \in a \circ b \cap e/d$. Then $(a,f,b) \in R$. If $d \neq e$, then f belongs to the line de.

Suppose $d{=}e$. By Theorem 5, we have $(b,c,d)\in R$, $(c,e,a)\in R$, whence $(b,c,a)\in R$, that means a belongs to the line bc, contrary to the hypothesis. Therefore $d \neq e$ and so, we obtain P6). ∎

9. Remark. The direct sum of two join spaces is a join space.

10. Theorem. *The postulate* P3) *is independent of the conditions of a join space definition.*

Proof. Let us define the following hyperoperation on $\mathbb{R}$: $\forall (a,b) \in \mathbb{R}^2$, $a \circ a = \{a\}$ and $a \circ b$ is the set of all real numbers between a and b.

Let $J = \mathbb{R} \times \mathbb{R}$. The element (x_1, x_2) of the cartesian plane J will be denoted by x.

Choose elements a, b, c, d in J, such that $\forall i \in \{1,2\}$, $a_i<b_i<c_i$, $c_1 = d_1$ and $c_2 < d_2$.

The line ab is composed of a, b, all the points which are above and on the right of b, all points which are below and on the left of a and all points which are simultaneously above and on the right of a and below and on the left of b, that is

$$ab = \{x = (x_1,x_2) \in J \mid [x_1 = a_1 \text{ and } x_2 = a_2] \text{ or }$$
$$[x_1 = b_1 \text{ and } x_2 = b_2] \text{ or } [b_1 < x_1 \text{ and } b_2 < x_2] \text{ or }$$
$$[a_1 < x_1 < b_1 \text{ and } a_1 < x_2 < b_2] \text{ or } [x_1 < a_1 \text{ and } x_2 < a_2]\}.$$

The line cd is the ordinary Euclidean vertical line cd, that is $cd =$ $= \{x = (x_1,x_2) \in J \mid c_1 = d_1 = x_1\}$.

We have $\{c, d\} \subset ab$, where $c \neq d$, but $a \notin cd$.

Hence P3) is not verified in J, which means postulate P3) is independent of the conditions of join space definition. ■

Some notions we shall use in the following:

Let $(J, \circ)$ be a join space, satisfying (τ) .

For $S \subset J$, then we denote by $< S >$ the least closed subhypergroup of $(J, \circ)$, which contains S and we call it the *closed subhypergroup generated by S*.

We say that S is a set of *generators* of $< S >$.

If A and B are closed subhypergroups of $(J, \circ)$ and B is a maximal proper subset of A, then we say that *A covers B*.

If $S \subset J$ and $\forall (a, b) \in S^2$, we have $a \circ b \subset S$, we say that S is *closed under "∘"* or, in geometrical language, S is *convex*.

11. Proposition. *If S_1 and S_2 are convex, then also $S_1 \cap S_2$, $S_1 \circ S_2$ and S_1/S_2 are convex.*

Proof. We have

$$(S_1 \circ S_2) \circ (S_1 \circ S_2) = (S_1 \circ S_1) \circ (S_2 \circ S_2) \subset S_1 \circ S_2 \text{ and}$$

$$\begin{aligned}(S_1/S_2)\circ(S_1/S_2) &\subset ((S_1/S_2)\circ S_1)/S_2 = \\ &= (S_1\circ(S_1/S_2))/S_2 \subset ((S_1\circ S_1)/S_2)/S_2 \subset (S_1/S_2)/S_2 = \\ &= S_1/(S_2\circ S_2) \subset S_1/S_2 \text{ (by Theorem 64, 5),4),3), p.12).}\end{aligned}$$

■

Now, let N be a closed subhypergroup of $(J, \circ)$ and $a \in J$. Then $N/(N/a)$, denoted $(a)_N$, is called the *coset of N determined by a*.

12. Theorem. *Let N be a closed subhypergroup of $(J, \circ)$. Then the cosets of N are closed under "∘", are mutually disjoint and cover J.*

Proof. By Proposition 11, $\forall a \in J$, $(a)_N$ is closed under "∘". We also have $a \in (a)_N$, since $\forall (a, b) \in J^2$, $b \in a/(a/b)$. We have to show only that the cosets are disjoint, that is if $a \in (b)_N$, then $(a)_N = (b)_N$. Since $a \in N/(N/b)$, it follows

$$N/a \subset N/(N/(N/b)) \subset (N \circ (N/b))/N \subset (N/b)/N =$$
$$= N/(b \circ N) = (N/N)/b = N/b, \text{ see Theorem 64, 3),4),5), p.12).}$$

On the other hand, from $a \in N/(N/b)$ it follows $b \in N/(N/a)$, so that $N/b \subset N/a$, by the above argument. Therefore $N/a = N/b$, whence $(a)_N = (b)_N$. ■

We shall denote by $J/\!/N$ the set of all cosets of N determined by elements of J. Define on $J/\!/N$ the hyperoperation:

$$(a)_N \otimes (b)_N = \{(x)_N \mid x \in a \circ b\}.$$

13. Remarks.

1. The hyperproduct "$\otimes$" of cosets in $J/\!/N$ is independent of the elements of J, which determine the cosets.

2. $J/\!/N$ has a unique identity element, namely N and if $n \in N$, then $(n)_N = N$. Indeed, if $n \in N$, then $(n)_N = N/(N/n) = N$.

14. Proposition. *For each element A of $J/\!/N$, there exists a unique element X such that $N \in A \otimes X$.*

Proof. Suppose $N \in A \otimes X$, where $A = (a)_N$ and $X = (x)_N$. Then $(a)_N \otimes (x)_N = \{(t)_N \mid t \in a \circ x\} \ni N$, that means there is $n \in N$, such that $n \in a \circ x$. Hence $a \in N/x$ so that $X = N/(N/x) \supset N/a$.

Since the cosets of N are disjoint, there exists at most one X such that $N \in A \otimes X$.

On the other hand, if we choose $x \in N/a$ then $X = (x)_N$ satisfies $N \in A \otimes X$. ■

X will be called the *inverse* of A and will be denoted by A'.

15. Corollaries.

1. *We have* $\forall A \in J/\!/N,\ (A')' = A$;
2. $(a)'_N = (a')_N$ *if and only if* $a \circ a' \cap N \neq \emptyset$ (*i.e.* $a' \in N/a$).

The *order* of $J/\!/N$ is the cardinality of the set $J/\!/N$.

16. Remark. If we restrict J to be an Euclidean space and N a point, then $J/\!/N$ is essentially the set of rays issuing from N.

Projecting the rays of $J/\!/N$ onto a hypersphere centered at N, we see that $J/\!/N$ is essentially a *spherical space*; we define the minor arc of a great circle joining two points as their "hyperproduct".

17. Definition. If A and B are closed subhypergroups of $(J, \circ)$ such that $B \subset A$ and the order of $A/\!/B$ is 3, then we say that B *separates* A.

Now, we introduce three new postulates, for a join space $(J, \circ)$ (in which condition (τ) holds), necessary to characterize a descriptive geometry.

J1) If $(a, b) \in J^2,\ a \neq b$, then $< a, b >$ covers a.

This is a consequence of the postulate "two points belong to a unique line".

18. Remarks.

1) A join space satisfying J1) is an exchange space.
2) J1) is independent of conditions of join space definition and of condition (τ).

Indeed, it is sufficient to consider $J = \mathbb{R}\times\mathbb{R}$, $\forall (a, b) \in \mathbb{R}^2$, $a\circ a=\{a\}$ and $a\circ b$ is the set of all real numbers between a and b.

As in Theorem 10, we consider a, b, c, d in J, such that $\forall i \in \{1, 2\}$, $a_i < b_i < c_i$, $c_1 = d_1$ and $c_2 < d_2$, where $x = (x_1, x_2)$, $\forall x \in J$.

We have $< c, b >= J$, $< c, d >$ is represented by the vertical "line" cd. Therefore $c \in < c, d > \subset < c, b >$ and $< c, d > \neq < c, b >$. Thus c makes J1) invalid in J, but all the conditions of a join space definition as well as (τ) are satisfied ■

19. Definition. A subset B of a closed subhypergroup A of an exchange space J is called a *basis* of A if it is independent and furthermore $< B > = A$.

Any closed subhypergroup of an exchange space has a basis.

Any two bases of A have the same cardinal number called the *dimension* of A, denoted by $d(A)$.

If B is another closed subhypergroup of $(J, \circ)$, such that $B \subset A$, then $d(B) \leq d(A)$.

If A and B are finite dimensional closed subhypergroups of $(J, \circ)$, such that $A \cap B \neq \emptyset$ then the *dimensional equality* holds:

$$d(< A, B >) + d(A \cap B) = d(A) + d(B).$$

If A covers B, then $d(A) = d(B) + 1$.

If $d(A) = n$ is finite, any independent set of n elements of A is a basis of A.

The following postulate J2) establishes that $(J, \circ)$ contains two closed subhypergroups A, B such that B separates A.

We may restate it, as follows:

J2) There exist A and B, closed subhypergroups of $(J, \circ)$ such that $B \subset A$ and $A /\!/ B$ has order 3.

J2) is verified in a descriptive geometry, since we can take A to be any line and B one of its points.

In order to avoid introducing the hypothesis $d(J) > 2$ in the following theorems, we postulate

J3) $d(J) > 2$

meaning J contains a set of three independent elements.

20. Theorem. *Let N be a closed subhypergroup of $(J,\circ)$ and $(a,b,a',b') \in J^4$, such that $a\circ a' \cap N \neq \emptyset \neq b\circ b' \cap N$. Then $< a,b,N > = N/(a\circ b) \cup N/(a\circ b') \cup N/(a'\circ b) \cup N/(a'\circ b') \cup N/a \cup \cup N/b \cup N/a' \cup N/b' \cup N$.*

Proof. First of all, notice that if A and B are closed subhypergroups of $(J,\circ)$ such that $A \cap B \neq \emptyset$ then $< A,B >= A/B$ (by Theorem 74, p.14). We have

$$\begin{aligned} < a,b,N > &= << a,N >, < b,N >> = < a,N > / < b,N > = \\ &= (N \cup N/a \cup N/a')/(N \cup N/b \cup N/b') = \\ &\text{(see [312], Theorem 10.6, Corollary 2)} \\ &= N/N \cup N/(N/b) \cup N/(N/b') \cup (N/a)/N \cup \\ &\cup (N/a)/(N/b) \cup (N/a)/(N/b') \cup (N/a')/N \cup \\ &\cup (N/a')/(N/b) \cup (N/a')/(N/b'). \end{aligned}$$

Notice that $(N/x)/(N/y) = (N/(N/y))/x = (N/y')/x = N/(x\circ y')$ (by Theorem 69, p.13). On the other hand, $N/N = N$, so we obtain the desired result. ∎

21. Corollary. *If N is a closed subhypergroup of $(J,\circ)$, then $< a,b,N > /\!\!/N = (a)_N \otimes (b_N) \cup (a)_N/(b)_N \cup (b)_N/(a)_N \cup ((a)_N \otimes \otimes (b)_N)' \cup (a)_N \cup (b)_N \cup (a)'_N \cup (b)'_N \cup N$.*

22. Theorem. *$\forall (a,b) \in J^2$, we have*

$$< a,b >= a\circ b \cup a/b \cup b/a \cup \{a,b\}.$$

Proof. The result is trivial if $a = b$ and thus suppose $a \neq b$.

By J3), there is an element x such that a,b,x are distinct and form an independent set.

According to Corollary 21, if $t \in < a,b >$ there are at most nine sets into which $(t)_x$ can fall. We shall consider these possibilities:

1) Suppose $(t)_x \in (a)_x \otimes (b)_x$. Then there is $t' \in a \circ b$, such that $(t)_x = (t')_x$. We show $t = t'$. We have $t' \in < x, t >$ so that $\{t, t'\} \subset < a, b > \cap < x, t >$. On the other hand, by the dimensional equality, we have $d(< a, b > \cap < x, t >) = d(< a, b >) + d(< x, t >) - d(<< a, b >, < x, t >>) = 2 + 2 - 3 = 1$, since $<< a, b >, < x, t >> = < a, b, x >$. Hence $< a, b > \cap < x, t >$ consists of a single element, so that $t = t'$ and $t \in a \circ b$.

2) Now, consider the possibility $(t)_x \in (a)_x / (b)_x$.

We have $(a)_x \in (t)_x \otimes (b)_x$, so that there is $a' \in b \circ t$, such that $(a)_x = (a')_x$. If $t = b$, then $a' = b$ and $(a)_x = (b)_x$, contrary to the independence of a, b, x. Hence $t \neq b$. By J1), from $< a, b > \supset < b, t > \ni b$, we obtain $< b, t > = < a, b >$, whence $a \in < a, x > \cap < b, t >$.

On the other hand, $a' \in < a, x > \cap < b, t >$. Applying the dimensional equality, as above, we obtain $a = a'$. Therefore, $a \in b \circ t$, so $t \in a/b$. Similarly, $(t)_x \in (b)_x/(a)_x$ implies $t \in b/a$. If $(t)_x = (a)_x$ we obtain $t = a$ and similarly, $(t)_x = (b)_x$ implies $t = b$.

3) Now, suppose $(t)_x \subset ((a)_x \otimes (b)_x)' = (a \circ b)_x$. Then there exists $\tilde{t} \in a \circ b$, such that $(t)_x = (\tilde{t})'_x$. From here it follows $t \circ \tilde{t} \ni x$. Since $\{t, \tilde{t}\} \subset < a, b >$, it follows $x \in t \circ \tilde{t} \subset < a, b >$, which is impossible since a, b, x are distinct and form an independent set.

In a similar way, we can show that the other three possibilities for $(t)_x$ vacuous. Therefore $t \in a \circ b \cup a/b \cup b/a \cup \{a, b\}$ and $< a, b > = a \circ b \cup a/b \cup b/a \cup \{a, b\}$. ∎

23. Remark. Postulate J3) is essential for the validity of the last theorem. We can show this, constructing the following example:

Let $J = ax \cup ay \cup az$, where ax, ay, az denote pairwise disjoint open intervals.

On J, we define the hyperoperation:

$$\forall (u, v) \in J^2, u \neq v, \text{ we have } u \circ u = u; \; u \circ v = au \cup av,$$

where au and av are open segments.

Then $(J, \circ)$ is a join space, in which (τ) holds; moreover, J1) and J2) hold.

Let $c \in ax$ and $b \in ay$. Then J3) is invalid since $< c, b >= J$ and $d(J) = 2$. The last theorem fails, because

$$c \circ b \cup c/b \cup b/c \cup \{c, b\} = ax \cup ay \neq < c, b > .$$

The following theorem is a characterization of descriptive geometries.

24. Theorem. *Descriptive geometries are characterized as join spaces satisfying* J1), J2), J3) *and* (τ).

Proof. Let (S, R) be a descriptive geometry. $\forall (x, y) \in S^2,\ x \neq y$, we define $x \circ x = \{x\}$ and $x \circ y = \{t \mid (x, t, y) \in R\}$. Then $(S, \circ)$ is a join space, which verifies J1), J2), J3) and (τ), as we have seen before.

Conversely, if $(J, \circ)$ is a join space satisfying(τ) then P1), P2), P4) and P6) hold, as we have seen before.

Recall that $(a, b, c) \in R \Longleftrightarrow a \neq c$ and $b \in a \circ c$.

We have to show that J1), J2) and J3) imply P3) and P5).

The line ab (where $a \neq b$) is the set $a \circ b \cup a/b \cup b/a \cup \{a, b\}$ and we have proved before that the line ab coincides with $< a, b >$.

We verify that: if $a \neq b$ and $\{c, d\} \subset < a, b >$, where $c \neq d$, then $< a, b >=< c, d >$. This follows from J1): for $c \neq a$ or b; suppose $c \neq a$. Then $< a, b > \supset < a, c > \ni a$ and by J1), it follows $< a, b > = \, = < a, c >$. Thus $d \in < a, c >$. Since $c \neq d$, we can show similarly $< a, c > = < c, d >$, so that $< a, b > = < c, d >$. This obviously implies P5).

We have to verify now only P3). By J3) there exist distinct elements a, b, c of J, which form an independent set. Suppose a, b, c are contained in a line, say the line $pq = < p, q >$. But then $< a, b > = < p, q > \ni c$, contrary to the independence of a, b, c. Therefore, a, b, c are not in the same line and so, P5) is verified. ■

§2. Spherical geometries and join spaces

25. Definition. An (abstract) *spherical geometry* is a system (S, R), where S is a set of elements called *points* and R is a ternary relation on S called *betweenness*, which satisfies the following postulates:

(i) if $(x, y, z) \in R$, then x, y, z are distinct;

(ii) if $(x, y, z) \in R$, then $(z, y, x) \in R$;

(iii) for any x, there exists a unique x' (called the *opposite* of x) such that $x' \neq x$ and the following implication holds:

$$(x, u, v) \in R \Longrightarrow (u, v, x') \in R;$$

(iv) if $y \neq x$ and $y \neq x'$, then there exists u such that $(x, u, y) \in R$;

(v) if "$\circ$" is defined in $(*)$ then $(x \circ y) \circ z = x \circ (y \circ z)$, whenever both members are defined.

26. Examples of spherical geometries:

1. Let S be an Euclidean n-sphere and R be defined as follows:

$$(x, y, z) \in R \Longleftrightarrow \begin{cases} x, z \text{ are distinct and nonopposite and} \\ y \text{ is an interior point of the minor arc} \\ \text{of a great circle with joins } x \text{ and } z. \end{cases}$$

Then (S, R) is a spherical geometry, called the *Euclidean* spherical geometry.

2. Let S be the set of rays emanating from a point of an ordered affine space of arbitrary (finite or infinite) dimension. We define

$$(x, y, z) \in R \Longleftrightarrow \begin{cases} x, z \text{ are distinct} \\ \text{the ray } y \text{ is interior to the angle} \\ \text{formed by the non–opposite rays } x \text{ and } z. \end{cases}$$

Then (S, R) is a spherical geometry (which includes the first, in the sense of isomorphism).

We can define on S the following partial hyperoperation: $\forall\, (x, y) \in S^2$, $y \neq x$, $y \neq x'$, we have

$$(*) \qquad x \circ y = \{t \mid (x, t, y) \ni R\}, \quad x \circ x = \{x\}.$$

Except in the trivial cases, it is impossible to extend this partial hyperoperation to an semihypergroup on S:

27. Theorem. *The partial hyperoperation* $(*)$ *for a spherical geometry on at least three points does not extend to a semihypergroup.*

Proof. Suppose such an extension of "$\circ$" possible in S.

I) First of all, we shall check the following equality:

$$\forall (x,y) \in S^2,\ y \neq x,\ y \neq x',$$
$$x\circ(x'\circ y) = x\circ y \cup \{y\} \cup x'\circ y.$$

Notice that the hyperproduct $x\circ(x'\circ y)$ can be considered. Indeed, since $y \neq x'$, if we suppose $x' \in x'\circ y$, then we have $(x', x', y) \in R$, a contradiction. Thus, $x' \notin x'\circ y$. Moreover, $x \notin x'\circ y$, otherwise $(x', x, y) \in R$, whence $(x, y, x) \in R$, a contradiction.

Therefore, we can consider the hyperproduct $x\circ(x'\circ y)$. Now, we verify:

$$u \in x\circ(x'\circ y) \Longleftrightarrow u \in x\circ y \cup \{y\} \cup x'\circ y.$$

"$\Longrightarrow$" There exists $v \in x'\circ y$, such that $u \in x\circ v$. Hence $(x', v, y) \in R$, whence $(v, y, x) \in R$ and so $(x, y, v) \in R$ and $(y, v, x') \in R$. It follows $x \neq v$ and $x' \neq v$. On the other hand, from $u \in x\circ v$ it follows $(x, u, v) \in R$. If $u = y$ we have $u = y \in x\circ y \cup \{y\} \cup x'\circ y$. Suppose $u \neq y$. Notice that if $(a, t, b) \in R$ and $(a, s, b) \in R$, $t \neq s$, then it can be easily verified that $(a, t, s) \in R$ or $(a, s, t) \in R$. Therefore, from $(x, u, v) \in R$, $(x, y, v) \in R$ and $u \neq y$ it results $(x, u, y) \in R$ or $(x, y, u) \in R$.

If $(x, u, y) \in R$, then $u \in x\circ y$.

If $(x, y, u) \in R$, then $(y, u, x') \in R$, whence $(x', u, y) \in R$ and so $u \in x'\circ y$.

Therefore, $u \in x\circ y \cup \{y\} \cup x'\circ y$.

"$\Longleftarrow$" Suppose $u \in x\circ y$. Then $(x, u, y) \in R$. Since $y \neq x$ and $y \neq x'$, there exists z, such that $(x', z, y) \in R$. Hence $(z, y, x) \in R$ and so $(x, y, z) \in R$. So we have $x' \neq z$ and $x \neq z$.

Using the associative law, the following implication can be obtained:

$$(a,b,c) \in R \text{ and } (a,c,d) \in R \Longrightarrow (a,b,d) \in R.$$

Hence, from $(x,u,y) \in R$, $(x,y,z) \in R$ it follows $(x,u,z) \in R$. Thus $u \in x{\circ}z$. From $(x',z,y) \in R$ it follows $z \in x'{\circ}y$. Therefore $u \in x{\circ}(x'{\circ}y)$.

Now, suppose $u = y$. Then we obtain $u \in x{\circ}z$ and $z \in x'{\circ}y$, with the same choice of z and so $u \in x{\circ}(x'{\circ}y)$.

Finally, suppose $u \in x'{\circ}y$. Then $(x',u,y) \in R$ and we have $u \neq x$, $u \neq x'$.

Now choose z such that $(x',z,u) \in R$. Then $(z,u,x) \in R$ and $(x,u,z) \in R$. From $(x',z,u) \in R$ and $(x',u,y) \in R$, we obtain $(x',z,y) \in R$, that means $z \in x'{\circ}y$. On the other hand, $u \in x{\circ}z$ and so $u \in x{\circ}(x'{\circ}z)$. Therefore $\forall\,(x,y) \in S^2$, $x \neq y \neq x'$, we have $x{\circ}(x'{\circ}y) = x{\circ}y \cup \{y\} \cup x'{\circ}y$.

II) Suppose $(x,p) \in S^2$, $x \neq p \neq x'$. From I) and the associative law we obtain $(x{\circ}x'){\circ}p = x{\circ}p \cup \{p\} \cup x'{\circ}p$. So, $p \in (x{\circ}x'){\circ}p$, that means there is $s \in x{\circ}x'$, such that $p \in s{\circ}p$. If $p \neq s$ and $p \neq s'$ then $p \in s{\circ}p$ implies $(s,p,p) \in R$, a contradiction.

Therefore $p = s$ or $p = s'$. It follows $p \in x{\circ}x'$ or $p' \in x{\circ}x'$. It is not restrictive to suppose $p \in x{\circ}x'$. Since $x \neq p \neq x'$, there exists $q \in S$, such that $(p',q,x) \in R$. Hence $(q,x,p) \in R$ and so $(p,x,q) \in R$, whence $x \in p{\circ}q$. Since $p \in x{\circ}x'$, we obtain $x \in (x{\circ}x'){\circ}q$. But $q \neq x$ and $q \neq x'$, and since $(x{\circ}x'){\circ}q = x{\circ}q \cup \{q\} \cup x'{\circ}q$, we obtain $x \in x{\circ}q \cup \{q\} \cup x'{\circ}q$. All the possibilities $x \in x{\circ}q$ (that is $(x,x,q) \in R$), $x = q$, $x \in x'{\circ}q$ (that is $(x',x,q) \in R$, whence $(x,q,x) \in R$) are false, and so the proof is complete. ∎

However, we can enlarge S by the adjunction of an "ideal point", which will play the role of an identity. In this manner we obtain a join space associated with the given spherical geometry.

Let $e \notin S$ and let $S' = S \cup \{e\}$. We extend the hyperoperation "$\circ$" as follows:

$$(**) \qquad \begin{cases} \forall x \in S, \ x \circ x' = \{x, x', e\}; \\ \forall y \in S', \ y \circ e = e \circ y = y. \end{cases}$$

Thus, we obtain a join space $(S', \circ)$ with identity "e".

Remark that the associative law holds for $(S', \circ)$.

Now, let us check the implication:

$$a/b \cap c/d \neq \emptyset \Longrightarrow a \circ d \cap b \circ c \neq \emptyset.$$

Notice that $\forall a \in S'$, a has a unique inverse a' and $\forall (a, b) \in S'^2$, $a/b = a \circ b'$. So, $a/b \cap c/d \neq \emptyset$ implies $a \circ b' \cap c \circ d' \neq \emptyset$, whence $\{a\} \cap b \circ (c \circ d') \neq \emptyset$, hence $a/d' \cap b \circ c \neq \emptyset$, that is $a \circ d \cap b \circ c \neq \emptyset$ (by Theorem 64, 2), p.12).

Therefore, $(S', \circ)$ is a join space, called the *associated join space* or a *spherical join space* of the spherical geometry (S, R). ■

28. Theorem. *A join space $(J, \circ)$ is the associated join space of a spherical geometry if and only if $(J, \circ)$ has an identity and $\forall a \in J$, we have $a \circ a = a$ and $\forall x \in J$, x distinct from the identity, $< x >$ has cardinality* 3.

Denote by $< x >$ the least closed subhypergroup of $(J, \circ)$ which contains $x \in J$.

Proof. "$\Longrightarrow$ Let $(S', \circ)$ be the associated join space of a spherical geometry (S, R).

We have only to check that if $x \in S'$, $x \neq e$, then the order of $< x >$ is 3. Any closed subhypergroup of $(S', \circ)$ which contains x, must contain $\{x, x', e\} = X$. The set X is the least closed subhypergroup of S' which contains x. Moreover $x \neq x' \neq e \neq x$, hence $< x >$ has order 3.

"$\Longleftarrow$" Let e be the identity of J and let $S = J - \{e\}$. We define the following ternary relation on S:

$$(x, y, z) \in R \Longleftrightarrow y \in x \circ z, \ z \notin \{x, x'\}$$

(where x' is the unique inverse of x).

Now, we show that $\forall x \in S$, $x \circ x' = \{x, x', e\}$. Since $e \in \in x \circ x'$ it follows that $x \in x \circ (x \circ x') = (x \circ x) \circ x' = x \circ x'$. Similarly, $x' \in x \circ x'$. Moreover $x \neq e \neq x' \neq x$. Indeed, if $x = x'$, then $x \circ x' = x \circ x = \{x\}$ and so $\{x, e\}$ is the least closed subhypergroup of $(J, \circ)$, which contains x and so $< x >$ has order 2, contrary to hypothesis.

Therefore $\{x, x', e\} \subset x \circ x'$. But $\{x, x', e\}$ is the least closed subhypergroup which contains x, so $x \circ x' \subset \{x, x', e\}$. Hence $x \circ x' = = \{x, x', e\}$.

Now, we shall verify that $\forall x \in S$, $x \circ e = x$, that is the identity e is a scalar one. Indeed, we have $x \circ e \subset \{x, x', e\}$, which is a subhypergroup. Suppose to the contrary $e \in x \circ e$. Then $x' \in x \circ e$, otherwise $x \circ e = \{x, e\}$ and so $\operatorname{ord} x = 2$, a contradiction.

Hence, $x' \in x \circ e$, whence $x \in e/x' \cap x'/e$. Since $(J, \circ)$ is a join space, it follows that $e = x'$, which is false.

We shall prove that (S, R) is a spherical geometry.

(i) if $(x, y, z) \in R$, then we have $z \notin \{x, x'\}$ and $y \in x \circ z$. Suppose to the contrary $y = x$. Then $x \in x \circ z$. On the other hand, $(J, \circ)$ is join space with a scalar identity, so it is a canonical hypergroup. From $x \in x \circ z$, we obtain $z \in x \circ x' = \{x, x', e\}$, which is false. So, $y \neq x$. Similarly, we obtain $y \neq z$.

(ii) if $(x, y, z) \in R$ then $y \in x \circ z = z \circ x$, whence $(z, y, x) \in R$.

(iii) if $(x, u, v) \in R$, then $u \in x \circ v$, whence $v \in u \circ x'$, hence $(u, v, x') \in R$.

(vi) if $y \notin \{x, x'\}$, there is $u \in x \circ y$, and so $(x, u, y) \in R$.

Therefore, (S, R) is a spherical geometry. Moreover, $(J, \circ)$ is the associated join space. ■

§3. Projective geometries and join spaces

29. Definition. A *projective geometry* is a system (S, T) where S is a set of elements called *points*, and T is a set of subsets of S, called *lines*, which satisfies the following properties:

(i) any line contains at least three points;

(ii) any two distinct points a, b are contained in a unique line, denoted by $L(a, b)$;

(iii) if a, b, c, d are distinct and $L(a, b) \cap L(c, d) \neq \emptyset$, then $L(a, c) \cap L(b, d) \neq \emptyset$.

We have already mentioned the connections between the projective join spaces and the analytic projective geometries.

Now, we wish to associate a join space with a given projective geometry.

Remember that in a projective join space $(L, \square)$ over a division ring R, we have

$$\begin{array}{ll} \forall a \in L, & a \square a = 0 \quad \text{if } R = \mathbb{Z}_2, \\ a \square a = \{a, 0\}, & \text{otherwise.} \end{array}$$

Moreover, if $R = \mathbb{Z}_2$, each line of the analytic projective geometry has exactly three points, otherwise each line has more than three points.

Therefore, when an abstract projective geometry (S, T) contains a line, we are able to tell how the hyperoperation of a point to itself should be defined.

If we consider a projective geometry with one point and no lines, we are not able to discriminate between the two choices of the hyperproduct of the point to itself.

According to these considerations, we shall associate with a projective geometry (S, T), a join space $(S', \circ)$ as follows: let $S' = S \cup \{e\}$, where "e" is the ideal point, which plays the role of 0^* $(e \notin S)$.

Case I. $T \neq \emptyset$.

1. If $(x,y) \in S^2$, $x \neq y$, then $x \circ y = L(x,y) - \{x,y\}$.
2. Let $x \in S$. If some line of T contains exactly three points, then $x \circ x = \{e\}$, otherwise $x \circ x = \{x,e\}$.
3. If $a \in S'$, $e \circ a = a \circ e = a$.

Case II. $T = \emptyset$.

1. if $S = \{a\}$, then we can define two hyperoperations on S' and for each of these, "e" is an identity, so we have $a \circ a = \{e\}$, while for the other $a \circ a = \{a,e\}$.
2. if $S = \emptyset$, we define $e \circ e = \{e\}$.

30. Theorem. $(S', \circ)$ *is a join space.*

Proof. First of all, notice that:

(α) $e \in x \circ y \Longleftrightarrow x = y$;

(β) Let $(x,y,z) \in S^3$. Then x,y,z are distinct and collinear if and only if $z \in x \circ y$ and $x \neq y$.

Now, let us check that

(γ) $\forall (x,y) \in S'^2$, $x/y = x \circ y$.

1. Indeed, if $(x,y) \in S^2$, $x \neq y$ and $z \in x \circ y$ then $z \in S$, so, by (β), the points x,y,z are distinct and collinear, whence y,z,x are distinct and collinear, so that $x \in y \circ z$ and $z \in x/y$.

Conversely, if $z \in x/y$, then $x \in y \circ z$ and $z \in S$. If $y = z$ then $x \in y \circ y \subset \{y,e\}$, whence $x = y$, contradiction. Thus $y \neq z$ and the steps can be retraced to yield $z \in x \circ y$.

2. If $(x,y) \in S^2$, $x = y$. Suppose $x \circ x = e$. Then

$$x/x = \{t \mid x \in x \circ t\} = e = x \circ x.$$

Suppose $x \circ x = \{x,e\}$. Then

$$x/x = \{t \mid x \in x \circ t\} = \{x,e\} = x \circ x.$$

3. The remaining cases $x = e$, and $y = e$ are easily disposed of. So, $\forall\,(x, y) \in S'^2$, we have $x/y = x \circ y$.

We have to check now that the following implication holds in S' :

$(\mu)\;\; x/y \cap z/t \neq \emptyset \Longrightarrow x \circ t \cap y \circ z \neq \emptyset.$

Since $x/y \cap z/t \neq \emptyset$ it results that there is $u \in x \circ y \cap z \circ t$.

Case 1. If x, y, z, t are distinct in S and noncollinear, then

$$(L(x, y) - \{x, y\}) \cap (L(z, t) - \{z, t\}) \neq \emptyset \text{ and } L(x, y) \cap L(z, t) \neq \emptyset.$$

By the definition of a projective geometry it follows $L(x, t) \cap L(y, z) \neq \emptyset$, so

$$(x \circ t \cup \{x, t\}) \cap (y \circ z \cup \{y, z\}) \neq \emptyset.$$

Suppose $y \in x \circ t$. It results $t \in y/x = x \circ y$ so that $L(x, y) \cap L(z, t) \ni t$. But $u \in L(x, y) \cap L(z, t)$ and $u \neq t$. Therefore, $L(x, y) = L(z, t)$, contrary to hypothesis. Thus $y \notin x \circ t$. Similarly, $z \notin x \circ t$, $x \notin y \circ z$ and $t \notin y \circ z$. Hence the only possibility is $x \circ t \cap y \circ z \neq \emptyset$.

Case 2. If x, y, z, t are distinct in S and collinear, then by the definition of a projective geometry, we have $L(x, t) = L(x, y) = L(y, z)$. Hence, there is $u \in L(x, t) \cap L(y, z)$. Then $u \notin \{x, y, z, t\}$ implies $u \in x \circ t \cap y \circ z$.

Case 3. x, y, z, t are not distinct and in S. Since the proof is based on $x \circ y \cap z \circ t \neq \emptyset$, it suffices to consider the situations $x = y$, $x = z$ and $x = t$.

The result is immediate for $x = z$.

If $x = y$, then we have two possibilities:

i) The result is clear for $z = t$.

ii) Thus let $z \neq t$. We have $u \in x \circ y = x \circ x \subset \{x, e\}$. Moreover $u \neq e$, otherwise $z = t$. Thus $u = x$ and by the definition of the hyperproduct in S', every line of the projective geometry (S, T)

contains at least four points. Since $x \in z \circ t$, points x, z, t are distinct and collinear.

Let $v \in L(x,z)$, $v \notin \{x,z,t\}$. Then $v \in x \circ t \cap x \circ z = x \circ t \cap y \circ z$.

If $x = t$. We may assume $x \notin \{y,z\}$. Then $u \in L(x,y) \cap L(x,z)$. Hence $u \neq x$ yields $L(x,y) = L(x,z)$. If $y = z$ then $x \circ t \cap y \circ z = x \circ x \cap y \circ y \ni e$. Suppose $y \neq z$. Then x, y, z, u are distinct and collinear. By a well–known theorem of projective geometry, all lines of (S,T) have the same cardinality and so contain at least four points. By the definition of the hyperproduct in S', we have $x \circ x = \{x, e\}$. Hence $x \circ t \cap y \circ z = x \circ x \cap y \circ z \ni x$.

Case 4. One of x, y, z, t is e. Say $x = e$. Then $x \circ y \cap z \circ t \neq \emptyset$ yields $y \in z \circ t$, so that $x \circ t \cap y \circ z = \{t\} \cap y/z \neq \emptyset$. The other possibilities are treated similarly.

Now, let us verify the associativity. Suppose $w \in (x \circ y) \circ z$, where $(x,y,z) \in S'^3$. Then $w/z \cap y/x = w/z \cap x \circ y \neq \emptyset$, whence, by (μ) it follows $w \circ x \cap z \circ y \neq \emptyset$. Then $w/x \cap y \circ z \neq \emptyset$ and $w \in x \circ (y \circ z)$. Thus $(x \circ y) \circ z \subseteq x \circ (y \circ z)$.

The reverse inclusion can be verified similarly and since the commutativity holds, $(S', \circ)$ is a join space. ■

31. Remark. "e" is an identity of $(S', \circ)$ and $\forall a \in S$, we have $< a >= \{a, e\}$ since $\{a, e\} \subseteq < a >$ and $\{a, e\}$ is linear. Thus, $< a >$ has cardinality 2, for any $a \in S$.

$(S', \circ)$ is called the *associated join space* of the projective geometry (S,T) or a *projective join space.*

Except of the choice of "e", $(S', \circ)$ is unique, except when S consists of a single point. In this case, there are only two associated join spaces.

32. Proposition. *If $(J, \circ)$ is a join space with identity e, such that $< x >$ has cardinality 2, for any $x \in J - \{e\}$, then $(J, \circ)$ is an exchange space, for which the following properties hold, for any $(x,y) \in (J - \{e\})^2$:*

(i) $< x >= \{x, e\}$;

(ii) $x^{-1} = x$;

(iii) $e \in x \circ x \subset \{x, e\}$;

(iv) $< x, y > = x \circ y \cup \{x, y, e\}$.

Proof. (i) follows directly from the hypothesis, whence (ii) results and then (iii) is immediate from (i) and (ii).

(iv) We have $< x, y > = << x >, < y >> = < x > / < y > = \\ = < x > \circ < y > = \{x, e\} \circ \{y, e\} = x \circ y \cup \{x, y, e\}$.

Recall now that an exchange space is a join space which satisfies the following conditions:

I) if $x \in < y >$ and x is not an identity then $< x > = < y >$.

II) if $z \in < x, y >$ and $z \notin < y >$ then $< z, y > = < x, y >$.

Moreover, if the given join space has a scalar identity then I) and II) are equivalent (Theorem 75, p.14). So, it sufficies to verify I).

Let $u \in < x >$, $u \neq e$. By (i), we have $< x > = \{x, e\}$. Thus $u = x$ and $< u > = < x >$. ∎

33. Lemma. *Let* $(J, \circ)$ *be a join space with identity* e *and such that* $< s >$ *has cardinality* 2, *for any* $s \neq e$. *Suppose there exist* a, b *in* J, *such that* $a \circ b$ *is a singleton and* $e \notin \{a, b\}$. *Then any hyperproduct of elements of* J *is a singleton and* $(J, \circ)$ *is a commutative group.*

Proof. First of all, we prove that there is $x \in J - \{e\}$, such that $x \circ x = e$. Suppose this is false. Let $y \in J$. By the above Proposition,

$$e \in y \circ y \subset \{y, e\}, \text{ whence } y \circ y = \{y, e\}.$$

Let $w = a \circ b$. Then

$$\{w, e\} = w \circ w = a \circ b \circ a \circ b = a \circ a \circ b \circ b = \{a, e\} \circ \{b, e\} = \\ = \{ab, a, b, e\} = \{w, a, b, e\}.$$

Since $e \notin \{a, b\}$ we have $a = b = w$ and then $a = a \circ a = \{a, e\}$, a contradiction.

Therefore, there exists $x \in J - \{e\}$, such that $xox = e$.

We shall prove more, that $\forall r \in J$, we have $ror = e$.

Suppose on the contrary $ror \neq e$. We have $roxox = roe = r$. Let $t \in rox$. Then $tox \subset roxox = r$, hence $tox = r$. Since $ror \neq e$, it results

$$\{r, e\} = ror = toxotox = totoxox = tot \subset \{t, e\}.$$

Hence $r = t$ and $rox = r$, whence $x \in r/r = \{r, e\}$ so that $x = r$, contradiction with $ror \neq e$.

Therefore, $\forall r \in J$, $ror = e$.

Finally, we prove that $\forall (u, v) \in J^2$, uov is a singleton. Let $r_1 \in uov \ni r_2$. Then

$$r_1 o r_2 \subset uovouov = uouovov = eoe = e,$$

whence $r_1 = r_2$ as desired. ■

34. Theorem. *Let* $(J, \circ)$ *be a join space with identity e, such that* $< t >$ *has cardinality* 2 *for* $\forall t \in J - \{e\}$. *Then* $(J, \circ)$ *is the associated join space of a projective geometry.*

Proof. Let $S = J - \{e\}$. The element of S will be called the *points*. Let $L(x, y) = < x, y > -\{e\}$, if $(x, y) \in S^2$, $x \neq y$. We call $L(x, y)$ a *line* and we denote the set of all lines by T.

We prove that (S, T) is a projective geometry and $(J, \circ)$ is the associated join space.

First of all, we shall check that $\forall (x, y) \in S^2$, $x \neq y$, we have

$$L(x, y) = xoy \cup \{x, y\}.$$

By the above Proposition, we have

$$L(x, y) = < x, y > -\{e\} = (xoy \cup \{x, y, e\}) - \{e\} = xoy \cup \{x, y\}.$$

Now, we verify that any line is a set of points, which contains at least three points. Let $(x, y) \in S^2$, $x \neq y$. We have $L(x, y) \subset S$.

Suppose $x \in x \circ y$. Then $y \in x/x \subset \{x, e\}$, which is false. So, $x \notin x \circ y$ and similarly, $y \notin x \circ y$. Therefore, $L(x, y)$ contains at least three points.

We verify that any two distinct points are contained in a unique line. Indeed, if $(a, b) \in S^2$, $a \neq b$, then $a \in L(a, b) \ni b$. Suppose that $a \in L(x, y) \ni b$. Then $\{a, b\} \subset < x, y >$ and $\{a, b\}$ is independent.

By the Exchange Theorem, we have that $< a, b > = < x, y >$, so that $L(a, b) = L(x, y)$.

Finally, we verify that if a, b, c, d are distinct and $L(a, b) \cap \cap L(c, d) \neq \emptyset$, then $L(a, c) \cap L(b, d) \neq \emptyset$. From $L(a, b) \cap L(c, d) \neq \emptyset$, it follows $(a \circ b \cup \{a, b\}) \cap (c \circ d \cup \{c, d\}) \neq \emptyset$. Suppose that $a \circ b \cap c \circ d \neq \emptyset$. Then $a \circ b^{-1} \cap d \circ c^{-1} \neq \emptyset$ and $a/b \cap d/c \neq \emptyset$. Hence $a \circ c \cap b \circ d \neq \emptyset$ and so $L(a, c) \cap L(b, d) \neq \emptyset$.

Now, suppose that $c \in a \circ b$. Then $b \in c/a = a \circ c$, whence $L(a, c) \cap L(b, d) \neq \emptyset$.

The remaining cases are symmetrical to $c \in a \circ c$.

Therefore (S, T) is a projective geometry.

The next step is to verify that $(J, \circ)$ is the associated join space of (S, T). Consider $S' = S \cup \{e\} = J$. If $J = \{e\}$ then $S = \emptyset$ and $T = \emptyset$ and $(J, \circ)$ is an associated join space of (S, T) by definition. If $J = \{u, e\}$, $u \neq e$, then $S = \{u\}$, $T = \emptyset$. We have $e \circ u = u \circ e = u$, $e \circ e = e$ and $e \in u \circ u$. Hence $u \circ u = e$ or $u \circ u = \{u, e\}$ and in both cases $(J, \circ)$ is an associated join space of (S, T).

Suppose now that J has at least 3 elements, that is S has at least two elements and $T \neq \emptyset$.

The associated join space $(S', \square)$ of (S, T) is in this case defined as follows:

for $(x, y) \in S^2,\ x \neq y,\ x \square y = L(x, y) - \{x, y\}$,
if $x \in S$ and if some line of T contains exactly three points, then $x \square x = e$; otherwise, $x \square x = \{x, e\}$;
if $x \in S'$, $e \square x = x \square e = x$.

Now, we have only to verify that

$$(S', \square) = (S', \circ).$$

For $\forall x \in S'$, $x \circ e = x \,\Box\, e$. For all $x \in S'$, $x \circ e = x \,\Box\, e$. If $(x, y) \in S^2$, $x \neq y$, we obtain as above that $\{x, y\} \cap x \circ y = \emptyset$. Then we have $x \,\Box\, y = L(x, y) - \{x, y\} = x \circ y$.

Finally, consider $x \in S$. Suppose that some line $L(u, v)$ of T contains exactly three points, so that $x \,\Box\, x = e$. Since $L(u, v) = = u \circ v \cup \{u, v\}$, we see that $u \circ v$ is a singleton. By the Lemma, it results that $x \circ x$ is a singleton. Hence $x \circ x = e$ and $x \,\Box\, x = x \circ x$ and the theorem is proved. ■

Now, we can characterize projective geometries in terms of join spaces as follows:

> "A join space $(J, \circ)$ is the associated join space of a projective geometry if and only if it has an identity e and $\forall x \in J - \{e\}$, $< x >$ has cardinality 2."

§4. Multivalued loops and projective geometries

In this paragraph, we prove that the associated join space of a finite projective geometry (S, T) with N points on each line $(N \geq 3)$ is isomorphic to a quotient of an ordinary loop modulo a special equivalence relation. The following results have been obtained by St. Comer.

Let $(A, \cdot, e)$ be a loop and ρ an equivalence relation on A. If $\widehat{x}, \widehat{y}$ and $\widehat{z}$ are equivalence classes, and $\widehat{z} \subseteq \widehat{x} \cdot \widehat{y}$, we say that $\widehat{z}$ is *$(\widehat{x}, \widehat{y})$–projective* if $\forall u \in \widehat{x}$, $\exists v \in \widehat{y}$ such that $u \cdot v \in \widehat{z}$ and $\forall v_1 \in \widehat{y}$, $\exists u_1 \in \widehat{x}$, such that $u_1 \cdot v_1 \in \widehat{z}$. We say that the equivalence relation ρ is *special* if $\{e\}$ is an equivalence class and every product $\widehat{x} \cdot \widehat{y}$ of equivalence classes is a union of $(\widehat{x}, \widehat{y})$–projective equivalence classes.

A quasihypergroup $(B, \cdot, e)$ (where $e \in B$) is called a *multivalued loop* if e is a scalar identity of B and $\forall a \in B$, there exist unique x, y in B such that $e \in ax \cap ya$.

It is easy to verify that a quotient of a loop $(A, \cdot, e)$ modulo a special equivalence relation ρ is a multivalued loop $(A/\rho, *\{e\})$, where $\widehat{z} \in \widehat{x} * \widehat{y}$ if and only if $\widehat{z} \subseteq \widehat{x} \cdot \widehat{y}$. Note that not every multivalued loop is isomorphic to a quotient of a loop modulo a special equivalence relation (see [43, Prop. 2]).

The construction of the corresponding loop and of a special equivalence relation

Let $(S', \circ)$ be an associated join space of the finite projective geometry (S, T) with N points on each line ($N \geq 3$). We shall construct a loop $(L, \cdot, q)$ and a special equivalence relation ρ on L, such that $(S', \circ)$ is isomorphic to $(L/\rho, *\{q\})$.

Case I: First, we consider a finite projective geometry (S, T), where each line contains N points, $N \geq 4$. Let us denote the points of S by $p_1, p_2, \ldots$ For each p_i, we choose a set A_i with exactly $N - 2$ elements, such that $\forall i \neq j$, $A_i \cap A_j = \emptyset$.

Let q be an element, such that $q \notin \bigcup\limits_i A_i$. Set $L = \{q\} \cup \bigcup\limits_i A_i$ and let ρ be the equivalence relation on L, for which $\{q\}$ and the A_i are the equivalence classes.

If $p_i \neq p_j$, then let $L(p_i, p_j) = \{p_{k_1}, \ldots, p_{k_N}\}$, where $k_1 < k_2 < \cdots < k_N$. Let $L(p_i, p_j)^*$ be obtained from $L(p_i, p_j)$ by permuting $(p_{k_1}, \ldots, p_{k_N})$ cyclically to start with p_i and then deleting p_i and p_j.

Let $L(p_i, p_j)^* = (p_{s^{ij}(1)} \ldots p_{s^{ij}(N-2)})$. If F is a finite subset of S and $p_i \in F$, we say that p_i has *rang n in F*, if p_i is the nth element of F, with respect to the linear ordering induced by the indices.

For $i \neq j$ and $p_m \in L(p_i, p_j) - \{p_i, p_j\}$ let $r^{ij}(m)$ be the rank of p_j in $L(p_i, p_j) - \{p_i, p_m\}$ and let $\widetilde{r}^i(j)$ be the rank of p_j in $L(p_i, p_j)$.

Notice that if p_i, p_j, p_m are three distinct collinear points, then $\widetilde{r}^i(j) = \widetilde{r}^m(j)$. We also find that

$$r^{ij}(m) = \begin{cases} \widetilde{r}^i(j), & \text{if } i > j < m \\ \widetilde{r}^i(j) - 1, & \text{if } i < j < m, \text{ or } m < j < i \\ \widetilde{r}^i(j) - 2, & \text{if } i < j > m. \end{cases}$$

In the following, we regard the second index k of $a_{i,k} \in A_i$ as an integer modulo $N-1$, hence sums and differences involving these indices are calculated modulo $N-1$.

Let us define the following operation on L:

$\forall a \in L,\ q \cdot a = a \cdot q = a$

$$\forall (a_{i,k}, a_{i,\ell}) \in A_i^2,\ a_{i,k} \cdot a_{i,\ell} = \begin{cases} a_{i,k+\ell}, & \text{if } k+\ell \not\equiv 0 (\operatorname{mod} N-1) \\ q, & \text{if } k+\ell \equiv 0 (\operatorname{mod} N-1) \end{cases}$$

$\forall i \neq j,\ a_{i,k} \cdot a_{j,\ell} = a_{m,n},$ where

$$(\gamma) \qquad m = \begin{cases} s^{ij}(k+\ell-1), & \text{if } i<j \\ s^{ij}(k+\ell), & \text{if } i>j \end{cases} \quad \text{and}$$

$$n = r^{ij}(m) + k - 1 (\operatorname{mod} N-1).$$

We verify that $(L, \cdot, q)$ is a loop and $(S', \circ) \simeq (L/\rho, *, \{q\})$.

Case II: As above, a similar loop construction for the case $N=3$ also yields a loop whereby the special corresponding equivalence relation is the identity. Therefore, this construction does not give us the desired isomorphism.

As above, order the set P of all points of (S,T) as $p_1, p_2, \ldots$ For each p_i, choose pairwise disjoint two–element sets $A_i = \{a_{i0}, a_{i1}\}$ and $q \notin \bigcup_i A_i$. Let $L = \{q\} \cup \bigcup_i A_i$ and ρ defined as in Case I.

We define the following operation on L:

$$\forall a \in L,\ a \cdot q = q \cdot a = a,$$

$$\forall (a_{ik}, a_{i\ell}) \in A_i^2,\ a_{ik} \cdot a_{i\ell} = \begin{cases} q, & \text{if } k \neq \ell \\ a_{i,k+\ell}, & \text{if } k = \ell \end{cases}$$

where $k+\ell$ is calculated mod 2) and $\forall i \neq j\ a_{ik} \cdot a_{j\ell} = a_{mn}$, where $n = k+\ell (\operatorname{mod} 2)$ and m is such that $p_m = L(p_i, p_k) - \{p_i, p_k\}$.

35. Theorem. *With the above constructions, we have:*

1) $(L, \cdot, q)$ *is a loop;*
2) ρ *is a special equivalence relation on* L*;*
3) $(S', \circ) \simeq (L/\rho, *\{q\})$.

Proof. We verify for the case $N \geq 4$. The case $N = 3$ is similar but simpler.

1) First, we check that in the equality $z = xy$, z and any of x and y determines uniquely the other.

It is immediate if $q \in \{x, y, z\}$.

Let $x = a_{ik}$, $y = a_{j\ell}$ and $z = a_{mn}$. If card$\{i, j, m\} \leq 2$, then $i = j = m$ and the conclusion is immediate.

Suppose $i \neq j \neq m \neq i$. We have two possibilities:

a) a_{ik} and a_{mn} are given. Notice that $r^{ij}(m)$ increases monotonically with j, so there exists a unique $p_j \in L(p_i, p_m) - \{p_i, p_m\}$, $r^{ij}(m) + k - 1 \equiv n (\operatorname{mod} N - 1)$. Now, we can obtain, in a similar way, the unique h such that $s^{ij}(h) = m$. If $i < j$, we can obtain ℓ uniquely from $h = k + \ell - 1 (\operatorname{mod} N - 1)$, and respectively, if $i > j$, from $h = k + \ell (\operatorname{mod} N - 1)$.

b) $a_{j\ell}$ and a_{mn} are given. First, suppose $j < m$. We have

$$r^{ij}(m) = \begin{cases} \widetilde{r}^{i}(j) - 1, & \text{if } i < j \\ \widetilde{r}^{i}(j), & \text{if } i > j. \end{cases}$$

We seek i and k, which satisfy the equalities: $r^{ij}(m) = n - k + 1$ and $[(s^{ij}(k + \ell - 1) = m$, if $i < j)$ or $(s^{ij}(k + \ell) = m$, if $i > j)]$.

Using the definition of $r^{ij}(m)$ and the fact that $\widetilde{r}^{i}(j) = \widetilde{r}^{m}(j)$, we obtain:

$$\begin{array}{lll} \text{if } i < j, & k = n - \widetilde{r}^{m}(j) + 2, & \text{respectively} \\ \text{if } i > j, & k = n - \widetilde{r}^{m}(j) + 1, & \end{array}$$

whence, $s^{ij}(n - \widetilde{r}^{m}(j) + \ell + 1) = m$, for $i < j$ and also for $i > j$.

In the last equality, j, ℓ, m and n are known and i is unknown and from this equality we obtain uniquely i.

With this value for i, we obtain a unique solution for k.

The case $j > m$ is similar.

2) It is sufficient to verify that for $i \neq j$, $A_i A_j$ is a union of (A_i, A_j)–projective equivalence classes A_m.

Let the multiplication rule be:

$$\mu : A_i \times A_j \to \bigcup_{m \notin \{i,j\}} A_m$$

From the equalities (γ), we obtain the map μ is an onto map.

Now, we have to only show that each A_m (for $m \notin \{i,j\}$) is (A_i, A_j)–projective, that is we have to check that $\forall\, a_{ik} \in A_i$, $\exists a_{j\ell} \in A_j$, such that $a_{ik}a_{j\ell} \in A_m$ and symmetrically. This can be easily obtained from (γ).

3) An isomorphism between $(S', \circ)$ and $(L/\rho, *, \{q\})$ is given by:

$$f : S' \to L/\rho,\ f(a) = \begin{cases} \{q\}, & \text{if } a = e \\ A_i, & \text{if } a = p_i \in S. \end{cases}$$

36. Remark. The paper [43] presentes a relationship between multivalued loops and representations of atomic structures of certain 3-dimensional cylindric algebras.

Chapter 2

Graphs and Hypergraphs

Since the middle of the last century, Graph Theory has been an important tool in different fields, like Geometry, Algebra, Number Theory, Topology, Optimization, Operations Research, Median Algebras and so on. To solve new combinatorial problems, it was necessary to generalize the concept of a Graph.

The notion of a "hypergraph" appeared around 1960 and one of the initial concerns was to extend some classical results of graph theory.

Hypergraph Theory is an useful tool for discrete optimization Problems.

A very good presentation of Graph and Hypergraph Theory is in C. Berge [442] and Harary [448].

In this chapter, we have presented some important connections between Graph, Hypergraph Theory and Hyperstructure Theory.

§1. Generalized graphs and hypergroups

The following results on generalized graphs and hypergroups have been obtained by M. Gionfriddo.

1. Definition. Let $V \subseteq G$, $V \neq \emptyset$, where G is a finite non–empty set and $f : G \to \mathcal{P}(G)$, such that:

(i) $G - V \neq \emptyset$;

(ii) $\forall x \in V,\ f(x) = \{x\}$;

(iii) $\forall y \in G - V,\ f(y) \in \mathcal{P}(V)$ and $|f(y)| = n+1$ for some $\in \mathbb{N}^*$.

The pair (G, f) is called a (*generalized*) *graph* on G of *dimension* n or an n–*graph*.

Every $x \in V$ is called a *vertex* of (G, f) and each $y \in G - V$ is called an *edge* of (G, f).

A *connected graph* is a graph (G, f) such that $\forall (x, y) \in V^2$, there exists $E = \{e_1, e_2, ..., e_h\} \subseteq G - V$, with $x \in f(e_1)$, $y \in f(e_h)$ and $\forall i \in \{1, 2, ..., h-1\} = I_{h-1}$, we have $f(e_i) \cap f(e_{i+1}) \neq \emptyset$.

Now, for a non-empty set M set

$$\mathcal{H}(M) = \left\{ f : M \to \mathcal{P}^*(M) \mid \bigcup_{x \in M} f(x) = M \right\}.$$

2. Theorem. *Define on* $\mathcal{H}(M)$ *the following hyperoperation* $*$:

$$\forall (h, k) \in \mathcal{H}(M)^2,$$

$$h * k = \left\{ \ell \in \mathcal{H}(M) \mid \forall x \in M,\ \ell(x) \subseteq \bigcup_{y \in k(x)} h(y) \right\}.$$

Then $(\mathcal{H}(M), *)$ *is a regular hypergroup.*

Proof. Let us verify first the associativity law, that is

$$\forall (h, k, \ell) \in \mathcal{H}(M)^3,\ h * (k * \ell) = (h * k) * \ell.$$

First of all, we check that $\forall x \in M$, $\forall (h, k, \ell) \in \mathcal{H}(M)^3$, we have

$$\bigcup_{t \in \ell(x)} \left(\bigcup_{y \in k(t)} h(y) \right) = \bigcup_{y \in \bigcup\limits_{t \in \ell(x)} k(t)} h(y).$$

On the other hand, we have:

$$u \in \bigcup_{t\in\ell(x)} \left(\bigcup_{y\in k(t)} h(y)\right) \Longrightarrow \left(\exists t_1 \in \ell(x) : u \in \bigcup_{y\in k(t_1)} h(y)\right) \Longrightarrow$$
$$\Longrightarrow u \in \bigcup_{y\in \bigcup\limits_{t\in\ell(x)} k(t)} h(y).$$

On the other hand,

$$u \in \bigcup_{y\in \bigcup\limits_{t\in\ell(x)} k(t)} h(y) \Longrightarrow (\exists y_1 \in \bigcup_{t\in\ell(x)} k(t) :$$
$$u \in h(y_1)) \Longrightarrow (\exists t_1 \in \ell(x) : y_1 \in k(t_1) \text{ and}$$
$$u \in h(y_1)) \Longrightarrow u \in \bigcup_{y\in k(t_1)} h(y) \Longrightarrow u \in \bigcup_{t\in\ell(x)} \left(\bigcup_{y\in k(t)} h(y)\right).$$

For any $(h,k) \in \mathcal{H}(M)^2$, we denote by $a_{h,k}$ the element of $h * k$, for which

$$\forall x \in M,\ a_{h,k}(x) = \bigcup_{y\in k(x)} h(y).$$

We have:

$$u \in h * (k * \ell) = \bigcup_{u\in k*\ell} h * u \Longrightarrow$$
$$\Longrightarrow (\exists u_1 \in k * \ell : u \in h * u_1) \Longrightarrow$$
$$\Longrightarrow \left(\exists u_1 \in k * \ell : \forall x \in M,\ u(x) \subseteq \bigcup_{y\in u_1(x)} h(y)\right) \Longrightarrow$$
$$\Longrightarrow \left(\forall x \in M,\ u(x) \subseteq \bigcup_{y\in \bigcup\limits_{t\in\ell(x)} k(t)} h(y) = \bigcup_{t\in\ell(x)} \left(\bigcup_{y\in k(t)} h(y)\right)\right) \Longrightarrow$$
$$\Longrightarrow \left(\forall x \in M,\ u(x) \subseteq \bigcup_{t\in\ell(x)} a_{h,k}(t)\right) \Longrightarrow$$
$$\Longrightarrow u \in a_{h,k} * \ell \Longrightarrow u \in (h * k) * \ell.$$

Conversely,

$$u \in (h * k) * \ell \Longrightarrow u \in \bigcup_{v \in h*k} v * \ell \Longrightarrow (\exists v_1 \in h * k : u \in v_1 * \ell) \Longrightarrow$$

$$\Longrightarrow \left(\exists v_1 \in h * k : \forall x \in M,\ u(x) \subseteq \bigcup_{t \in \ell(x)} v_1(t) \subseteq \bigcup_{t \in \ell(x)} \left(\bigcup_{y \in k(t)} h(y) \right) = \right.$$

$$\left. = \bigcup_{y \in \underset{t \in \ell(x)}{\cup} k(t)} h(y) \right) \Longrightarrow \left(\forall x \in M,\ u(x) \subseteq \bigcup_{y \in a_{k,\ell}(x)} h(y) \right) \Longrightarrow$$

$$\Longrightarrow u \in h * a_{k,\ell} \Longrightarrow u \in \bigcup_{v \in k * \ell} h * v = h * (k * \ell).$$

Therefore, $\forall\, (h, k, \ell) \in \mathcal{H}(M)^3$,

$$h * (k * \ell) = (h * k) * \ell.$$

Now, we shall prove that $\mathcal{H}(M)$ has at least an identity and any element has an inverse.

Let $\mathcal{I} = \{i \in \mathcal{H}(M) \mid \forall x \in M,\ x \in i(x)\}$. If $i \in \mathcal{I}$, then $\forall\, h \in \mathcal{H}(M)$, $\forall\, x \in M$,

$$h(x) \subseteq \left(\bigcup_{y \in h(x)} i(y) \right) \cap \left(\bigcup_{y \in i(x)} h(y) \right),$$

hence $h \in h * i \cap i * h$.

Therefore, $\mathcal{I}$ is the set of identities of $\mathcal{H}(M)$.

Now, let $h \in \mathcal{H}(M)$ and $\mathcal{I}_h = \{k \in \mathcal{H}(M) \mid \forall x \in M,\ k(x) = \{y \in M \mid x \in h(y)\}\}$. If $k \in \mathcal{I}_h$, then

$$\forall x \in M,\ x \in \bigcap_{y \in k(x)} h(y) \cap \bigcap_{y \in h(x)} k(y).$$

If $i \in \mathcal{I}$ is such that $i(x) = \{x\}$, $\forall x \in M$, then $i \in h * k \cap k * h$. Therefore

$$\forall\, h \in \mathcal{H}(M),\ \exists \mathcal{I}_h \in \mathcal{P}^*(\mathcal{H}(M)) : \forall k \in \mathcal{I}_h,\ h * k \cap k * h \cap \mathcal{I} \neq \emptyset,$$

that is any element $h \in \mathcal{H}(M)$ has an inverse.

From this, we also obtain the reproductibility of $(\mathcal{H}(M), *)$.

Indeed, $\forall\, i \in \mathcal{I}$, $\forall\, h \in \mathcal{H}(M)$, we have $h \in i * h$. Moreover, for any $k \in \mathcal{H}(M)$, $\exists i \in \mathcal{I}$, $\exists k'$ inverse of k, such that $i \in k * k'$. Then, there is $u \in k' * h$, such that $h \in k * u$.

Similarly, there is $v \in \mathcal{H}(M)$ such that $h \in v * k$. ∎

3. Theorem.

(i) *For every* $M \neq \emptyset$, *the hypergroup* $(\mathcal{H}(M), *)$ *is left–reversible;*

(ii) *For* $|M| > 1$, $(\mathcal{H}(M), *)$ *is not right–reversible.*

Proof. i) Let $(\ell, h, k) \in \mathcal{H}(M)^3$. If $\ell \in h * k$ and $u \in \mathcal{H}(M)$ is such that $\forall\, x \in M$, $u(x) = M$, then

$$\forall\, x \in M,\ k(x) \subseteq \bigcup_{y \in \ell(x)} u(y) = M,$$

hence $k \in u * \ell$.

ii) We shall prove that there exist $(h, k, \ell) \in \mathcal{H}(M)^3$ and $x_0 \in M$, such that $\ell \in h * k$ and for every inverse v of k, $h(x_0) \not\subseteq \bigcup\limits_{y \in v(x_0)} \ell(y)$.

Let a, b be two distinct elements of M. Let $(h, k, \ell) \in \mathcal{H}(M)^3$ be such that:

$h(a) = \{b\}$, $h(b) = a$ and $\forall\, x \in M - \{a, b\}$, $h(x) = M$;
$k(a) = \{b\}$ and $\forall\, x \in M - \{a\}$, $k(x) = M$ and
$\forall\, x \in M$, $\ell(x) = \{a\}$.

Since $\bigcup\limits_{x \in M} h(x) = M$ and $f \in h * k$ if and only if ($f(a) = \{a\}$ and $\forall\, x \in M - \{a\}$, $f(x) \subseteq M$), it follows that $\ell \in h * k$.

Moreover, since for every inverse v of k,

$$\ell * v = \left\{ g \in \mathcal{H}(M) \mid \forall\, x \in M,\ g(x) \subseteq \bigcup_{y \in v(x)} \ell(y) = \{a\} \right\}$$
$$\text{and } h(a) = \{b\},$$

we have $h \notin \ell * v$. ∎

4. Definition. A *subgraph* of (G, f) is a graph $(\widetilde{G}, \widetilde{f})$ such that

$$\widetilde{G} \subseteq G \text{ and } \widetilde{f} = f/G.$$

5. Definition. Let (G_1, f_1) and (G_2, f_2) be two graphs. The map $\psi : G_1 \to \mathcal{P}^*(G_2)$ is called a *generalized multihomomorphism* (or, simply a GMH) from (G_1, f_1) to (G_2, f_2) if

(i) $\forall x \in G_1, \ \bigcup\limits_{y \in f_1(x)} \psi(y) = \bigcup\limits_{y \in \psi(x)} f_2(y);$

(ii) $\forall y \in G_1 - V_1$, if $\widetilde{f_1(y)} = f_1(y) \cup \{y\}$ and $Y = \bigcup\limits_{t \in \widetilde{f_1(y)}} \psi(t)$, then $(Y, f_2/Y)$ is a connected subgraph of (G_2, f_2).

6. Definition. Let (G_1, f_1) and (G_2, f_2) be two graphs. The map $\varphi : V_1 \to \mathcal{P}^*(V_2)$ (where $\forall i \in \{1, 2\}$, V_i is the set of vertex of (G_i, f_i)) is called an *Ore multihomomorphism* (or simply an OMH) from (G_1, f_1) to (G_2, f_2) if $\forall X \subseteq V_1$, $X \neq \emptyset$, such that $\exists y \in G_1 - V_1$ with $f_1(y) = X$, the set $\bigcup\limits_{t \in X} \varphi(t)$ is the set of vertices of a connected subgraph of (G_2, f_2).

Let (G, f) be a connected graph. Let $H_G = \{\psi \mid \psi$ is a GMH in (G, f), $\bigcup\limits_{x \in G} \psi(x) = G$, $\forall y \in G - V$, $\left|\bigcup\limits_{t \in \widetilde{f(y)}} \psi(t)\right| > 1\}$ and $\circ : H_G^2 \to \mathcal{P}^*(H_G)$ defined as follows:

$$\forall (h, k) \in H_G^2, \ h \circ k = \left\{ \ell \in H_G \mid \forall x \in G, \ \ell(x) \subseteq \bigcup_{y \in k(x)} h(y) \right\}.$$

Let $K_G = \{\varphi \mid \varphi$ is an OMG in (G, f), $\bigcup\limits_{x \in V} \varphi(x) = V$, $\forall y \in G - V$, $\left|\bigcup\limits_{t \in f(y)} \varphi(t)\right| > 1\}$ and $* : K_G^2 \to \mathcal{P}^*(K_G)$ defined as follows:

$$\forall (h, k) \in K_G^2, \ h * k = \left\{ \ell \in K_G \mid \forall x \in V, \ \ell(x) \subseteq \bigcup_{y \in k(x)} h(y) \right\}.$$

7. Theorem.

(i) $(H_G, \circ)$ *is a regular hypergroup;*

(ii) $(K_G, *)$ *is a regular hypergroup.*

Proof. i) For any $(h,k) \in H_G^2$, let $a_{h,k}$ be the map defined as follows:

$$\forall\, x \in G,\ a_{h,k}(x) = \bigcup_{y \in k(x)} h(y).$$

Since $\forall\, y \in G - V$, $\left|\bigcup_{t \in \widetilde{f(y)}} k(t)\right| > 1$, we obtain that $a_{h,k} \in H_G$. Moreover, we have $a_{h,k} \in h \circ k$. So, $\forall\,(h,k) \in H_G^2$, we have $h \circ k \neq 0$.

Now, let us verify the associativity law.

We have $\forall\,(h,k,\ell) \in H_G^3$,

$$u \in (h \circ k) \circ \ell \Longrightarrow u \in h \circ a_{k,\ell} \Longrightarrow u \in h \circ (k \circ \ell) \text{ and}$$
$$v \in h \circ (k \circ \ell) \Longrightarrow v \in a_{h,k} \circ \ell \Longrightarrow v \in (h \circ k) \circ \ell.$$

Finally, let us notice that if $i \in \mathcal{H}(G)$ is such that $\forall\, x \in G$, we have that $i(x) = x$, then i is an identity GMH and $i \in H_G$.

If $\mu \in \mathcal{H}(G)$ is such that $\forall\, x \in V$, $\mu(x) = V$ and $\forall\, x \in G - V$, $\mu(x) = G$, then $\mu \in H_G$ and for any $\psi \in H_G$, we have $i \in \psi \circ \mu \cap \mu \circ \psi$.

ii) Similarly, for any $(h,k) \in K_G$, let $b_{h,k}$ be the map defined as follows:

$$\forall\, x \in V,\ b_{h,k}(x) = \bigcup_{y \in k(x)} h(y).$$

Since $\forall\, y \in G - V$, $\left|\bigcup_{t \in f(y)} k(t)\right| > 1$ it follows that $b_{h,k} \in K_G$ and $b_{h,k} \in h * k$.

Now, let us verify the associativity law. We have

$$u \in (h * k) * \ell \Longrightarrow u \in h * b_{k,\ell} \Longrightarrow u \in h * (k * \ell)$$
$$v \in h * (k * \ell) \Longrightarrow v \in b_{h,k} * \ell \Longrightarrow v \in (h * k) * \ell.$$

Finally, $\forall x \in V$, if $i(x) = \{x\}$, then $i \in K_G$ and $\forall h \in K_G$, $h \in i * h \cap h * i$. If $\forall x \in V$, $\eta(x) = V$, then $\eta \in K_G$ and $\forall h \in K_G$, $i \in \eta * h \cap h * \eta$.

8. Theorem. *There exists a homomorphism from* $(H_G, \circ)$ *to* $(K_G, *)$.

Proof. Let $F : H_G \to K_G$ defined as follows: $\forall \psi \in H_G$, $F(\psi) = = \psi/V$. We have $\psi/V \in K_G$.

For any $(h, k) \in H_G^2$, if $\varphi \in F(h \circ k)$, then there exists $\psi \in H_G$ such that $F(\psi) = \psi/V = \varphi$ and $\forall x \in G$, $\psi(x) \subset \bigcup_{y \in k(x)} h(y)$. For any $x \in V$, we have

$$\varphi(x) = \psi(x) \subseteq \bigcup_{y \in k(x)} h(y) = \bigcup_{y \in (k/V)(x)} (h/V)(y),$$

whence $\varphi \in F(h) * F(k)$. ∎

§2. Chromatic quasi–canonical hypergroups

The quasi–canonical hypergroups were utilised by St. Comer to establish connections with edge–coloured graphs.

Let $\mathcal{C}$ be a non-empty set of colours and ε an involution of $\mathcal{C}$, that means $\mathcal{E} \circ \mathcal{E} = 1_{\mathcal{C}}$.

Let V be a set of vertex. A pair $(x, y) \in V^2$ with $x \neq y$ is called an *edge*. For any $a \in \mathcal{C}$, let C_a be a binary relation on V.

A system $\mathcal{V} =< V, C_a >_{a \in \mathcal{C}}$ is called a *colour scheme* if the following conditions are satisfied:

1° $\{C_a \mid a \in \mathcal{C}\}$ is a partition of $\{(x, y) \in V^2 \mid x \neq y\}$,

2° $\forall a \in \mathcal{C}$, $C_{\mathcal{E}(a)} = \{(y, x) \mid (x, y) \in C_a\}$;

3° $\forall a \in \mathcal{C}$, $\forall x \in V$, $\exists y \in V : (x,y) \in C_a$, that means each vertex has an edge of each colour emanating from it;

4° if $a, b, c \in \mathcal{C}$, then the following implication holds:

$$C_c \cap C_a \circ C_b \neq \emptyset \Longrightarrow C_c \subset C_a \circ C_b,$$

where "$\circ$" is the composition of relations.

9. Remark. $\forall a \in \mathcal{C}$, C_a is thought of as the set of directed edges with *colour* a in the complete directed graph with no loops on the set V.

The involution $\mathcal{E}$ guarantees that the colour assigned to an edge (y,x) depends only on the colour assigned to the reverse directed edge (x,y) and not on the particular (x,y), so we can say that colours a and $\mathcal{E}(a)$ are *paired.*

If $\forall a \in \mathcal{C}$, $\mathcal{E}(a) = a$, that means the colours are self–paired, then the colours schemes can be pictured by colouring the edges of undirected graphs.

A *partial colour scheme* is a system $\mathcal{V} =< V, C_a >_{a\in\mathcal{C}}$, which satisfies only the conditions 1° and 2°.

Notice that two special cases of the notion of colour scheme were been widely studied:

1. *homogeneous coherent configurations* (see D.G. Higman [449]), which are studied via matrix algebra, because of so called *intersection numbers.* An intersection number is the number of (a,b)-paths from x to y, where $(x,y) \in C_c$. In a homogeneous coherent configuration, a such number is independent of the choice of $(x,y) \in C_c$.

2. *association schemes*, which are homogeneous coherent configurations with $\mathcal{E}(a) = a$, for all $a \in \mathcal{C}$. Associative schemes have a large literature. We mention only Bose and Mesner [30].

Some of the important association schemes are those associated with distance–transitive and strongly regular graphs (Biggs [444], Cameron and Van Lint [33]).

Let us associate now a quasi–canonical hypergroup with a colour scheme $\mathcal{V} =< V, C_a >_{a\in\mathcal{C}}$.

Let $e \notin \mathcal{C}$. We shall consider the following *colour algebra* on $\mathcal{V}$:

$$\mathcal{A}_{\mathcal{V}} =< \mathcal{C} \cup \{e\}, \square, ^{-1}, e >,$$

where the inverse is defined by $a^{-1} = \mathcal{E}(a)$, for $a \in \mathcal{C}$ and $e^{-1} = e$. The product is defined by: $a \square e = e \square a = a$, for $a \in \mathcal{C} \cup \{e\}$, $\forall (a,b) \in \mathcal{C}^2$, $b \neq a^{-1}$, $a \square b = \{c \in \mathcal{C} \mid C_c \subset C_a \circ C_b\}$ and $\forall a \in \mathcal{C}$, $a \square a^{-1} = \{c \in \mathcal{C} \mid C_c \subset C_a \circ C_{a^{-1}}\} \cup \{e\}$. It results the following

10. Proposition. *$\mathcal{A}_{\mathcal{V}}$ is a quasi–canonical hypergroup with the unit e.*

11. Definition. A quasi–canonical hypergroup is called *chromatic* if it is isomorphic to $\mathcal{A}_{\mathcal{V}}$.

In the following, we shall present an important example of chromatic quasi–canonical hypergroup.

12. Definition. Let ρ be an equivalence relation on a quasi–canonical hypergroup $(H, \square)$.

1. ρ is called a *full conjugation* on H if the following implications hold:
   $x\rho y \Longrightarrow x^{-1}\rho y^{-1}$;
   $z \in x \square y$ and $z\rho z' \Longrightarrow \exists (x', y') \in H^2$,
   such that $x'\rho x$, $y'\rho y$ and $z' \in x' \square y'$.

2. ρ is called a *special conjugation* if 1) holds and, moreover, $x\rho e$ implies $x = e$.

13. Theorem. (see Comer [46]) *Let $(H, \square)$ be a quasi–canonical hypergroup and ρ an equivalence relation on H. Then*

ρ is a full conjugation on H if and only if $(\{\rho_x | x \in H\}, \cdot)$ is a quasi–canonical hypergroup, called a (double) quotient of H and it is denoted by $H /\!/ \rho$. Notice that "$\cdot$" is the induced operation on the set of ρ-classes (that is $\rho_z \in \rho_x \cdot \rho_y \Longleftrightarrow \exists x', \exists y' : x\rho x', y\rho y', z \in x' \,\square\, y'$)

Let us denote by Q^2(Group) the set of all quasi–canonical hypergroups isomorphic to a double quotient of a group.

14. Examples The following are full conjugations on a group G:

1. any congruence relation ρ on G is a full conjugation and $G /\!/ \rho$ is just the usual quotient group;

2. if H is a subgroup of G and $\rho_H \subset G \times G$ is defined as follows:

$$x\rho_H y \Longleftrightarrow HxH = HyH,$$

   then ρ_H is a full conjugation on G.

3. if K is a group of automorphisms of G and ρ is defined as follows:

$$x\rho y \Longleftrightarrow \exists \sigma \in K : y = \sigma(x),$$

   then ρ is a special conjugation on G.

Utumi [396] used special conjugations of groups to obtain important examples of cogroups.

We have:

15. Proposition. *If G is a group and ρ is a full conjugation on G, then ρ_e is a subgroup of G.*

We point out that the double quotient of groups are related to chromatic quasi–canonical hypergroups.

16. Theorem. *Every quasi–canonical hypergroup in Q^2(Group) is chromatic.*

Proof. Let ρ be a full conjugation on a group G. Then $\rho_e = H$ is a subgroup of G.

Set $\mathcal{C} = \{\rho_x \mid x \in G,\ \rho_x \neq H\}$, $V = \{Hx \mid x \in G\}$, $\forall\, a \in \mathcal{C}$, $\mathcal{E}(a) = a^{-1}$ and $C_a = \{(Hx, Hy) \in V^2 \mid xy^{-1} \in a\}$.

It can be easily verified that $\mathcal{V} =< V, C_a >_{a\in\mathcal{C}}$ is a colour scheme, which is usually called the *regular colour scheme* representation of $G/\!\!/\rho$. In order to verify the implication: $C_c \cap C_a \circ C_b \neq \emptyset \Longrightarrow C_c \subset C_a \circ C_b$, we show that $c \in a \cdot b$ (in $G/\!\!/\rho$) $\Longleftrightarrow C_c \subseteq C_a \circ C_b$ for any $a, b, c \in \mathcal{C}$.

Indeed, if $c \in a \cdot b$ (in $G/\!\!/\rho$) and $(Hu, Hv) \in C_c$ then there exist $r \in a$ and $s \in b$, such that $uv^{-1} = rs$. Denote sv by z. Then $(Hu, Hz) \in C_a$, since $uz^{-1} = uv^{-1}s^{-1} = r \in a$ and $(Hz, Hv) \in C_b$, since $zv^{-1} = s \in b$. So, $(Hu, Hv) \in C_a \circ C_b$ and hence $C_c \subset C_a \circ C_b$. Conversely, if $C_c \subset C_a \circ C_b$ and $x \in c$, then we have $(Hx, H) \in C_c$, so there exists $z \in b$ such that $(Hx, Hz) \in C_a$, that is $xz^{-1} \in a$, whence $x = (xz^{-1})z \in a \cdot b$. Therefore, $c \subseteq a \cdot b$ in $G/\!\!/\rho$. Let $< \mathcal{C} \cup \{e_0\}, \square, ^{-1}, e_0 >$ the quasi–canonical hypergroup associated with the colour scheme $\mathcal{V} =< V, C_a >_{a\in\mathcal{C}}$.

Finally, we have only to notice that $\varphi : G/\!\!/\rho \longrightarrow \mathcal{C} \cup \{e_0\}$, $\varphi(\rho_x) = \rho_x$ and $\varphi(\rho_e) = e_0$ (e_0 is the identity of $\mathcal{C} \cup \{e_0\}$) is an isomorphism. ∎

§3. Hypergroups induced by paths of a direct graph

The following results on graphs and hypergroups are due to I.G. Rosenberg.

These hyperoperations have also been considered by P. Corsini.

17. Definition. We say that $G = (V, E)$ is a *directed* (simple and loopless) *graph* if V is a nonvoid set and E a binary areflexive relation on V (i.e., $E \subseteq V^2 = V \times V$ and $(v, v) \in \rho$ for no $v \in V$). For $(x, y) \in V^2$ a *path from x to y* or an *$x - y$ path*, is a finite sequence $< z_0, ..., z_m >$ over V, such that

(i) $x = z_0,\ y = z_m$

(ii) for all $0 \leq i < j \leq m,\ z_i = z_j \Longrightarrow i = 0,\ j = m,$

(iii) $(z_i, z_{i+1}) \in E$ for all $i \in \{0, ..., m-1\}$

For every $x \in V$ we consider $\langle x \rangle$ is an $x - x$ path.

We assume throughout that G is *connected* in the sense that for any $(x, y) \in V^2$, there is at least one $x - y$ path.

Let $\circ_1 : V^2 \longrightarrow \mathcal{P}^*(V)$ be defined as follows: $\forall\, x \in V,\ x \circ_1 x = \{x\}$ and $\forall\,(x,y) \in V^2,\ x \neq y,\ x \circ_1 y$ is the set of all vertices on all $x - y$ paths, that is $u \in x \circ_1 y$ if there is a $x - y$ path $< z_0, ..., z_m >$ and there is $0 \leq i \leq m$ such that $u = z_i$.

Let $\circ_2 : V^2 \longrightarrow \mathcal{P}^*(V)$ be defined as follows: $\forall\, x \in V,\ x \circ_2 x$ is the set of all vertices on all $x - x$ paths and $\forall\,(x,y) \in V^2,\ x \neq y,\ x \circ_2 y = x \circ_1 y$.

18. Definition. We say that a vertex set B *separates* a vertex set A from a vertex set C (in that order) if every path starting from A and ending in C meets B. If A is a singleton $\{a\}$ we say that B *separates* a from C and similarly B *separates* A from c whenever $C = \{c\}$.

Let us introduce the following property (α) of $(V, \circ_i)$ (where $i \in \{1,2\}$), which consists in two parts: (α_1) and (α_2). There are:

(α_1) If $< z_0, ..., z_m >$ and $< w_0, ..., w_n >$ are paths and $0 \leq i < k \leq m$ and $0 < j < r < n$ are such that the following conditions hold:

(1) $z_0 \neq z_m$ and $w_0 \neq w_n$ if $i = 1$;

(2) $z_k = w_j,\ z_i = w_n$ while the sets $\{z_{i+1}, ..., z_{k-1}\}$ and $\{w_{j+1}, ..., w_{n-1}\}$ are disjoint;

(3) $\{w_{r+1}, ..., w_{n-1}, z_i, ..., z_m\}$ separates the set $w_0 \circ z_0$ from w_r;

(4) each of the sets $\{z_0, ..., z_k, w_{j+1}, ..., w_{r-1}\}$ and $\{w_0, ..., w_{r-1}\}$ separates w_r from z_m;

then there is $y \in z_0 \circ w_0$ such that w_r is on a $y - z_m$ path.

(α_2) If $< z_0, ..., z_m >$ and $< w_0, ..., w_n >$ are paths and $0 \le k < i \le m$ and $0 < r < j < n$ are such that the following conditions hold:

(1) $z_0 \neq z_m$ and $w_0 \neq w_n$ if $i = 1$;

(2) $z_i = w_0, \ z_k = w_j$ while the sets $\{z_{k+1}, ..., z_{i-1}\}$ and $\{w_1, ..., w_{j-1}\}$ are disjoint;

(3) $\{z_0, ..., z_i, w_1, ..., w_{r-1}\}$ separates w_r from $z_m \circ w_n$;

(4) both sets $\{w_{r+1}, ...w_n\}$ and $\{w_{r+1}, ..., w_j, z_{k+1}, ..., z_m\}$ separates z_0 from w_r,

then there is $y \in z_m \circ w_0$ such that w_r is on a $z_0 - y$ path.

We say that G satisfies (α) if G satisfies both (α_1) and (α_2).

19. Theorem. *The following statements are equivalent for a directed connected graph G and the associated hypergroupoid* $< V, \circ_i >$ $(i \in \{1, 2\})$:

(1) $(V, \circ_i)$ *is a semihypergroup;*

(2) $(V, \circ_i)$ *is a hypergroup;*

(3) *G satisfies* (α).

Proof. We shall denote $(V, \circ_i)$ by $(V, \circ)$.

(1)$\Longrightarrow$(2). Let $(V, \circ)$ be a semihypergroup and $(x, y) \in V^2$ be aribtrary. Notice that $y \in (x \circ y) \cap (y \circ x)$ and consequently $V \subseteq (x \circ V) \cap (V \circ x)$. Clearly, $x \circ V \subseteq V \supset V \circ x$ and so $\forall x \in V$, $x \circ V = V = V \circ x$. Therefore, $(V, \circ)$ is a hypergroup.

(2)$\Longrightarrow$(3). Let $(V, \circ)$ be a hypergroup. To prove (α_1) let $\langle z_0, ..., z_m \rangle$ and $\langle w_0, ..., w_n \rangle$ be two paths satisfying the conditions (1) and (2) of (α_1). Clearly, $w_n = z_i \in z_0 \circ z_m$ and $w_r \in w_0 \circ (z_0 \circ z_m)$.

Since $(V, \circ)$ is a hypergroup, $w_0 \circ (z_0 \circ z_m) = (w_0 \circ z_0) \circ z_m$. Thus $w_r \in (w_0 \circ z_0) \circ z_m$ and so there is $y \in w_0 \circ z_0$ such that $w_r \in y \circ z_m$ proving the conclusion of (α_1). Next suppose that the condition (2) from (α_2) holds. Then $w_0 = z_i \in z_0 \circ z_m$; hence $w_r \in (z_0 \circ z_m) \circ w_n = z_0 \circ (z_m \circ w_n)$ which is the conclusion of (α_2).

(3)$\Longrightarrow$(1). Let (α) hold.

I. Let $(x, y, z) \in V^3$. We shall verify that

$$(*) \qquad (x \circ y) \circ z \supset x \circ (y \circ z)$$

1) First, suppose that either $y \neq z$ or $(V, \circ) = (V, \circ_2)$ that is $y \circ y$ is the set of all $y - y$ paths. Let $u \in x \circ (y \circ z)$ be arbitrary. Then there exist an $y - z$ path $< z_0, ..., z_m >$, $0 \leq t \leq m$ and an $x - z_t$ path $< w_0, ..., w_v >$ such that $u = w_r$ for some $0 \leq r \leq v$. Denote by q the least index such that $w_q \in \{z_0, ..., z_m\} = Z$ and let $w_q = z_h$.

1° First, suppose that $r \leq q$. We claim that $< w_0, ..., w_q, z_{h+1}, ..., z_m >$ is a path. Indeed, $< w_0, ..., w_q >$ and $< z_h, ..., z_m >$ are paths and $\{w_0, ...w_q\}$ is disjoint from $\{z_{h+1}, ..., z_m\}$ on account of the minimality of q. Thus $u = w_r \in x \circ z \subseteq (x \circ y) \circ z$ so, in this case, the inclusion $(*)$ holds.

2° Let us consider now $r > q$. If $u \in Z$ then $u \in y \circ z \subseteq (x \circ y) \circ z$ and the inclusion $(*)$ holds. Thus we may assume that $u \notin Z$. Then there are $q \leq j < r < n \leq v$ and $0 \leq i < k \leq m$ such that

$$\text{(X)} \qquad \{z_i, ..., z_k\} \cap \{w_j, ..., w_n\} = \{z_i, z_k\} = \{w_j, w_n\}.$$

We have two cases:

a) Let $w_j = z_i$ and $w_n = z_k$. Then $\langle z_0, \ldots, z_i, w_{n-1}, z_k, \ldots, z_m \rangle$ is an $y - z$ path, hence $u \in y \circ z \subseteq (x \circ y) \circ z$.

b) Thus let $w_j = z_k$ and $w_n = z_i$. We try to take advantage of the following three paths.

b$_1$) The $u - z$ path $\lambda = \langle w_r, ..., w_{n-1}, z_i, ..., z_m \rangle$. If the set $Y = \{w_{r+1}, ..., w_{n-1}, z_i, ..., z_m\}$ does not separate $X = x \circ y$ from $u = w_r$, then there is $t \in x \circ y$ such that the vertex w_r is on a $t - w_r$ path μ sharing only w_r with λ, hence $u \in t \circ z \subset (x \circ y) \circ z$ and the inclusion $(*)$ holds. Thus we can assume that Y separates $x \circ y$ from u.

b$_2$) We try to use the $y - u$ path $\langle z_0, ..., z_k, w_{j+1}, ..., w_{r-}\rangle$. If the set $U = \{z_0, ..., z_k, w_{j+1}, ..., w_{r-1}\}$ does not separate w_r from z_m, then $u = w_r \in y \circ z \subseteq (x \circ y) \circ z$. Thus we can assume that U separates w_r from z_m.

b$_3$) Finally we try to use the $x - u$ path $\langle w_0, ..., w_r \rangle$. If $V = \{w_0, ..., w_{r-1}\}$ does not separate w_r from z_m, then $u = w_r \in x \circ z \subseteq (x \circ y) \circ z$. Thus we may assume that V separates w_r from z_m.

b$_4$) In the remaining case the condition (α_1) assures that $u \in (x \circ y) \circ z$.

2) Let $(V, \circ) = (V, \circ_1)$ and $y = z$. Then $x \circ (y \circ y) = x \circ y \subset \subset (x \circ y) \circ y$ as required.

II. Let $(x, y, z) \in V^3$. We shall prove that

$$(**) \qquad (x \circ y) \circ z \subseteq x \circ (y \circ z)$$

1) First, suppose that $y \neq z$ or $(V, \circ) = (V, \circ_2)$. Let $u \in (x \circ y) \circ z$. Then there are an $x - y$ path $\langle z_0, ..., z_m \rangle$, $0 \leq t \leq m$ and a $z_t - z$ path $\langle w_0, ..., w_v \rangle$ such that there is $0 \leq r \leq v$ for which $u = w_r$. Denote by q the greatest index such that $w_q \in Z = \{z_0, ..., z_m\}$ and let $w_q = z_h$.

1° First, suppose that $r \geq q$. Since $\langle z_0, ..., z_h, w_{q+1}, ..., w_v \rangle$ is an $x - z$ path, we obtain $u = w_r \in x \circ z \subseteq x \circ (y \circ z)$ and the inclusion $(**)$ is proved. Thus let $r < q$. If $u \in Z$ then again

$u \in x \circ y \subseteq x \circ (y \circ z)$. Thus we may assume that $u \notin Z$. Then there are $0 \leq j < r < n \leq q$ and $0 \leq i < k \leq m$ such that (X) holds.

a) Let $w_j = z_i$ and $w_n = z_k$. Then $\langle z_0, ..., z_i, w_{j+1}, ..., w_{n-1}, z_k, ..., z_m \rangle$ is an $x - z$ path and so $u = w_r \in x \circ z \subset \subset x \circ (y \circ z)$.

b) Thus let $w_j = z_k$ and $w_n = z_i$. We try to use the following three paths.

b$_1$) The path $\lambda = \langle z_0, ..., z_k, w_1, ..., w_r \rangle$. If the set $Y = \{z_0, ..., z_k, w_1, ..., w_{r-1}\}$ does not separate w_r from $z_m \circ w_n$, then there is $w \in z_m \circ w_v = y \circ z$ such that w_r is on a $z_0 - w$ path; consequently $u = w_r \in \in x \circ (y \circ z)$. Thus we may assume that Y separates w_r from $z_m \circ w_m$.

b$_2$) We try to use the path $\langle w_r, ..., w_n \rangle$. If $Y = \{w_{r+1}, ..., w_n\}$ does not separate z_0 from w_r, we have $w_r \in x \circ z \subseteq x \circ (y \circ z)$. Thus we may assume that Y separates z_0 from w_r.

b$_3$) Finally we try to use the path $\langle w_1, ..., w_j, z_{i+1}, ..., z_m \rangle$. If $A = \{w_r, ..., w_j, z_{i+1}, ..., z_m\}$ does not separate z_0 from w_r, then $u = w_r \in y \circ z \subseteq (y \circ z) \circ x$. In the remaining case the condition (α_2) yields $u \in (y \circ z) \circ x$.

2) Finally let $y = z$ and $(V, \circ) = (V, \circ_1)$. Then $(y \circ y) \circ x = = y \circ x \subseteq y \circ (y \circ x)$. This concludes the proof. ■

§4. Hypergraphs and hypergroups

We consider a general hypergraph Γ, and prove that it is always possible to construct from it a sequence of quasi–hypergroups $Q_0(\Gamma)$, $Q_1(\Gamma), ...,$ such that if $Q_k(\Gamma) = Q_{k+1}(\Gamma)$ for some k, then there exists $s \leq k$ such that $Q_s(\Gamma)$ is a join space. Conversely to any hypergroupoid Q satisfying (1), (2) and (3) of the proposition below, it is associated a hypergraph $\Gamma(Q)$ such that $Q_0(\Gamma(Q)) = Q$.

The following results have been obtained by P. Corsini.

20. Definition. Let $\Gamma = \langle H; \{A_i\}_i \rangle$ be a hypergraph, i.e. $\forall i$, $A_i \in P(H) - \{\emptyset\}$; $\bigcup\limits_i A_i = H$; $\forall x \in H$. Set $E(x) = \bigcup\limits_{x \in A_i} A_i$. The hypergroupoid $H_\Gamma = (H; \circ)$ where the hyperoperation is defined by

$$\forall (x, y) \in H^2, \quad x \circ y = E(x) \cup E(y)$$

is called a *hypergraph hypergroupoid* or an *h.g. hypergroupoid.*

21. Theorem. *The hypergroupoid* H_Γ *satisfies for each* $(x, y) \in H^2$:

(1) $x \circ y = x \circ x \cup y \circ y$;

(2) $x \in x \circ x$; *and*

(3) $y \in x \circ x \Longleftrightarrow x \in y \circ y$.

22. Theorem. *A hypergroupoid* H_Γ *satisfying* (1), (2), (3) *of the Theorem* 21 *also satisfies*

(4) $x \circ y \supset \{x, y\}$,

(5) $x \circ y = y \circ x$,

(6) $x \circ H = H$,

(7) $\langle H; \{x \circ x\}_{x \in H} \rangle$ *is a hypergraph,*

(8) $(x \circ x) \circ x = \bigcup\limits_{x \in z \circ z} z \circ z$,

(9) $(x \circ x) \circ (x \circ x) = x \circ x \circ x$.

Proof. It is enough to prove (8) and (9).

(8) We have $(x \circ x) \circ x = \bigcup_{z \in x \circ x} z \circ x$. Then from (1), $x \circ x \circ x = = \bigcup_{z \in x \circ x} (z \circ z \cup x \circ x)$; now from (2) it follows $x \circ x \circ x = \bigcup_{z \in x \circ x} z \circ z$, and finally from (3) we obtain (8).

(9) We have:

$$\begin{aligned}(x \circ x) \circ (x \circ x) &= \bigcup_{\{a,b\} \subset x \circ x} a \circ b = \bigcup_{\{a,b\} \subset x \circ x} ((a \circ a) \cup (b \circ b)) = \\ &= \bigcup_{a \in x \circ x} a \circ a = \bigcup_{x \in a \circ a} a \circ a = x \circ x \circ x.\end{aligned}$$

■

23. Remark. It is clear from (5) and (6) of Theorem 22, that an h.g. hypergroupoid is a commutative quasihypergroup.

24. Theorem. *A hypergroupoid $(H; \circ)$ satisfying* (1), (2) *and* (3) *of Theorem* 21 *is a hypergroup if and only if the following condition is valid:*

$$(\tau) \qquad \forall (a,c) \in H^2, \quad c \circ c \circ c - c \circ c \subset a \circ a \circ a.$$

Proof. We prove the implication "$\Longleftarrow$". For (1) it is enough to verify the associativity. We have:

$$\begin{aligned}\forall (a,b,c) \in H^3,\ (a \circ b) \circ c &= (a \circ a \cup b \circ b) \circ c = (a \circ a) \circ c \cup (b \circ b) \circ c, \\ a \circ (b \circ c) &= (b \circ c) \circ a = (b \circ b) \circ a \cup (c \circ c) \circ a.\end{aligned}$$

Moreover,

$$\begin{aligned}(a \circ a) \circ c &= \bigcup_{u \in a \circ a} u \circ c = \bigcup_{u \in a \circ a} ((u \circ u) \cup (c \circ c)) = \\ &= c \circ c \cup \left(\bigcup_{u \in a \circ a} u \circ u \right) = c \circ c \cup a \circ a \circ a \text{ (by Theorem 22, (8))}.\end{aligned}$$

Then we also have $(b{\circ}b){\circ}c = b{\circ}b{\circ}b \cup c{\circ}c$.

Therefore $(a{\circ}b){\circ}c = a{\circ}a{\circ}a \cup b{\circ}b{\circ}b \cup c{\circ}c$ and moreover $a{\circ}(b{\circ}c) = (b{\circ}c){\circ}a = b{\circ}b{\circ}b \cup c{\circ}c{\circ}c \cup a{\circ}a$.

Set $P = a{\circ}a{\circ}a \cup c{\circ}c$, $Q = c{\circ}c{\circ}c \cup a{\circ}a$.

It is clear that $(a{\circ}b){\circ}c = b{\circ}b{\circ}b \cup P$, $a{\circ}(b{\circ}c) = b{\circ}b{\circ}b \cup Q$ and also $P = (a{\circ}a{\circ}a - a{\circ}a) \cup a{\circ}a \cup c{\circ}c$. By the hypothesis (τ) we have $a{\circ}a{\circ}a - a{\circ}a \subset c{\circ}c{\circ}c$. Since $c{\circ}c \subset c{\circ}c{\circ}c$, it follows $P \subset Q$. In a similar way the inverse inclusion is proved and then the implication follows.

We prove the implication "$\Longrightarrow$". From the associativity it follows: $\forall\,(a,c) \in H^2$, $(a{\circ}a){\circ}c = a{\circ}(a{\circ}c)$.

From above we have also: $(a{\circ}a){\circ}c = c{\circ}c \cup a{\circ}a{\circ}a$, $a{\circ}(a{\circ}c) = \bigcup_{v\in a\circ c} a{\circ}v = \bigcup_{v\in a\circ c} (a{\circ}a \cup v{\circ}v) = a{\circ}a \cup \left(\bigcup_{v\in a\circ a} v{\circ}v\right) \cup \left(\bigcup_{v\in c\circ c} v{\circ}v\right) = a{\circ}a{\circ}a \cup c{\circ}c{\circ}c$ (by Theorem 22, (9)), consequently $c{\circ}c{\circ}c - c{\circ}c \subset a \circ a \circ a$. ■

25. Corollary. *If a hypergroupoid satisfies* (1), (2) *and* (3) *of Theorem* 21 *and the condition:*

$$(\tau') \qquad \forall x, \quad x \circ x \circ x = x \circ x,$$

then it is hypergroup.

26. Definition. An associative h.g.-quasihypergroup is called a h.g.-hypergroup.

27. Theorem. *If the hypergroup* $H_\Gamma = (H; \circ)$ *satisfies* (1), (2), (3) *of Theorem* 21*, then it is a join space.*

Proof. It is sufficient to prove that the following implication is satisfied:

$$x/y \cap z/w \neq \emptyset \Longrightarrow x \circ w \cap y{\circ}z \neq \emptyset,$$

where $x/y = \{z \mid x \in z{\circ}y\}$. We have:

$$u \in x/y \cap z/w \Longleftrightarrow [x \in u{\circ}y \text{ and } z \in u{\circ}w].$$

Moreover, $x \in u\circ y \iff x \in u\circ u \cup y\circ y$ and $z \in u\circ w \iff z \in u\circ u \cup w\circ w$. Four cases are possible:

(1) If $x \in u\circ u$, $z \in u\circ u$, then $u \in x\circ x \cap z\circ z$ and therefore $u \in x\circ w \cap y\circ z$.

(2) If $x \in u\circ u$, $z \in w\circ w$, it follows $w \in z\circ z$, hence $w \in x\circ w \cap y\circ z$.

(3) If $x \in y\circ y$, $z \in u\circ u$, then $y \in x\circ x$, it follows $y \in x\circ w \cap y\circ z$.

(4) If $x \in y\circ y$, $z \in w\circ w$, then $w \in z\circ z$ it follows $w \in x\circ w \cap y\circ z$.

■

28. Theorem. *Let $(H;\circ)$ be a quasihypergroup satisfying* (1), (2), (3) *of Theorem* 21. *Then there is a hypergraph Γ such that $(H;\circ)$ is the h.g.–quasihypergroup associated with Γ.*

Proof. Let Γ be the hypergraph $\langle H; \{x,y\}_{x\in H, y\in x\circ x}\rangle$.
Then for all z in H, we have:

$$E(z) = \bigcup_{z\in x\circ x} \{x,z\} = \bigcup_{x\in z\circ z} \{x,z\} = z\circ z.$$

Then for all (x,y) in H^2, $x\circ y = x\circ x \cup y\circ y = E(x)\cup E(y)$, so the quasihypergroup $< H;\circ >$ is the h.g.–hypergroupoid associated with the hypergraph Γ. ■

29. Theorem. *Let $(H;\circ)$ be a hypergroup satisfying* (1), (2), (3) *of Theorem* 21, *$H_0 = (H;\circ_0)$, $H_1 = (H;\circ_1), \ldots, H_k = (H;\circ_k), \ldots$ the sequence of the hypergroupoids obtained by setting $\forall (x,y) \in H^2$, $x \circ_0 y = x\circ y$, $\forall k \geq 0$, $x \circ_{k+1} x = x \circ_k x \circ_k x$, $x \circ_{k+1} y = x \circ_{k+1} x \cup y \circ_{k+1} y$. Then $\forall k \geq 0$,*

(α) *The hyperoperation $\circ_k$ satisfies* (1), (2), (3) *of Theorem* 21.

(β) $((x \circ_k x \circ_k x) \circ_k (x \circ_k x \circ_k x)) \circ_k (x \circ_k x \circ_k x) = x \circ_{k+2} x.$

Proof. (α) We prove (α) by induction on k. Let us suppose that $\circ_k$ satisfies (1), (2), (3). We prove that the same happens for $\circ_{k+1}$.

(1) is satisfied by definition.

(2) $x \in x \circ_k x \subset x \circ_k x \circ_k x = x \circ_{k+1} x$ by the inductive hypothesis.

(3) if $y \in x \circ_{k+1} x = x \circ_k x \circ_k x$, then, by the inductive hypothesis, there is $z \in x \circ_k x$ such that $y \in z \circ_k x = z \circ_k z \cup x \circ_k x$. If $y \in x \circ_k x$, then, by the inductive hypothesis $x \in y \circ_k y \subset y \circ_{k+1} y$. If not, we have $y \in z \circ_k z$ from which $z \in y \circ_k y$, but we also have $x \in z \circ_k z$ and therefore, by Theorem 22, (8), $x \in (y \circ_k y) \circ_k (y \circ_k y) = y \circ_k y \circ_k y = y \circ_{k+1} y$.

(β) For any $X \subset H$

$$X \circ X = \bigcup_{(y,z)\in X\times X} y \circ z = \bigcup_{(y,z)\in X\times X} (y \circ y \cup z \circ z) = \bigcup_{x\in X} x \circ x.$$

Then, for any $X \subset H$, by Theorem 22, (9)

$$\begin{aligned} X \circ X \circ X &\subset (X \circ X) \circ (X \circ X) = \\ &= \bigcup_{y \in X\circ X} y \circ y = \bigcup_{x\in X}\left(\bigcup_{y\in x\circ x} y\circ y\right) = \\ &= \bigcup_{x\in X} (x\circ x)\circ(x\circ x) = \\ &= \bigcup_{x\in X} x\circ x\circ x \subset X\circ X\circ X, \end{aligned}$$

we have

$$X\circ X\circ X = \bigcup_{x\in X} x\circ x\circ x.$$

Hence specifying $X = x \circ_k x \circ_k x = x \circ_{k+1} x$, by Theorem 22, (8)

$$\begin{aligned} X \circ_k X \circ_k X &= \bigcup_{z\in X} z \circ_k z \circ_k z = \\ &= \bigcup_{x \in x \circ_{k+1} x} z \circ_{k+1} z = (x \circ_{k+1} x) \circ_{k+1} (x \circ_{k+1} x) = \\ &= x \circ_{k+1} x \circ_{k+1} x = x \circ_{k+2} x. \end{aligned}$$

■

30. Theorem. *Every hypergraph* $\Gamma = < H; \{A_i\} >$ *determines a sequence of quasi–hypergroups* $Q_0 = (H; \circ_0)$, $Q_1 = (H; \circ_1), \ldots,$ $Q_m = (H; \circ_m), \ldots$ *such that* $\forall\, k \geq 1$, Q_k *is an enlargement of* Q_{k-1}. *If there exists* s *such that* $Q_s = Q_{s+1}$, *then* Q_{P_Γ} *is a hypergroup for some integer* P_Γ.

Proof. Let $\forall\, x \in H$, $E_0(x) = E(x)$, $E_{k+1}(x) = \bigcup_{y \in E_k(x)} E_k(y)$. We have a sequence of hypergraphs $\Gamma_k = \{E_k(x) \mid x \in H\}$ and of the associated quasihypergroups $H_k = (H; \circ_k)$, where $\forall\, x \in H$, $\forall\, k \geq 0$, $x \circ_k x = E_k(x)$. From Theorem 22, (8), and Theorem 29, (α), for

$$\forall\, k \geq 0, \quad \forall\, x \in H, \quad x \circ_{k+1} x = E_{k+1}(x) = x \circ_k x \circ_k x, \quad \text{and}$$
$$\forall\, (x, y) \in H^2, \quad x \circ_k y = x \circ_k x \cup y \circ_k y.$$

It is clear that $x \circ_{k+1} x \supset x \circ_k x$.

Set $m(x) = \min\{k \mid x \circ_{k+1} x = x \circ_k x\}$.

We have that $\forall\, s \geq m(x)$, $x \circ_s x = x \circ_{m(x)} x$. To see this, it is enough to prove the following implication: $x \circ_{k+1} x = x \circ_k x \Longrightarrow x \circ_{k+2} x = x \circ_{k+1} x$. Indeed, applying in turns Theorem 29, (β)) and Theorem 22, (9)

$$\begin{aligned}
&x \circ_{k+2} x = \\
&= (x \circ_k x \circ_k x) \circ_k (x \circ_k x \circ_k x) \circ_k (x \circ_k x \circ_k x) = \\
&= (x \circ_{k+1} x) \circ_k (x \circ_{k+1} x) \circ_k (x \circ_{k+1} x) = \\
&= (x \circ_k x) \circ_k (x \circ_k x) \circ_k (x \circ_k x) = \\
&= x \circ_k x \circ_k x) \circ_k (x \circ_k x) = (x \circ_{k+1} x) \circ_k (x \circ_k x) = \\
&= (x \circ_k x) \circ_k (x \circ_k x) = x \circ_k x \circ_k x = x \circ_{k+1} x.
\end{aligned}$$

Now let

$$P_\gamma = \max\{m(x) \mid x \in H\}.$$

It is clear that in $(H; \circ_{P_\Gamma})$, $\forall\, y \in H$, $y \circ_{P_\Gamma} y \circ_{P_\Gamma} y = y \circ_{P_\Gamma} y$ and therefore, by Corollary 25, the hypergroupoid $(H; \circ_{P_\Gamma})$ is a hyper–group. ■

31. Definition. Denoting by S the class of semihypergroups, set

$$a_\Gamma = \min\{s \in N^* \mid Q_s \in S\}.$$

32. Examples.

(1) If the edges are disjoint, i.e. $i \neq j \Longrightarrow A_i \cap A_j \neq \emptyset$, then (τ') is clearly satisfied and therefore the hyperproduct defined in (0) is associative.

(2) Let $\Gamma = \{\{1\}, \{1,2\}, \{3,4\}\}$. Also in this case H_Γ satisfies (τ') and therefore it is associative.

(3) Let $\Gamma' = \{\{1,2\}, \{2,3\}\}$. We have $1{\circ}1 = \{1,2\} \neq 1{\circ}1{\circ}1 = \{1,2,3\}$. Then $H_{\Gamma'}$ does not satisfy (τ'), but is satisfies (τ), and therefore by Theorem 24,

$H_{\Gamma'}$ is a hypergroup.

(4) Let $\Gamma'' = \{\{1,2\}, \{2,3\}, \{3,4\}, \{4,5\}\}$. It is $(1{\circ}2){\circ}3 = (1,2,3){\circ}3 = \{1,2,3,4\}$; $1{\circ}(2{\circ}3) = \{1\}{\circ}\{1,2,3,4\} = \{1,2,3,4,5\}$; and therefore $H_{\Gamma''}$ is not associative. Remark: $(1{\circ}1){\circ}(1{\circ}1) = 1{\circ}1{\circ}1 = \{1,2,3\}$, $((1{\circ}1){\circ}1){\circ}1 = \{1,2,3,4\}$.

(5) Let $\Gamma = \{\{1,2\}, \{2,3\}, \{3,4\}, \{4,5\}, \{5,6\}, \{6,7\}, \{7,8\}\}$. We can check that $4 \circ_2 4 = 5 \circ_2 5 = H$, $3 \circ_2 3 = H - \{8\}$, $\forall x \in H$, $x \circ_3 x = H$. Then in $< H; \circ_2 >$ the condition (τ) is satisfied and therefore, by Theorem 24, $< H; \circ_2 >$ is a hypergroup, different from the total hypergroup. Then it is clear that $2 = a_\Gamma < P_\Gamma = 3$.

33. Theorem. *Let* Γ *be a connected finite hypergraph, let* $x_0, x_1, ..., x_n$ *form a trail from* x_0 *to* x_n*, that is,* $A_{i_1}, ..., A_{i_n}$ *exist such that* $\{x_{k-1}, x_k\} \subset A_{i_k}$ *for all* $k \in 1, 2, ..., n$ *and let* $Q_0 = H_\Gamma, ..., Q_{P_\Gamma}$*, be the sequence of quasi–hypergroups associated with* Γ*. Then*

(1) $\forall i,\ \forall k,\ \forall s : 0 \leq s \leq 2^k,\ x_{i+s} \in x_i \circ_k x_i.$

(2) $\{x_i \mid 0 \le i \le n\} \subset x_0 \circ_{g(n)} x_0$, *where* $g(n) = \min\{m \in N \mid m \ge \log_2(n)\}$.

(3) *There is* m *such that* $(H;\, \circ_m\,)$ *is a total hypergroup. If* $\tau_\Gamma = \min\{m \mid (H;\, \circ_m\,)$ *is total* $\}$, *we have:* $a_\Gamma \le P_\Gamma = \tau_\Gamma \le g(\delta(\Gamma))$, *where* $\delta(\Gamma)$ *is the diameter of* Γ.

Proof. (1) For $k = 0$ obviously $\{x_i, x_{i+1}\} \subset x_i \circ_0 x_i$. Let us suppose: $\forall\, i, \forall\, s : 0 \le s \le 2^k$, $x_{i+s} \in x_i \circ_k x_i$, by induction we have:

for $0 \le t \le 2^k$,

$$x_{i+s+t} \in x_{i+s} \circ_k x_{i+s} \subset x_{i+s} \circ_k x_i \subset x_i \circ_k x_i \circ_k x_i = x_i \circ_{k+1} x_i$$

that is, $\forall\, r : 0 \le r \le 2^{k+1}, x_{i+r} \in x_i \circ_{k+1} x_i$.

(2) follows directly from (1).

(3) The first statement is a consequence of (2) and of the hypothesis of connectivity. For the second, it is enough to remark that $\forall\,(x,y) \in H^2$; there is a path c from x to y of length $d(x,y) \le \delta(\Gamma)$. Then $\forall\, y \in H$, $y \in x \circ_{g(d(x,y))} x \subset x \circ_{g(\delta(\Gamma))} x$ and therefore $\forall\, x \in H$, $x \circ_{g(\delta(\Gamma))} x = H$. ■

34. Theorem. *Let* Γ *be a connected finite hypergraph. Then we have:* $a_\Gamma = P_\Gamma - 1$.

Proof. It is clear that $a_\Gamma \le P_\Gamma - 1$. Indeed, if we let $k = P_\Gamma - 1$, we have $\forall\, x \in H$, $x \circ_k x \circ_k x = H$ and therefore

$$\forall\, v \in H,\ v \circ_k v \circ_k v - v \circ_k v \subset x \circ_k x \circ_k x,$$

whence, from Theorem 24, $< H;\, \circ_k\, >$ is a hypergroup. Let us prove now that $a_\Gamma \ge P_\Gamma - 1$ that is, if $\circ_k$ is associative, $k \ge P_\Gamma - 1$.

From Theorem 24, we have

$$\forall\,(x,y) \in H \times H,\ y \circ_k y \circ_k y \supset x \circ_k x \circ_k x - x \circ_k x.$$

In order to prove that, $\forall\, z \in H$, $z \in y \circ_k y \circ_k y$ we remark two cases can occur:

(1) $z \circ_k z \neq z \circ_k z \circ_k z$, then $x \in z \circ_{k+1} z$ exists such that $x \notin z \circ_k z$; it follows from (3) Theorem 21, $z \notin x \circ_k x$ but $x \in z \circ_{k+1} z$ implies $z \in x \circ_{k+1} x$, therefore $z \in x \circ_k x \circ_k x - x \circ_k x \subset y \circ_k y \circ_k y$.

(2) $z \circ_k z = z \circ_k z \circ_k z$. Since Γ is connected, $z \circ_k z \circ_k z = H$. If $z \notin y \circ_{k+1} y$, it follows $y \notin z \circ_{k+1} z$, a contradiction, then $z \in y \circ_{k+1} y$.

Finally, $y \circ_{k+1} y = H$. Since that is true for $\forall y \in H$, it follows $k+1 \geq P_\Gamma$, i.e. $a_\Gamma \geq P_\Gamma - 1$. ■

35. Definition. Let $\Gamma = \{H; \{A_j\}_j\}$ be a hypergraph and let x, y be points of H. We set xRy if and only if either $x = y$ or a trail exists form x to y, in other words R is the least equivalence relation which contains the relation R' defined by Γ, i.e.

$$xR'y \iff \exists j : \{x, y\} \subset A_j.$$

$\forall x \in H$, let $R(x)$ be the equivalence class mod R, determined by x.

36. Definition. If Γ is any finite hypergraph, and $C_1, C_2, ..., C_q$ are the connected components of Γ, set $\tau_\Gamma = \max\{\tau_C \mid 1 \leq i \leq q\}$, $\forall x \in H$, let $\Gamma(x)$ be the connected component of Γ to which x belongs.

37. Theorem. *Let Γ be a finite hypergraph. Then $\forall x \in H$, we have $x \circ_{\tau_{\Gamma(x)}} x = R(x)$.*

Proof. We prove the theorem by induction. It is clear that $\forall x$, $x \circ_0 x \subset R(x)$. Let us suppose $x \circ_{k-1} x \subset R(x)$.

We have $x \circ_k x = x \circ_{k-1} x \circ_{k-1} x$, thus, by Theorem 22, (8), if $z \in x \circ_k x$, there is $y \in x \circ_{k-1} x$ such that $z \in y \circ_{k-1} y$, therefore $zRyRx$ and then $\forall k$, $x \circ_k x \subset R(x)$ whence the theorem. ■

38. Theorem. *Let Γ be a finite hypergraph. Then, $\forall x \in H$, we have $x \circ_{\tau_\Gamma} x = R(x)$.*

Proof. It is immediate from Definition 36 and Theorem 37. ■

§5. On the hypergroup H_Γ associated with a hypergraph Γ

In the previous paragraph, we have seen that, given a hypergraph Γ on a set H, that is a family $\Gamma = \{A_i\}_{i \in I}$ of non empty subsets A_i of H such that $\bigcup_{i \in I} A_i = H$, we can associate with Γ, a hypergroupoid defined by

$$\forall x \in H,\ x \circ_1 x = \bigcup_{A_i \ni x} A_i,\ \forall (x,y),\ x \circ_1 y = x \circ_1 x \cup y \circ_1 y$$

and a sequence of hypergroupoids $((H;\, \circ_k\,))_{k \in \mathbb{N}^*}$, where for $k > 1$

$$x \circ_k x = x \circ_{k-1} x \circ_{k-1} x,\ x \circ_k y = x \circ_k x \cup y \circ_k y.$$

Now, we consider the case $\Gamma = \{(1,2),(2,3),...,(n-1,n)\}$, then the case when Γ is connected and finally when Γ is a tree.

The following results have been obtained by P. Corsini.

Let "$\circ_i$" be the hyperoperation defined $\forall x \in H,\ x \circ_i x = \{y \mid d(x,y) \leq i\}$ (where $d(x,y)$ is the graphic distance between x and y, that is the length of the shortest path between x and y),

$$\forall (x,y) \in H^2,\ x \circ_i y = x \circ_i x \cup y \circ_i y.$$

Let δ be the *diameter* (if it exists) of Γ, that is

$$\delta = \max\{d(x,y) \mid \{x,y\} \subset H\}.$$

If we set $xD_ky \iff x \in y \circ_k y$ and $xR_iy \iff d(x,y) \leq i$, we have $D_k = R_{2^{k-1}}$. So, $x \circ_k y = x \circ_{2^{k-1}} y$. Set $I(n) = \{1,2,...,n\}$.

39. Proposition. *$\forall s \in I(n)$, we have*

$$D_k(s) = s \circ_k s = \{x \mid \min\{n, s+2^{k-1}\} \geq x \geq \max\{1, s-2^{k-1}\}\}.$$

Proof. We prove the Proposition by induction on k. Set $k = 1$. We have

$$\forall s \in I(n) - \{1, n\},$$
$$s \circ_1 s = \{s-1, s, s+1\},\ 1 \circ_1 1 = \{1, 2\},\ n \circ_1 n = \{n, n-1\}.$$

It follows $s \circ_1 s \circ_1 s = \{s-2, s-1, s, s+1, s+2\}$ if $s - 2 \geq 1$, $s + 2 \leq n$. Moreover,

$$\forall s \in I(n),\ s \circ_1 s \circ_1 s = \{x \mid \min\{n, s+2\} \geq x \geq \max\{1, s-2\}\}.$$

Set $\forall \{\alpha, \beta\} \subset I(n)$, $I_1^n(\alpha, \beta) = \{x \mid \min\{n, \alpha\} \geq x \geq \max\{1, \beta\}\}$ and $T_k(s) = s \circ_k s \circ_k s$. Now, set by inductive hypothesis:

$$T_{k-1}(s) = I_1^n(s + 2^{k-1}, s - 2^{k-1}).$$

Then by [[74], Th. 5], we have

$$T_k(s) = (T_{k-1}(s) \circ_{k-1} T_{k-1}(s)) \circ_{k-1} T_{k-1}(s),$$

whence

$$T_k(s) =$$
$$=(I_1^n(s+2^{k-1}, s-2^{k-1}) \circ_{k-1} I_1^n(s+2^{k-1}, s-2^{k-1}) \circ_{k-1} I_1^n(s+2^{k-1}, s-2^{k-1})$$

It follows

$$T_k(s) = I_1^n(s+2^{k-1}+2^{k-2}, s-2^{k-1}-2^{k-2}) \circ_{k-1} I_1^n(s+2^{k-1}, s-2^{k-1}).$$

Finally,

$$T_k(s) = I_1^n(s + 2^{k-1} + 2^{k-2} + 2^{k-2}, s - 2^{k-1} - 2^{k-2} - 2^{k-2}),$$

so we have $T_k(s) = I_1^n(s + 2^k, s - 2^k)$. ∎

40. Corollary. $\forall s \in H$, $\forall k \geq 1$, *we have* $s \circ_k s = s \circ_{2^{k-1}} s$.

Proof. Immediate.

Let us suppose now that Γ is connected.

41. Theorem.

a) *If* R_i *has not outer elements (see Def. 5, Chapter 3), then* $(H;\, \circ_i)$ *is a join space. Let* Γ *be a tree and let* $(H;\, \circ_i)$ *be a hypergroup. Then*

b) R_i *has not outer elements.*

c) *We have* $\delta \leq 2i$.

Proof. a) The hypothesis implies the condition 4 of Theorem 8, Chapter 3, to be vacuous. The conditions 1,2,3 of Theorem 8, Chapter 3, are satisfied because R_i is reflexive. Therefore $(H;\, \circ_i)$ is a hypergroup. From Theorem 3 [38], follows that $< H;\, \circ_i >$ is a join space.

b) Suppose to the contrary that x is outer. Then there exists $h \in H$ such that $(h, x) \notin R_i^2$. Since Γ is a tree we have $R_i^2 = R_{2i}$. Therefore $(h, x) \notin R_{2i}$ whence $d(h, x) > 2i$. Let π be the path between h and x. Let p be the element of this path at a distance $2i$ from x. Then $(p, x) \in R_i^2 - R_i$, hence R_i does not satisfy 4 of Theorem 8, Chapter 3, and so $(H;\, \circ_i)$ is not a hypergroup.

c) It follows from the following remarks:

1. If R is a relation on H, then R is transitive if and only if $\forall\, a \in H$, we have $a \circ_R a \circ_R a = a \circ_R a$.

2. If R is a symmetric nontransitive relation on H, such that $R \subset R^2$, then H_R is a hypergroup if and only if $\forall\, x \in H$, we have $x \circ_R x \circ_R x = H$. ∎

42. Lemma. *Let* $< H; \Gamma >$ *be a connected graph,* $i \in \mathbb{N}^*$ *and* $< H;\, \circ_i >$ *the associated hypergroupoid. Then* $\forall\, (a, b, c) \in H^3$, *we have*

$$(a \circ_i b \circ_i c) = K'(a, b, c) = \\ = \{\lambda \mid d(a, \lambda) \leq 2i\} \cup \{\mu \mid d(b, \mu) \leq 2i\} \cup \{\nu \mid d(c, \nu) \leq i\}$$

$$a \circ_i (b \circ_i c) = K''(a, b, c) = \\ = \{x \mid d(a, x) \leq i\} \cup \{y \mid d(b, y) \leq 2i\} \cup \{z \mid d(c, z) \leq 2i\}.$$

Proof. We have $(a \circ_i b) \circ_i c \subset K'(a,b,c)$. Indeed,

$$\begin{aligned}(a \circ_i b) \circ_i c &= (\{y \mid d(a,y) \leq i\} \cup \{z \mid d(b,z) \leq i\} \circ_i c = \\ &= \{u \mid d(u,y) \leq i, d(y,a) \leq i\} \cup \\ &\cup \{v \mid d(v,z) \leq i, d(z,b) \leq i\} \cup \{w \mid d(w,c) \leq i\}.\end{aligned}$$

So,

$$(a \circ_i b) \circ_i c \subset K'(a,b,c) =$$
$$= \{\lambda \mid d(\lambda,a) \leq 2i\} \cup \{\mu \mid d(\mu,b) \leq 2i\} \cup \{w \mid d(w,c) \leq i\}.$$

Let us see now that also $(a \circ_i b) \circ_i c \supset K'(a,b,c)$.

Let $x \in \{\lambda | d(\lambda,a) \leq 2i\}$. Then there is $q \leq 2i$ such that $d(x,a)=q$.

If $q \leq i$, then $x \in a \circ_i a \subset (a \circ_i b) \circ_i c$.

If $q > i$, there is a path π between a and x, there are $t \leq i$ and $w \in \pi$ such that

$$d(a,w) = t, \; d(w,x) \leq p = q - t \leq i.$$

So $w \in a \circ_i a$ and $x \in w \circ_i w \subset w \circ_i c$, proving $x \in (a \circ_i a) \circ_i c \subset$ $\subset (a \circ_i b) \circ_i c$.

Analogously, one sees that $\{\mu \mid d(\mu,b) \leq 2i\} \subset (a \circ_i b) \circ_i c$; hence

$$K' \subset (a \circ_i b) \circ_i c.$$

In a similar way, it can be proved that

$$\begin{aligned}&a \circ_i (b \circ_i c) = K''(a,b,c) = \\ &\quad = \{x \mid d(x,a) \leq i\} \cup \{y \mid d(y,b) \leq 2i\} \cup \{z \mid d(z,c) \leq 2i\}.\end{aligned}$$

■

43. Theorem. *Let* $< H;\Gamma >$ *be a connected graph of finite diameter* δ. *Then* $\delta \leq 2i$ *if and only if* $\forall (a,b,c), (a \circ_i b) \circ_i c = H =$ $= a \circ_i (b \circ_i c)$.

Proof. Set $\delta \leq 2i$. Then it follows that

$$\forall q \in H, \; \{x \mid d(qx) \leq 2i\} = H,$$

therefore $\forall\,(a,b,c)$

$$\begin{aligned}(a\circ_i b)\circ_i c &=\\ &=\{x \mid d(ax)\le 2i\}\cup\{y \mid d(by)\le 2i\}\cup\{z \mid d(az)\le i\}=H=\\ &=\{x \mid d(ax)\le i\}\cup\{y \mid d(by)\le 2i\}\cup\{z \mid d(az)\le 2i\}.\end{aligned}$$

Whence $\circ_i$ is trivially associative since $\forall\,(a,b,c)$

$$(a\circ_i b)\circ_i c = H = a\circ_i (b\circ_i c)$$

For the converse, set $\forall\,(a,b,c)\in H^3$, $(a\circ_i b)\circ_i c = H$. Then $\forall\, a$ we have

$$H=(a\circ_i a)\circ_i a = K''(a,a,a)=\{x \mid d(ax)\le 2i\}$$

therefore $\delta\le 2i$. ■

§6. Other hyperstructures associated with hypergraphs

In this paragraph a new type of hypergroups associated with hypergraphs is defined. Some properties of their subhypergroups are studied. This is a generalization of hypergroups associated with graphs, given by P. Corsini. These results are obtained by V. and L. Leoreanu.

Let $\langle H,(A_i)_{i\in J}\rangle$ be a hypergraph, that is $\bigcup\limits_{i\in J} A_i = H$ and $\forall\, i\in J,\ A_i\ne\emptyset$.

Let x,y be two different points of H. We say that there is a *trail* between x and y if there is $\{x=x_0,x_1,\ldots,x_n=y\}\subset H$ and $\{j_1,j_2,\ldots,j_n\}\subset J$, such that $\forall\, i\in\{0,1,\ldots,n-1\}$, $\exists j_{i+1}\in J$, so that we have $\{x_i,x_{i+1}\}\subset A_{j_{i+1}}$ and $i\ne i' \Longrightarrow \{x_i,x_{i+1}\}\ne\{x_{i'},x_{i'+1}\}$. The trail is a *path* if the vertices are different.

We shall denote by $\gamma(x,y)$ the set of all the paths between x and y in $\langle H,(A_i)_{i\in J}\rangle$.

If $\pi \in \gamma(x,y)$, $\pi : x = x_0, x_1, ..., x_n = y$ and $e_k^{\pi} = [x_{k-1}, x_k]$ is the k–edge of π, counting from x, set $\alpha_{\pi} = \{j \in J \mid \exists k : e_k^{\pi} \subset A_j\}$ and $\alpha_{x,y} = \bigcup_{\pi \in \gamma(x,y)} \alpha_{\pi}$.

Let us suppose that $\langle H, (A_i)_{i \in J} \rangle$ is a connected hypergraph.

Let us define the hyperoperation on the set H as follows:

$$x \circ y = \begin{cases} \bigcup_{j \in \alpha_{x,y}} A_j, & \text{if } x \neq y \\ \{x\}, & \text{if } x = y \end{cases}$$

44. Remarks.

1. "$\circ$" is a commutative hyperoperation.

2. $\{x, y\} \subset x \circ y$, so that $\langle H, \circ \rangle$ is a quasihypergroup.

45. Proposition. *For any x, y in H, we have*

$$(x \circ x) \circ y = x \circ (x \circ y) = x \circ y.$$

Proof. It is sufficiently to check that for any distinct x, y in H, we have $x \circ (x \circ y) = x \circ y$.

Let s be arbitrary in $x \circ (x \circ y)$; then there is $t \in x \circ y$, such that $s \in x \circ t$. We need to prove $s \in x \circ y$.

If $t = x$, then $s = x \in x \circ y$.

If $t = y$, then $s \in x \circ y$.

Let us suppose $t \notin \{x, y\}$. Since $t \in x \circ y$, it follows that there are $\pi_1 \in \gamma(x,y)$ and $h \in \alpha_{\pi_1}$ such that $t \in A_h$.

$\pi_1 : x = \alpha_0, \alpha_1, ..., \alpha_{n-1}, \alpha_n = y$

Since $s \in x \circ t$, it results that there is $\pi_2 \in \gamma(x,t)$ and there is $k \in \alpha_{\pi_2}$, such that $s \in A_k$.

$\pi_2 : x = \beta_0, \beta_1, ..., \beta_{m-1}, \beta_m = t$

Let

$$J_1 = \{0, 1, 2, ..., k-1\}, \quad J_2 = \{k, k+1, ..., m\}$$
$$I_1 = \{0, 1, 2, ..., h-1\}, \quad I_2 = \{h, h+1, ..., n\}$$

We shall consider the following situations:

I. If there is $(i,j) \in (I_1 \cup I_2) \times J_2$ such that $\alpha_i = \beta_j$ and there is not $(i,j) \in (I_1 \cup I_2) \times J_1$, such that $\alpha_i = \beta_j$, set

$$\rho = \min\{j \in J_2 \mid \exists I_1 \cup I_2 : \alpha_i = \beta_j\} \text{ and } \beta_p = \alpha_{\bar{p}}.$$

Then the trail

$$x = \beta_0, \beta_1, ..., \beta_p = \alpha_{\bar{p}}, \alpha_{\bar{p}+1}, ..., \alpha_n = y$$

is a path and $s \in x \circ y$.

II. If there is no $(i,j) \in (I_1 \cup I_2) \times J_2$ such that $\alpha_i = \beta_j$ and there is $(i,j) \in (I_1 \cup I_2) \times J_1$, such that $\alpha = \beta_j$, set

$$p' = \max\{j \in J_1 \mid \exists i \in I_1 \cup I_2 : \alpha_i = \beta_j\} \text{ and } \beta_{p'} = \alpha_{\bar{p}'}$$

We have two possibilities:

II.1. If $p' = \max\{j \in J_1 \mid \exists i \in I_1 : \alpha_i = \beta_j\}$ then $\bar{p}' \in I_1$ and the trail:

$$x = \alpha_0, \alpha_1, ..., \alpha_{\bar{p}'} = \beta_{p'}, \beta_{p'+1}, ..., \beta_m = t,\ \alpha_{h+1}, \alpha_{h+2}, ..., \alpha_n$$

is a path and $s \in x \circ y$.

II.2 If $p' = \max\{j \in J_1 \mid \exists i \in I_2 : \alpha_i = \beta_j\}$ then $\bar{p}' \in I_2$ and the trail:

$$x = \alpha_0, \alpha_1, ..., \alpha_{h-1}, t = \beta_m, \beta_{m-1}, ..., \beta_{p'} = \alpha_{\bar{p}'}, \alpha_{\bar{p}'+1}, ..., \alpha_n = y$$

is a path and $s \in x \circ y$.

III. If there is no $(i,j) \in (I_1 \cup I_2) \times (J_1 \cup J_2)$, such that $\alpha_i = \beta_j$, then the trail

$$x = \beta_0, \beta_1, ..., \beta_m = t, \alpha_{h+1}, \alpha_{h+2}, ..., \alpha_n = y$$

is a path and $s \in x \circ y$.

IV. If there are $(i,j) \in (I_1 \cup I_2) \times J_2$ and $(i',j') \in (I_1 \cup I_2) \times J_1$, such that $\alpha_i = \beta_j$ and $\alpha_{i'} = \beta_{j'}$, set

$$\begin{array}{rclll} p & = & \min\{j \in J_2 \mid \exists i \in I_1 \cup I_2 : \alpha_i = \beta_j\}, & \beta_p = \alpha_{\bar{p}} & \text{and} \\ p' & = & \max\{j' \in J_1 \mid \exists i' \in I_1 \cup I_2 : \alpha_{i'} = \beta_{j'}\}, & \beta_{p'} = \alpha_{\bar{p}'}. & \end{array}$$

We shall consider the following possibilities:

IV.1. If $p' = \max\{j \in J_1 \mid \exists i \in I_1 : \alpha_i = \beta_j\}$ and $\bar{p} < \bar{p}'$, then the trail:

$$x = \alpha_0, \alpha_1, ..., \alpha_{\bar{p}} = \beta_p, \beta_{p-1}, ..., \beta_{p'} = \alpha_{\bar{p}'}, \alpha_{\bar{p}'+1}, ..., \alpha_n = y$$

is a path and $s \in x \circ y$.

IV.2. If $p' = \max\{j \in J_1 \mid \exists i \in I_1 : \alpha_i = \beta_j\}$ and $\bar{p}' < \bar{p}$, then the trail:

$$x = \alpha_0, \alpha_1, ..., \alpha_{\bar{p}'} = \beta_{p'}, \beta_{p'+1}, ..., \beta_p = \alpha_{\bar{p}}, \alpha_{\bar{p}+1}, ..., \alpha_n = y$$

is a path and $s \in x \circ y$.

IV.3. If $p' = \max\{j \in J_1 \mid \exists i \in I_1 : \alpha_i = \beta_j\}$ and $\bar{p} < \bar{p}'$, we consider the same path as at IV.1, and we have that $s \in x \circ y$.

IV.4. If $p' = \max\{j \in J_1 \mid \exists i \in I_1 : \alpha_i = \beta_j\}$ and $\bar{p}' < \bar{p}$, we consider the same path as at IV.2, and we have that $s \in x \circ y$. ∎

46. Theorem. *$\langle H, \circ \rangle$ is a regular reversible hypergroup.*

Proof. First, we verify the associativity. It remains to check that:

$\forall\, (x,y,z) \in H^3;\ x \neq y \neq z \neq x,$ we have $(x \circ y) \circ z = x \circ (y \circ z)$.

We show that:

$$(*) \qquad \forall\, (x,y,z) \in H^3,\ (x \circ y) \circ z = x \circ z \cup x \circ y.$$

By $(*)$ and by Remark 20,

$$\forall\, (x,y,z) \in H^3, (x \circ y) \circ z \subset x \circ (y \circ z).$$

Therefore,

$$x\circ(y\circ z)=(y\circ z)\circ x\subset y\circ(z\circ x)=(z\circ x)\circ y\subset z\circ(x\circ y)=(x\circ y)\circ z.$$

Hence, if $(*)$ holds, we have:

$$\forall\,(x,y,z)\in H^3,\ (x\circ y)\circ z=x\circ(y\circ z).$$

It sufficis to verify that

$$\forall\,(x,y,z)\in H^3,\ x\neq y\neq z\neq x,\ \text{we have } (x\circ y)\circ z\subseteq x\circ z\cup x\circ y.$$

Let s be an arbitrary element of $x\circ y$ and w an arbitrary element of $s\circ z$. We need to prove that $w\in x\circ z\cup x\circ y$.

If $s=z$, then $w\in z\circ z=\{z\}$, so $w=z\in x\circ z\cup x\circ y$.

Next, we consider $s\neq z$.

Since $s\in x\circ y$, there are $\pi_1\in\gamma(x,y)$ and $k\in J$, such that $s\in A_k$

$$\pi_1: x=\beta_0,\beta_1,...,\beta_{k-1},\beta_k,...,\beta_m=y.$$

Since $w\in s\circ z$, there are $\pi_2\in\gamma(s,z)$ and $h\in J$, such that $w\in A_h$

$$\pi_2: x=\alpha_0,\alpha_1,...,\alpha_{h-1},\alpha_h,...,\alpha_n=z.$$

Set

$$J_1=\{0,1,...,k-1\},\quad J_2=\{k,k+1,...,m\}$$
$$I_1=\{0,1,...,h-1\},\quad I_2=\{h,h+1,...,n\}$$

We consider the following situations:

I. If there are $(i,j)\in I_1\times(J_1\cup J_2)$ and $(i',j')\in I_2\times(J_1\cup J_2)$, such that $\alpha_i=\beta_j$ and $\alpha_{i'}=\beta_{j'}$, then set

$$p_1=\min\{i\in I_2\mid \exists j\in J_1\cup J_2:\alpha_i=\beta_j\},\qquad \alpha_{p_1}=\beta_{\bar{p}_1},$$
$$p_2=\min\{i'\in I_1'\mid \exists j'\in J_1\cup J_2:\alpha_{i'}=\beta_{j'}\},\quad \alpha_{p_2}=\beta_{\bar{p}_2}.$$

We have the following cases:

I.1. If $p_1=\min\{i\in I_2\mid \exists j\in J_1:\alpha_i=\beta_j\}$ and $\bar{p}_1<\bar{p}_2$, then the trail:

$$x=\beta_0,\beta_1,...,\beta_{\bar{p}_1}=\alpha_{p_1},\alpha_{p_1-1},...,\alpha_{p_2}=\beta_{\bar{p}_2},\beta_{\bar{p}_2+1},...,\beta_m=y$$

is a path and $w\in x\circ y$.

I.2. If $p_1 = \min\{i \in I_2 \mid \exists j \in J_1 : \alpha_i = \beta_j\}$ and $\bar{p}_2 < \bar{p}_1$, then the trail:

$$x = \beta_0, \beta_1, ..., \beta_{\bar{p}_2} = \alpha_{p_2}, \alpha_{p_2+1}, ..., \alpha_{p_1} = \beta_{\bar{p}_1}, \beta_{\bar{p}_1+1}, ..., \beta_m = y$$

is a path and $w \in x \circ y$.

I.3. If $p_1 = \min\{i \in I_2 \mid \exists j \in J_2 : \alpha_i = \beta_j\}$ and $\bar{p}_2 < \bar{p}_1$, then the trail:

$$x = \beta_0, \beta_1, ..., \beta_{\bar{p}_2} = \alpha_{p_2}, \alpha_{p_2+1}, ..., \alpha_{p_1} = \beta_{\bar{p}_1}, \beta_{\bar{p}_1+1}, ..., \beta_m = y$$

is a path and $w \in x \circ y$.

I.4. If $p_1 = \min\{i \in I_2 \mid \exists j \in J_2 : \alpha_i = \beta_j\}$ and $\bar{p}_1 < \bar{p}_2$, then the trail:

$$x = \beta_0, \beta_1, ..., \beta_{\bar{p}_1} = \alpha_{p_1}, \alpha_{p_1-1}, ..., \alpha_{p_1} = \beta_{\bar{p}_2}, \beta_{\bar{p}_2+1}, ..., \beta_m = y$$

is a path and $w \in x \circ y$.

II. If there is $(i, j) \in I_1 \times (J_1 \cup J_2)$ such that $\alpha_i = \beta_j$ and there is no $(i, j) \in I_2 \times (J_1 \cup J_2)$, such that $\alpha_i = \beta_j$, let

$$p = \max\{i \in I_1 \mid \exists j \in J_1 \cup J_2 : \alpha_i = \beta_j\}, \ \alpha_p = \beta_{\bar{p}}.$$

Then the trail:

$$x = \beta_0, \beta_1, ..., \beta_{\bar{p}} = \alpha_p, \alpha_{p+1}, ..., \alpha_n = z$$

is a path and $w \in x \circ z$.

III. If there is no $(i, j) \in I_1 \times (J_1 \cup J_2)$ such that $\alpha_i = \beta_j$ and there is $(i, j) \in I_2 \times (J_1 \cup J_2)$, such that $\alpha_i = \beta_j$, let

$$p = \min\{i \in I_2 \mid \exists j \in J_1 \cup J_2 : \alpha_i = \beta_j\}, \ \alpha_p = \beta_{\bar{p}}.$$

We have the following cases:

III.1. If $p = \min\{i \in I_2 \mid \exists j \in J_1 : \alpha_i = \beta_j\}$, then $\bar{p} \in J_1$. The trail:

$$x = \beta_0, \beta_1, .., \beta_{\bar{p}} = \alpha_p, \alpha_{p-1}, ..., \alpha_0 = s, \ \beta_{k+1}, \beta_{k+2}, ..., \beta_m = y$$

is a path and $w \in x \circ y$.

III.2. If $p = \min\{i \in I_2 \mid \exists j \in J_2 : \alpha_i = \beta_j\}$, then $\bar{p} \in J_2$. The trail:

$$x = \beta_0, \beta_1, .., \beta_{k-1}, s = \alpha_0, \alpha_1, ..., \alpha_p = \beta_{\bar{p}}, \beta_{\bar{p}+1}, ..., \beta_m = y$$

is a path and $w \in x \circ y$.

IV. If there is no $(i, j) \in (I_1 \cup I_2) \times (J_1 \cup J_2)$, such that $\alpha_i = \beta_j$, then the trail:

$$x = \beta_0, \beta_1, ..., \beta_{k-1}, s = \alpha_0, \alpha_1, ..., \alpha_n = z$$

is a path and $w \in x \circ z$.

Therefore, $\langle H, \circ \rangle$ is a hypergroup.

Since for any $(a, x) \in H^2$, we have $a \in x \circ a = a \circ x$, it follows that any element of H is an identity of H and H is the set of inverses of an arbitrary element of H. Therefore, H is a regular hypergroup.

Let $(a, b, c) \in H^3$, such that $a \in b \circ c$; there is $c' = b$ inverse of c, be such that $b \in a \circ c'$ and there is $b' = c$ inverse of b such that $c \in b' \circ a$, whence it follows that H is a regular reversible hypergroup.

47. Remark. $(H, \circ)$ is a join space if and only if $\langle H, (A_i)_{i \in J} \rangle$ is a tree.

Indeed, if there is at least one $i_0 \in J$, such that $|A_{i_0}| \geq 3$, this means there are y, a, b in A_{i_0}, $y \neq a \neq b \neq y$; then we can consider $x \in H$, $x \notin \{a, b\}$, such that there is $i \in J : \{x, y\} \subset A_i$.

We have: $a \in A_{i_0} \cup A_i \subset x \circ b$ and $b \in A_{i_0} \cup A_i \subset x \circ a$, so that $x \in a/b \cap b/a$, but $a \circ a \cap b \circ b = \{a\} \cap \{b\} = \emptyset$. Therefore,

$\langle H, (A_i)_{i\in J}\rangle$ is a graph. But the only type of connected graph, for which the associated hypergroup is a join space is a tree (see [71]). ■

We present the following results on subhypergroups of $(H, \circ)$.

48. Proposition.

(i) *For any* $n \in \mathbb{N}^*$ *and for any* $(x_1, ..., x_n) \in H^n$, *the set* $\prod_{j=1}^{n} x_j$ *is a subhypergroup of* H.

(ii) *Any finite subhypergroup of* H *can be written as a hyperproduct of elements of* H.

(iii) *There are hypergraphs, whose hypergroups have subhypergroups, that are not hyperproducts.*

(iv) *The only closed subhypergroup of* H *is* H.

Proof. (i) Let $S = \prod_{j=1}^{n} x_j$ and a an arbitrary element of S. We need to prove that

$$a \circ S = S.$$

Indeed, $\forall\, s \in S,\ s \in a \circ s$, so $S \subset a \circ S$.

Let $t \in S$. Then, since for any $x \in H$, we have $x \circ x = x$ and by the associativity and the commutativity, it results:

$$a \circ t \subset S \circ S = \prod_{j=1}^{n} x_j \circ \prod_{j=1}^{n} x_j = \prod_{j=1}^{n} x_j = S.$$

(ii) Let $S = \{x_1, x_2, ..., x_n\}$be a finite subhypergroup of H. Then $S = \prod_{j=1}^{n} x_j$.

(iii) We can consider the following examples:

1°. Let $H = \mathbb{N}$ be the graph, for which there is an edge between i and j, where $\{i, j\} \subset \mathbb{N}$, if i and j are consecutive numbers. Then

$$i \circ j = \{k \in \mathbb{N} \mid \min\{i, j\} \leq k \leq \max\{i, j\}\}$$

It results that, for any $\{i_1, i_2, ..., i_s\} \subset \mathbb{N}$, $s \in \mathbb{N}$, $s \geq 2$,

$$\prod_{j=1}^{s} i_j = \{k \in \mathbb{N} \mid \min\{i_1, i_2, ..., i_s\} \leq k \leq \max\{i_1, i_2, ..., i_s\}$$

is a finite set.

For $i_0 \in \mathbb{N}$, the set $S = \{j \in \mathbb{N} \mid j \geq i_0\}$is an infinite subhypergroup of H and S is not a hyperproduct.

2°. Let $\langle H, (A_i)_{i \in \mathbb{N}} \rangle$ be a hypergraph (that is $\forall\, i$, $A_i \neq \emptyset$ and $\bigcup\limits_{i \in I} A_i = H$) such that: for any $i \in \mathbb{N}$, A_i is the smallest subset of $\mathbb{Q}_+$, containing i and $i + 1$ and such that if $\{x, y\} \subset A_i$, then

$$\frac{x + y}{2} \in A_i.$$

For $x \in \mathbb{R}$, the number $[x]$ is the greatest integer not exceeding x.

Then, for any $\{x, y\} \subset H$, we have

$$x \circ y = \begin{cases} \bigcup\limits_{\min\{[x],[y]\} \leq k \leq \max\{[x],[y]\}} A_k & , \text{ if } x \neq y \\ \\ x & , \text{ if } x = y \end{cases}$$

whence for any $m \in \mathbb{N}$, $m \geq 2$, and for any different elements $x_1, x_2, ..., x_m$ of H, we have:

$$\prod_{j=1}^{m} x_j = \bigcup_{\min\{[x_1],[x_2],...,[x_m]\} \leq k \leq \max\{[x_1],[x_2],...,[x_m]\}} A_k,$$

that is a bounded set.

But, for any $j_0 \in \mathbb{N}$, $S = \{j \in H \mid j \geq j_0\}$ is an unbounded subhypergroup of H, so S can not be written as a hyperproduct.

(iv) Let S be a subhypergroup of H, $S \neq H$ and let $x \in S$ and $y \in H - S$. We have $x \in y \circ x$, so S is not a closed subhypergroup.

Therefore, H has no proper closed or invertible or ultraclosed or complete part subhypergroup. ■

Chapter 3

Binary Relations

The first connection between a hyperstructure and a binary relation is implicit in Nieminen [300], who associated a hypergroup with a connected simple graph.

In the same direction, albeit with different hyperoperations associated with graphs, went the papers by Corsini ([74], [79]) and Rosenberg ([326]) and, in the following, by V. Leoreanu and L. Leoreanu ([238]).

Later, Chvalina ([38]) found a correspondence between partially ordered sets and hypergroups. Rosenberg ([326]) generalized Chvalina definition, associating with any binary relation a hypergroupoid.

Rosenberg hypergroup was studied by Corsini ([79]) and then, by Corsini and Leoreanu ([88]), who considered hypergroups associated with union, intersection, product, Cartesian product, direct limit of relations, as we have seen before.

There are still open problems on this subject. One of them is to find necessary and sufficient conditions for the hypergroupoids associated with union, intersection, product etc, to be hypergroups. Recently, Spartalis, De Salvo and Lo Faro have obtained new results on hyperstructures associated with binary relations.

§1. Quasi-order hypergroups

Quasi-order hypergroups have been introduced and studied by Jan Chvalina.

1. Definition. Let $(H, \cdot)$ be a hypergroupoid. We say that H is a *quasi–order hypergroup* (that is a *hypergroup determined by a quasi–order*) if $\forall\,(a, b) \in H^2$, $a \in a^3 \subseteq a^2$ and $a \cdot b = a^2 \cup b^2$. Moreover, if the following implication holds:

$$a^2 = b^2 \Longrightarrow a = b$$

for any $(a, b) \in H^2$, then $(H, \cdot)$ is called an *order hypergroup*.

2. Proposition. *A hypergroupoid $(H, \cdot)$ is a (quasi)–order hypergroup if and only if there exists a (quasi)–order ρ on the set H, such that*

$$\forall\,(a, b) \in H^2, \quad a \cdot b = \rho(a) \cup \rho(b).$$

Proof. "$\Longrightarrow$" Let $(H, \cdot)$ be a quasi–order hypergroup. Let us define on H, the following binary relation:

$$a\rho b \Longleftrightarrow b \in a^2.$$

ρ is reflexive, since $\forall\, a \in H$, we have $a \in a^3 \subseteq a^2$.

If $a\rho d$ and $d\rho b$, then $d \in a^2$ and $b \in d^2 \subseteq a^4 = a^2$ (since $a^3 = a^2$), so that $a\rho b$, that means ρ is transitive.

Thus, ρ is a quasi–order on H and

$$\forall\,(a, b) \in H^2,\ a \cdot b = a^2 \cup b^2 = \rho(a) \cup \rho(b).$$

Now, let $(H, \cdot)$ be an order hypergroup. The conditions $a\rho b$ and $b\rho a$ imply $a \in b^2$ and $b \in a^2$, whence $a^2 \subseteq b^4 = b^2$, $b^2 \subseteq a^4 = a^2$, that means $a^2 = b^2$. Since $(H, \cdot)$ is an order hypergroup, we obtain $a = b$, so that ρ is an order.

"$\Longleftarrow$" Let (H, ρ) be a quasi–ordered set. If we define on H the hyperoperation $a \cdot b = \rho(a) \cup \rho(b)$, then $(H, \cdot)$ is a hypergroup satisfying $a \in a^2 = a^3$ and $a^2 = \rho(a)$, for any $a \in H$.

Moreover, if ρ is antisymmetric and if we have $a^2 = b^2$ (for $(a,b) \in H^2$) then $\rho(a) = \rho(b)$, that means $a\rho b$ and $b\rho a$, so we obtain $a = b$. ■

3. Notations. For any $(a,b) \in H^2$, we denote

$$L_\rho(a,b) = \rho^{-1}(a) \cap \rho^{-1}(b) \quad \text{and} \quad U_\rho(a,b) = \rho(a) \cap \rho(b).$$

4. Theorem. *Let $(H,\cdot)$ be a quasi–order hypergroup and ρ the associated quasi–order on H. The following conditions are equivalent:*

(i) *$(H,\cdot)$ is a join space;*

(ii) *for $\forall\,(a,b) \in H^2$, such that $a \cdot b \subseteq c^2$ for a suitable element $c \in H$, there exists an element $d \in H$, such that $d^2 \subseteq a^2 \cap b^2$;*

(iii) *for $\forall\,(a,b) \in H^2$, such that $L_\rho(a,b) \neq \emptyset$, we also have $U_\rho(a,b) \neq \emptyset$.*

Proof. (i)$\Longrightarrow$(ii) Let $(H,\cdot)$ be a join space and $(a,b) \in H^2$, such that $\exists c \in H : a \cdot b \subseteq c^2$. We have $\rho(a) \cup \rho(b) = a^2 \cup b^2 = a \cdot b \subseteq c^2 = = \rho(c)$ so $a \in \rho(a) \subseteq \rho(c)$, $b \in \rho(b) \subseteq \rho(c)$. Hence $a \in \rho(b) \cup \rho(c) = bc$, $b \in \rho(a) \cup \rho(c) = a \cdot c$, whence $c \in a/b \cap b/a$. Therefore, $a^2 \cap b^2 \neq \emptyset$ and for any $d \in a^2 \cap b^2$, we have $d^2 = \rho(d) \subseteq \rho(a^2 \cap b^2) = \rho(\rho(a) \cap \rho(b)) \subseteq \subseteq \rho^2(a) \cap \rho^2(b) \subseteq \rho(a) \cap \rho(b) = a^2 \cap b^2$, so we obtain (ii).

(ii)$\Longrightarrow$(iii) Let $(a,b) \in H^2$, such that $L_\rho(a,b) \neq \emptyset$. Then there is $c \in L_\rho(a,b) = L_\rho(a,a) \cap L_\rho(b,b) = \rho^{-1}(a) \cap \rho^{-1}(b)$; hence $c\rho a$ and $c\rho b$. Since $a \in \rho(c)$ it results $\rho(a) \subset \rho^2(c)$ and we have $\rho^2(c) \subseteq \rho(c)$ so $\rho(a) \subseteq \rho(c)$ and similarly, $\rho(b) \subseteq \rho(c)$. Hence $a \cdot b = \rho(a) \cup \rho(b) \subseteq \subseteq \rho(c) = c^2$. By hypothesis, there exists $d \in H$, such that $d^2 \subseteq a^2 \cap b^2$.

On the other hand, $a^2 \cap b^2 = \rho(a) \cap \rho(b) = U_\rho(a,a) \cap U_\rho(b,b) = = U_\rho(a,b)$. It results $U_\rho(a,b) \neq \emptyset$.

(iii)$\Longrightarrow$(i) We have to verify the following implication:

$$a/b \cap c/d \neq \emptyset \Longrightarrow a \cdot d \cap b \cdot c \neq \emptyset.$$

Let $x \in a/b \cap c/d$. It results $a \in xb = \rho(x) \cup \rho(b)$ and $c \in xd = \rho(x) \cup \rho(d)$. We have the following possibilities:

1° $x \in \rho^{-1}(a) \cap \rho^{-1}(c) = L_\rho(a,c)$. From (iii) it results $U_\rho(a,c) = \rho(a) \cap \rho(c) \neq \emptyset$. Therefore, $ad \cap bc = (\rho(a) \cup \rho(d)) \cap (\rho(b) \cup \rho(c)) = (\rho(a) \cap \rho(b)) \cup (\rho(d) \cap \rho(b)) \cup (\rho(a) \cap \rho(c)) \cup (\rho(d) \cap \rho(c)) \neq \emptyset$.

2° $a \in \rho(b)$. Then $\rho(a) \subset \rho^2(b) \subset \rho(b)$, so $\rho(a) \cap \rho(b) \neq \emptyset$, whence $ad \cap bc \neq \emptyset$.

3° Similarly, if $c \in \rho(d)$, then $\rho(c) \subset \rho^2(d) \subset \rho(d)$ so $\rho(c) \cap \rho(d) \neq \emptyset$, whence $ad \cap bc \neq \emptyset$.

In all the situations, we obtain $ad \cap bc \neq \emptyset$. Therefore, $(H, \cdot)$ is a join space. ■

§2. Hypergroups associated with binary relations

I.G. Rosenberg associates a hypergroupoid H_R with every binary relation R on a set H and with full domain, in this manner:

$$\forall (x,y) \in H^2,\ x o y = \{z \in H \mid (x,z) \in R \text{ or } (y,z) \in R\}.$$

He characterizes all R such that the hypergroupoid $H_R = (H, \circ)$ is a semihypergroup, hypergroup and join space.

Let $R \subset H \times H$ and for all $(x,y) \in H^2$, set

$$x o x = \{y \in H \mid (x,y) \in R\},\ x o y = x o x \cup y o y \text{ and } H_R = < H; \circ > .$$

5. Definition. We say that $x \in H$ is an *outer element* of R if $\exists h \in H$, such that $(h,x) \notin R^2$ and an *inner element* of R otherwise.

First of all, we have the following:

6. Lemma. *H_R is a hypergroupoid if and only if H is the domain of R.* ∎

7. Theorem. *Let R be a binary relation on H with full domain. Then H_R is a semihypergroup if and only if $R \subseteq R^2$ and the following implication is satisfied:*

(α) $(a, x) \in R^2 \Longrightarrow (a, x) \in R$
whenever x is an outer element of R.

Proof. First notice that for H_R the associative law for "$\circ$" becomes

$$a \circ a \cup \left(\bigcup_{u \in b \circ b \cup c \circ c} u \circ u \right) = \left(\bigcup_{v \in a \circ a \cup b \circ b} v \circ v \right) \cup c \circ c$$

which can be expressed as follows: For all $(a, b, c, x) \in H^4$

$$(\beta) \qquad \begin{aligned} &(a, x) \in R \text{ or } (b, x) \in R^2 \text{ or } (c, x) \in R^2 \Longleftrightarrow \\ &\Longleftrightarrow (a, x) \in R^2 \text{ or } (b, x) \in R^2 \text{ or } (c, x) \in R. \end{aligned}$$

($\Longrightarrow$) Let H_R be a semihypergroup. Assume to the contrary that $R \not\subseteq R^2$. Then there exists $(b, x) \in R - R^2$. Consider (β) for $a = x$ and $c = b$. Then the right–hand side of (β) is clearly satisfied on account of $(c, x) = (b, x) \in R$. On the left–hand side $(b, x) = (c, x) \notin R^2$ and so $(x, x) = (a, x) \in R$. Now $(b, x) \in R$ and $(x, x) \in R$ yield the contradiction $(b, x) \in R^2$. Thus $R \subseteq R^2$. To prove (α) suppose to the contrary that there exist an outer element x of R and $a \in H$ such that $(a, x) \in R^2 - R$. By the definition of an outer element clearly $(b, x) \notin R^2$ for some $b \in H$. Set $c = b$ in (β). In view of $(a, x) \in R^2$ the right–hand side of (β) holds while the left–hand side is invalid on account of $(a, x) \notin R$ and $(b, x) \notin R^2$. This contradiction proves the validity of (α).

($\Longleftarrow$) Let $R \subseteq R^2$ and $(a, x) \in R^2 \Longrightarrow (a, x) \in R$ provided x is an outer element of R. Let $(a, b, c, x) \in H^4$. If $(b, x) \in R^2$ then both sides of (β) are satisfied. Thus let $(b, x) \notin R^2$. Then x outer and (α) yield $(a, x) \in R^2 \Longrightarrow (a, x) \in R$. Notice that in view of

$R \subseteq R^2$ we have $(a, x) \in R^2 \Longleftrightarrow (a, x) \in R$. By the same taken $(c, x) \in R^2 \Longleftrightarrow (c, x) \in R$; together with $(b, x) \notin R^2$ this proves (β). ∎

The above Theorem can be reformulated for hypergroups in the following manner:

8. Theorem. *Let R be a binary relation. Then H_R is a hypergroup if and only if*

1) *R has full domain;*

2) *R has full range;*

3) *$R \subseteq R^2$, and*

4) *$(a, x) \in R^2 \Longrightarrow (a, x) \in R$*
whenever x is an outer element of R.

9. Proposition. *Let $< H; \circ >$ be a semihypergroup. There is a binary relation R on H, such that $< H; \circ >$ is of the form H_R if and only if $\forall\, (a, b) \in H^2$, the following conditions are satisfied:*

(1°) $a \circ b = a^2 \cup b^2$;

(2°) $a^2 \subseteq (a^2)^2$, *and*

(3°) $(a^2)^2 \cap (H - (b^2)^2) \subseteq a^2$. $\qquad (\gamma)$

Proof. ($\Longrightarrow$) Let $H_R =< H; \circ >$ be a semihypergroup. Notice that

$$(z, t) \in R \Longleftrightarrow t \in z^2, \quad (z, t) \in R^2 \Longleftrightarrow t \in (z^2)^2.$$

Now (1°) follows from the definition of H_R and (2°) is a translation of $R \subseteq R^2$. To prove (3°) let x belong to the left side of (γ). Then $(a, x) \in R^2$ and $(b, x) \notin R^2$ and therefore x is an outer element of R. From (α) in Theorem 7 we obtain $(a, x) \in R$ which means $x \in a^2$.

($\Longleftarrow$) Let $< H; \circ >$ satisfy (1°)–(3°). Set

$$R = \{(a, b) \mid a \in H,\ b \in a^2\} \qquad (\delta)$$

Now (1°) means $a \circ b = a \circ a \cup b \circ b$ for all $a, b \in H$. As a^2 is nonvoid for each $a \in H$, clearly the domain of R is $\mathbb{D}_R = H$. It can be easily verified that (2°) translates into $R \subseteq R^2$. To prove (α) let $(a, x) \in R^2$ where x is an outer element of R. Then $(b, x) \notin R^2$ for some $b \in H$. From (δ) we obtain $x \in (a^2)^2$ and $x \notin (b^2)^2$. Now (γ) yields $x \in a^2$ and $(a, x) \in R$ by (δ). ∎

§3. Hypergroups associated with union, intersection, direct product, direct limit of relations

As we have seen in the previous paragraph, with any binary relation R on a set H, a partial hypergroupoid $H_R = < H; \circ >$ is associated, as follows:

$$\forall (x, z) \in H^2,\ x \circ x = \{y \in H \mid (x, y) \in R\},\ x \circ z = x \circ x \cup z \circ z.$$

Let

$$\mathbb{D}(R) = \{x \in H \mid \exists y \in H : (x, y) \in R\},$$
$$\mathbb{R}(R) = \{x \in H \mid \exists z \in H : (z, x) \in R\},$$

for all $k \geq 2$,

$$R^k = \{(a_1, a_{k+1}) \in H^2 \mid \exists (a_2, ..., a_k) \in H^{k-1} : (a_1, a_2) \in R, ..., (a_k, a_{k+1}) \in R\}.$$

x is called an outer element for R if $\exists h \in H : (h, x) \notin R^2$.

Rosenberg found conditions on R, such that H_R is a hypergroup or a join space (see [326]). Let us recall Theorem 8, §2:

H_R is a hypergroup if and only if:

1. $H = \mathbb{D}(R)$;
2. $H = \mathbb{R}(R)$;
3. $R \subset R^2$;
4. *if x is an outer element for R, then* $\forall a \in H,\ (a,x) \in R^2 \Longrightarrow (a,x) \in R$.

If H_R is a hypergroup, then it is called the *Rosenberg hypergroup.*

In this paragraph, the hypergroup H_R associated by Rosenberg with a binary relation R, is analysed especially in the case R is symmetric, and conditions are found on relations R_i so that the hypergroupoid associated with the union, intersection, direct product, direct limit of the R_i is a hypergroup.

Let $< H_R; \circ >$ be the hypergroup associated to a binary relation R satisfying the conditions 1-4 of Theorem 8.

Set $P = \{x \in H \mid x \circ x \not\ni x\}$ and $K = \{e \in H \mid e \circ e \supset P\}$.

10. Theorem. *H_R is regular if and only if $K \neq \emptyset$.*

Proof. Let us prove the two implications:

"$\Longrightarrow$" Let e be an identity of the regular hypergroup H_R.

If $P = \emptyset$, clearly $e \circ e \supset P$.

If $P \neq \emptyset$, then $\forall x \in P$, we have $e \circ x = e \circ e \cup x \circ x$. Since $x \notin x \circ x$, it follows that $e \circ e \ni x$, therefore $e \circ e \supset P$, whence $K \neq \emptyset$.

"$\Longleftarrow$" If $P = \emptyset$, then $\forall x \in H,\ x \circ x \ni x$, whence $\forall (x,y) \in H^2$, we have: $x \circ y = x \circ x \cup y \circ y \supset \{x,y\}$ so $H = I_H$ and $\forall x \in H$, $H = i(x)$ (the set of inverses of x).

Now, let us suppose $P \neq \emptyset$. Then if $e \in H$ is such that $e \circ e \supset P$, we have

$$\forall x \in P,\ e \circ x = e \circ e \cup x \circ x \supset e \circ e \ni x.$$

$$\forall y \in H - P,\ e \circ y = e \circ e \cup y \circ y \supset y \circ y \ni y.$$

Therefore I_H is not empty, since $I_H \supset K$. On the other side, if $e \in I_H$, we have $\forall z \in H,\ e \circ z = e \circ e \cup z \circ z \supset e \circ e \ni e$, whence $e \in i(z)$ and so H_R is regular. ∎

11. Remark.

1. $K \cap P = \emptyset$.
2. $K = I_H$.

Proof.

1. If $e \in K \cap P$, then $e \in P$ implies $e \circ e \not\ni e$, but $e \in K$ implies $e \circ e \supset P \ni e$, a contradiction.

2. Let $e \in K$. Then $e \circ e \supseteq P$ and for every $h \in P$, $h \in e \circ e \subseteq h \circ e$. For $h \in H - P$, clearly $h \in h \circ h \subseteq h \circ e$. This proves $e \in I_H$ and $K \subseteq I_H$. We prove the inverse inclusion. Let $e \in I_H$, and $x \in P$. Then $e \circ x = e \circ e \cup x \circ x \ni x$. Since $x \circ x \not\ni x$ we obtain $x \in e \circ e$ hence $e \in K$. ∎

For an equivalence relation θ on H denote by H/θ the set of blocks (or equivalence classes) of θ.

12. Theorem. *If H_R is a hypergroup then*

(i) *R^2 is transitive,*

(ii) *if, moreover, R is symmetric, then R^2 is an equivalence relation on H,*

(iii) *if R is symmetric and $|H/R^2| > 1$ then R is an equivalence relation on H.*

Proof. (i) Suppose to the contrary that there exist $(x, y) \in R^2 \ni \ni (y, z)$ such that $(x, z) \notin R^2$. Then z is outer and so $(y, z) \in R$. Since $(x, y) \in R^2$, there exists $a \in H$ such that $(x, a) \in R \ni (a, y)$. Now $(a, z) \in R^2$ shows $(a, z) \in R$. Thus $(x, z) \in R^2$, a contradiction.

(ii) Let R be symmetric. Let $x \in H$. We have $(x, y) \in R$ for some $y \in H$ (since the domain of R is H, according to 1, Theorem 8) and by symmetry $(x, y) \in R \ni (y, x)$ whence $(x, x) \in R^2$, proving the reflexivity of R^2. It is clear that R^2 is symmetric and so by (i) the relation R^2 is an equivalence relation on H.

(iii) Let R be symmetric and $|H/R^2| > 1$. Then each $h \in H$ is outer and so $R^2 \subset R \subset R^2$ by 3 and 4 of Theorem 8, proving $R = R^2$. ■

13. Theorem. *If $K \neq \emptyset$ and R is symmetric, then H_R is a regular reversible hypergroup.*

Proof. By Theorem 10, we know that H_R is regular, so only the reversibility has to be proved. For any $a \in H$, set $U_a = a \circ a$.

Let $a \in b \circ c$. Since $b \circ c = U_b \cup U_c$ we can suppose $a \in U_b$ whence $(b, a) \in R$. Hence $(a, b) \in R$, and so $b \in U_a$. It follows that for all $x \in H$, we have $b \in U_a \cup U_x = a \circ x$; thus if c' is any inverse of c, then $b \in a \circ c'$. So H_R is reversible on one side.

Let us remember now that $\forall e \in I_H$, e is an inverse of every element of H.

I. If $c \notin c \circ c$, then $c \in e \circ e$, whence $c \in e \circ e \cup U_a = e \circ a$ and $e \in i(b)$.

II. If $c \in c \circ c$ let us distinguish two cases:

1. $|H/R^2|>1$. In this case, $\forall x \in H$, we have $i(x) = I_H = H$ hence $c \in i(b)$, so $c \in c \circ a$, where $c \in i(b)$. Therefore, if $|H/R^2| > 1$, H is reversible.
2. Let us suppose now $|H/R^2|=1$, whence $R^2=H^2$. Let e be an identity, since eR^2c, there is $d \in H$ such that $(e, d) \in R \ni (d, c)$. It follows $c \in d \circ d$ and $d \in e \circ e$ from which $e \in d \circ d$ and $c \in d \circ d$. Therefore we have $d \circ b = U_d \cup U_b = d \circ d \cup U_b$. Since $d \circ d \ni e$, it follows $d \circ b \ni e$, so $d \in i(b)$, but we have also $d \circ d \ni c$, then $d \circ a = d \circ d \cup a \circ a \ni c$. Therefore, we can conclude that H_R is reversible on both sides. ■

Operations on $\mathcal{R}(H)$ and the corresponding H_R

Let R, S be binary relations on H, satisfying the conditions 1–3 of Theorem 8. Then also $R \cup S$ satisfies 1-3, but generally, as the following examples show, $H_{R \cup S}$ is not a hypergroup even if both H_R and H_S are.

I. Let $H = \{1,2,3,4\}$, $I_H = \{(x,x) \mid x \in H\}$. $R = I_H \cup \{(1,2)\}$, $S = I_H \cup \{(2,3)\}$. Clearly, H_R and H_S are hypergroups and we have: $R^2 = R$, $S^2 = S$.
$(R \cup S)^2 = R^2 \cup S^2 \cup RS \cup SR = R \cup S \cup \{(1,3)\} \underset{\neq}{\supset} R \cup S$.
Hence $(4,3) \notin (R \cup S)^2$, so 3 is outer for $R \cup S$, but $(1,3) \in (R \cup S)^2 - R \cup S$. Therefore, 1, 2, 3 of Theorem 8 are satisfied, but 4 is not, so $H_{R\cup S}$ is not a hypergroup.

II. If we suppose $RS \cup SR \subset R^2 \cup S^2$, and $R^2 = R$, $S^2 = S$, then 4 is satisfied by $R \cup S$ and therefore $H_{R\cup S}$ is a hypergroup.

14. Remark. The condition $RS \cup SR \subset R^2 \cup S^2$ (that is $(R \cup S)^2 = R^2 \cup S^2$) is not necessary for $H_{R\cup S}$ to be a hypergroup as we see in III:

III. Set $H = \{1,2,3\}$. $R = I_H \cup \{(1,2)\}$, $S = I_H \cup \{(2,3)\}$.
We have $R^2 = R$, $S^2 = S$ and $(R \cup S)^2 = R^2 \cup S^2 \cup \{(1,3)\}$ so $(R \cup S)^2 \neq R^2 \cup S^2$ but $R \cup S$ satisfies the condition 4.

15. Remark. Neither of $R^2{=}R$, $S^2{=}S$, nor both H_R, H_S be hypergroups is necessary for $H_{R\cup S}$ to be a hypergroup as one sees in IV:

IV. Set $H = \{1,2,3\}$.
$R=\{(1,2),(2,1),(2,2),(3,3)\}$, $S=\{(2,3),(3,2),(1,1),(3,3)\}$, so $R^2 = I_H \cup \{(1,2),(2,1)\} \underset{\neq}{\supset} R$, $S^2 = I_H \cup \{(2,3),(3,2)\} \underset{\neq}{\supset} S$ and $(R \cup S)^2 = H \times H \underset{\neq}{\supset} R^2 \cup S^2$ whence all the conditions 1–4 are satisfied by $R \cup S$.

Let us remark also that H_R, H_S do not satisfy 4. Indeed:
$(3,1) \notin R^2$ implies that 1 is outer, but $(1,1) \notin R^2 - R$.
$(1,2) \notin S^2$ implies that 2 is outer, but $(2,2) \notin S^2 - S$.

16. Theorem. *Let R and S be reflexive and transitive relations on H (that is, quasi–orders). Then $H_{R\cap S}$ is a hypergroup.*

Proof. Indeed, $R \cap S$ is quasi–order and so it satisfies 1–4 of Theorem 8. ∎

17. Corollary. *If H_R and H_S are hypergroups, R and S are symmetric, and $|H/R^2| > 1 < |H/S^2|$, then $H_{R\cap S}$ is a hypergroup.*

Proof. It follows directly from Theorem 12 (iii) and Theorem 16. ∎

18. Theorem. *Let R, S be relations on H such that*

(α) $\mathbb{D}(R) = \mathbb{R}(R) = H = \mathbb{D}(S) = \mathbb{R}(S)$

(β) $R^2 = R,\ S^2 = S,\ RS = SR.$

Then H_{RS} is a hypergroup.

Proof. Indeed, we have $(RS)\ (RS) = R(RS)S = R^2S^2 = RS$, whence the conditions 3, 4 of Theorem 8 are satisfied. Moreover, 1, 2 of Theorem 8 follow from (α). ∎

19. Corollary. *If R and S are equivalence relations on H such that $RS = SR$, then H_{RS} is a hypergroup.*

Proof. It follows from Theorem 18.

20. Theorem. *Let H_1, H_2 be non empty sets. Le R_i be a binary relation on H_i $(i = 1, 2)$ and $(H_i)_{R_i} =< H_i; \circ_i >$ be the hypergroupoid associated with R_i. Let $H = H_1 \times H_2$ and let H be endowed with the hyperoperation $(x_1, x_2) \circ (y_1, y_2) = (x_1 \circ_1 y_1, x_2 \circ_2 y_2)$ and let $R_1 \times R_2$ be the binary relation on H defined as follows:*

$((a_1, x_1), (a_2, x_2)) \in R_1 \times R_2$ *if and only if* $(a_1, a_2) \in R_1,\ (x_1, x_2) \in R_2.$

Then

a) $H_{R_1 \times R_2} = (H_1)_{R_1} \times (H_2)_{R_2}.$

b) $H_{R_1 \times R_2}$ *is a hypergroup if and only if for* $j \in \{1, 2\}$

(i) $(H_j)_{R_j}$ *is a hypergroup, and*

(ii) $R_j^2 \neq H_j^2 \Longrightarrow R_{3-j} = R_{3-j}^2$.

Proof. (a) Direct verification.

(b) ($\Longrightarrow$) Let $H_{R_1 \times R_2}$ be a hypergroup. We prove (i) for $j = 1$. By 1 and 2 we get $\mathbb{D}(R_1 \times R_2) = H = \mathbb{R}(R_1 \times R_2)$ proving 1 and 2 for R_1. Next, by 3

$$R_1 \times R_2 \subseteq (R_1 \times R_2)^2 = R_1^2 \times R_2^2$$

and so R_1 satisfies 3. To prove 4 let z_1 be an outer element of R_1 and $(a_1, z_1) \in R_1^2$. Since $\mathbb{D}(R_2) = H_2$, there exists $(a_2, z_2) \in R_2^2$. Clearly, (z_1, z_2) is outer for $R_1 \times R_2$ and from $((a_1, a_2), (z_1, z_2)) \in R_1^2 \times R_2^2$ and 4 we obtain $((a_1, a_2), (z_1, z_2)) \in R_1 \times R_2$ and $(a_1, z_1) \in R_1$. Thus R_1 satisfies 4 and $(H_1)_{R_1}$ is a hypergroup. The same proof shows that $(H_2)_{R_2}$ is a hypergroup.

To prove (ii) let $j = 1$ and $R_1^2 \neq H_1^2$. Choose $(z, x_1) \in H_1^2 - R_1^2$.

To prove $R_2^2 \subset R_2$ let $(y_1, y_2) \in R_2^2$. As $\mathbb{R}(R_1) = H_1 = \mathbb{D}(R_1)$, we have $(a, x_1) \in R_1$ for some a and $(b, a) \in R_1$ for some b. Then $(b, x_1) \in R_1^2$ and $(y_1, y_2) \in R_2^2$ show $((b, y_1), (x_1, y_2)) \in R_1^2 \times R_2^2 = (R_1 \times R_2)^2$. Here (x_1, y_2) is outer for $R_1 \times R_2$ due to $(z, x_1) \notin R_1^2$. By 4 clearly $((b, y_1), (x_1, y_2)) \in R_1 \times R_2$ and $(y_1, y_2) \in R_2$ proving $R_2^2 \subset R_2$. We showed above that $R_2 \subset R_2^2$. Together $R_2^2 = R_2$. The same proof works for $j = 2$.

($\Longleftarrow$) Let (i) and (ii) hold. It is easy to see that $H_{R_1 \times R_2}$ satisfies the conditions 1–3. To prove 4 let $z = (z_1, z_2)$ be an outer element of $R_1 \times R_2$. Then for some $j \in \{1, 2\}$ the element z_j is an outer element of R_j. Then $R_j^2 \neq H_j^2$ and from (ii) we see that $R_{3-j} = R_{3-j}^2$. Let $a = (a_1, a_2)$ satisfy $(a, z) \in R_1^2 \times R_2^2$. Since $(H_j)_{R_j}$ satisfies 4 we obtain $(a_j, z_j) \in R_j$. Moreover, since we have $R_{3-j} = R_{3-j}^2$, it results $(a, z) \in R_1 \times R_2$. ∎

Now, let us recall some definitions. We call *model* a pair $< H; R >$, that is a set H endowed with a binary relation R.

If $< H'; R' >$ is another model, we say that a function $f : H \to H'$ is a *homomorphism* of *models*, and we write $f \in \mathrm{Hom}(H, H')$, if the following implication is satisfied: $(x,y) \in R \Longrightarrow (f(x), f(y)) \in R'$.

We say that a family of models $\{< H_i, R_i >\}_{i\in I}$ is *direct* if it satisfies the following conditions:

(i) $< I; \leq)$ is a direct partially ordered set.

(ii) $\forall (i,j) \in I^2,\ i \neq j \Longleftrightarrow H_i \cap H_j = \emptyset$.

(iii) for any $(i,j) \in I^2$, if $i \leq j$, a homomorphism of models $\varphi^i_j : H_i \to H_j$ is defined, such that if $i \leq j \leq k$, we have $\varphi^j_k \varphi^i_j = \varphi^i_k$ and $\forall i \in I,\ \varphi^i_i = \mathrm{Id}\,(H_i)$.

Set $H = \bigcup\limits_{i\in I} H_i$ and let us define in H the following binary relation:

$$\forall (x_i, y_j) \in H_i \times H_j,\ x_i \sim y_j \Longleftrightarrow \exists k \in I,\ k \geq i, k \geq j,$$
$$\text{such that } \varphi^i_k(x_i) = \varphi^j_k(y_j).$$

The relation $\sim$ is an equivalence relation on H.

We shall denote $\varphi^i_j(x_i)$ by x_j. The direct limit $\overline{H} = \varinjlim (H_i)_{i\in I}$ is the quotient $H/\sim$ endowed with the binary relation $\overline{R}$

$$(\bar{x}, \bar{z}) \in \bar{R} \Longleftrightarrow \exists\, q \in I,\ \exists x_q \in \bar{x} \cap H_q,\ \exists z_q \in \bar{z} \cap H_q$$

such that $(x_q, z_q) \in R_q$.

21. Theorem. *Let $K = \{< H_i, R_i >\}_{i\in I}$ be a direct family of models. If $\forall i \in I$, there is $k \in I$, $k \geq i$, such that $(H_k)_{R_k}$ is a hypergroup, then $(\overline{H}_{\overline{R}})$ is a hypergroup.*

Proof. To prove 1 of Theorem 8 for $\overline{R}$, let $\bar{x} \in \overline{H}$ be arbitrary. Choose $x \in \bar{x}$ Then $x \in H_i$ for some $i \in I$. There exists $k \geq i$ such that $(H_k)_{R_k}$ is a hypergroup. Clearly $x_k = \varphi^i_k(x) \in \bar{x} \cap H_k$ (due to $\varphi^i_k(x) = x_k = \varphi^i_i(x_k)$). From $\mathbb{D}(R_k) = H_k$ we obtain that $(x_k, y) \in R_k$ for some $y \in H_k$. Clearly $(\bar{x}, \bar{y}) \in \overline{R}$ proving 1 for $\overline{R}$.

The proof of 2 is similar.

To prove 3 let $(\bar{x}, \bar{y}) \in \overline{R}$. Then there exist $i \in I$ and $(x_i, y_i) \in \in (\bar{x} \times \bar{y}) \cap R_i$. By assumption $(H_k)_{R_k}$ is a hypergroup for some $k \geq i$. Set $x_k = \varphi_k^i(x_i)$ and $y_k = \varphi_k^i(y_i)$. Notice that $x_k \in \bar{x}$, $y_k \in \bar{y}$ and $(x_k, y_k) \in R_k$ because φ_k^i is a homomorphism of models. Applying 3 to R_k we obtain $(x_k, y_k) \in R_k^2$ and so $(x_k, u), (u, y_k) \in R_k$ for some $u \in H_k$. Finally, $(\bar{x}, \bar{u}), (\bar{u}, \bar{y}) \in \overline{R}$ proving $\overline{R} \subseteq \overline{R}^2$.

To prove 4 let $\bar{z}$ be an outer element of $\overline{R}$. Then $(\bar{a}, \bar{z}) \notin \overline{R}^2$ for some $\bar{a} \in \overline{H}$. Let $\bar{b} \in \overline{H}$ satisfy $(\bar{b}, \bar{z}) \in \overline{R}^2$, whence $(\bar{b}, \bar{u}) \in \overline{R} \ni (\bar{u}, \bar{z})$ for some $\bar{u} \in \overline{H}$. Then, from (i) and (iii), we obtain $(b, u) \in R_q \ni (u, z)$ for some $q \in I$, $b \in \bar{b} \cap H_q$, $z \in \bar{z} \cap H_q$ and $u \in \bar{u} \cap H_q$. Choose $a \in \bar{a}$. Then $a \in H_r$ for some $r \in I$. There exists $i \in I$ such that $i \geq q$, $i \geq r$. By the hypothesis $H' = (H_k)_{R_k}$ is a hypergroup for some $k \in I$, $k \geq i$. Set $a' = \varphi_k^r(a)$ and $z' = \varphi_k^q(z)$. We show that z' is an outer element of R_k. Indeed, $a' \in \bar{a} \cap H_k$ and $z' \in \bar{z} \cap H_k$ satisfy $(a', z') \notin R_k^2$ since otherwise we would have $(\bar{a}, \bar{z}) \in \overline{R}^2$.

Set $b' = \varphi_k^q(b)$ and $u' = \varphi_k^q(u)$. Then $(b', u') \in R_k \ni (u', z')$ because φ_k^q is a homomorphism. Thus, $(b', z') \in R_k^2$. Now the hypergroup H' satisfies 4 and so $(b', z') \in R_k$. This implies $(\bar{b}, \bar{z}) \in \overline{R}$ proving 4 for $\overline{H}$.

Hypergraphs, relations and H_R

Denote by $\mathcal{H}(H)$ the set of hypergraphs on H, that is of families $K = \{A_i^K\}_{i \in I_K}$ where I_K is nonempty, $\forall\, i \in I_K$, $A_i^K \in \mathcal{P}^*(H)$ and $\bigcup\limits_{i \in I_K} A_i^K = H$.

Denote by $SR(H)$ the set of reflexive and symmetric binary relations on H. For any $K \in \mathcal{H}(H)$, define the relation $R_K = \Psi(K)$ as follows:

$$\forall\, (x, y) \in H^2,\ xR_Ky \text{ if and only if } \exists i \in I_K : \{x, y\} \subset A_i^K.$$

Clearly, $R_K \in SR(H)$ and Ψ is a function $\Psi : \mathcal{H}(H) \to SR(H)$. Ψ is surjective but not injective. Set $\Psi^{-1}(\Psi(K)) = Q_K$.

Let now $\leq$ be the partial order on Q_K defined on $\mathcal{H}(H)$: $K_1 \leq K_2$ if and only if $\forall i \in I_{K_1}$, $\exists j \in I_{K_2}$ such that $A_i^{K_1} \subseteq A_j^{K_2}$.

Let $\mathcal{O} : SR(H) \to \mathcal{P}^*(\mathcal{H}(H))$ be the function defined by setting $\forall R \in SR(H)$, $\mathcal{O}(R) = \{K \in \mathcal{H}(H) \mid \Psi(K) = R\}$. Clearly, $\mathcal{O}(R_K) = Q_K$. For H infinite we assume the axiom of choice.

22. Theorem. *Let $R \in \mathcal{H}(H)$. Then $\mathcal{O}(R_K)$ is an interval of the order $\leq$, that is $\mathcal{O}(R)$ has a least element $\mu(R) = \{\{x,y\} \mid (x,y) \in R\}$ and a greatest element $M(R)$ which is the set of inclusion maximal subsets B of H such that $B \times B \subset R$.*

Proof.

(1) $\forall (x,y) \in H^2$, if xRy, then there exists j such that $\{x,y\} \subset A_j^K$; therefore $\forall K \in \mathcal{O}(R)$, $\mu(R) \leq K$.

(2) Let $K \in \mathcal{O}(R)$. $\forall i \in I_K$ we have clearly $A_i^K \times A_i^K \subset R$ and there exists $B \in \mathcal{P}^*(H)$ such that $A_i^K \subset B$, $B \times B \subset R$ and $B \times B \subset P \times P \subset R$ implies $P = B$. So $K \leq M(R)$. ■

23. Definition. Let ${}_R\Psi_m$ and ${}_R\Psi_M$ be the restrictions of Ψ to the least and greatest hypergraphs of $\mathcal{O}(R)$, respectively. Let Ψ_m and Ψ_M be respectively the functions

$$\Psi_m : \{\mu(R) \mid R \in SR(H)\} \longrightarrow SR(H)$$
$$\Psi_M : \{M(R) \mid R \in SR(H)\} \longrightarrow SR(H)$$

defined $\forall R \in SR(H)$,

$$\Psi_m(\mu(R)) = {}_R\Psi_m(\mu(R)),$$
$$\Psi_M(M(R)) = {}_R\Psi_M(M(R)).$$

24. Proposition. *We have $\Psi_m \mu = I_{SR(H)} = \Psi_M M$, whence μ, M are injective, Ψ_m, Ψ_M surjective.*

Proof. It is enough to remark that if $R = R_K = \Psi(K)$, we have $\mu(R) \in \Psi^{-1}(\Psi(K)) \ni M(R)$, whence $\Psi_m \mu(R_K) \in \Psi(\Psi^{-1}(\Psi(K))) \ni$ $\ni \Psi_M M(R_K)$ from which $\Psi_m \mu(R_K) = \Psi_M M(R_K) = \Psi(K) = R_K$.■

Other topics

Let R be a binary relation on H. Let $H(k) = < H; \circ_k >$ be the succession of hypergroupoids defined recursively as follows:

$$\begin{aligned} &\forall (x,y) \in H^2, \ x \circ_1 y = x \circ y, \\ &\forall k \geq 1, \forall x \in H, \ x \circ_{k+1} x = \bigcup_{y \in x \circ_k x} y \circ_1 y, \\ &\forall (x,y) \in H^2, \ x \circ_{k+1} y = x \circ_{k+1} x \cup y \circ_{k+1} y. \end{aligned}$$

We have clearly $z \in x \circ_k x$ if and only if $xR^k z$.

Let us denote $C_t(R)$ the transitive closure of the relation R.

25. Theorem. *Let R be a reflexive relation on H. Then the extension $< H; \bar{\circ} >$ of H_R defined by setting*

$$\forall x \in H, \quad x \bar{\circ} x = \bigcup_{k \geq 1} x \circ_k x$$

$$\forall (x,y) \in H^2, \quad x \bar{\circ} y = x \bar{\circ} x \cup y \bar{\circ} y$$

is a hypergroup.

Proof. It is enough to remark that $< H; \bar{\circ} > = H_{\overline{R}}$ where $\overline{R} = C_t(R) = \bigcup_{k \geq 1} R^k$ satisfies $\overline{R}^2 = \overline{R}$ whence the conditions 1–4 of Theorem 8. ■

26. Corollary. *Let R be a reflexive relation on H and let $|H| = n$. Then the hypergroupoid $< H; \circ_{n-1} >$ is a hypergroup.*

Proof. The hypothesis implies that $R^{n-1} = C_t(R)$. ■

27. Theorem. *Let R be a relation on H and K a subhypergroup of H_R. If $K \neq H$ then K is not closed.*

Proof. Indeed, if $(a, b, x) \in H^3$ is such that $a \in K$, $b \in U_a$, $x \in H - K$, we have $b \in a \circ x = a \circ a \cup x \circ x$. So $\{b, a\} \subset K$ but $x \notin K$ whence K is not closed. ■

28. Theorem. *Let R be a symmetric relation on H and H_R a hypergroup.*

1. *If $|H/R^2| = 1$, then H_R has not proper subsemihypergroups.*

2. *If $|H/R^2| > 1$, then every subsemihypergroup of H_R is a subhypergroup of H_R.*

Proof. It is enough to remark that $\forall\,(a,b) \in H^2$, aR^2b if and only if $\exists x \in H$ such that $(a,x) \in R$, $(x,b) \in R$ whence $b \in a^4$. It follows that $\forall\, a \in H$, $R^2(a) \subset < a >$, where $< a >$ is the subsemihypergroup generated by a. ■

29. Theorem. *Let H_R be a hypergroup and suppose R to be symmetric. Then R is regular. If $|H/R^2| > 1$, then R is an equivalence relation whence H_R/R is a hypergroup.*

Proof. Let xRy and $z \in H$. We have: $x{\circ}z = U_x{\cup}U_z$, $y{\circ}z = U_y{\cup}U_z$. Set $q \in x{\circ}z$.

1. Let $R^2 = R$. Then if $q \in z{\circ}z$, we have $q \;\bar{\bar{R}}\; (z{\circ}z)$; if $q \in x{\circ}x$, we have $q \;\bar{\bar{R}}\; (y{\circ}y)$ whence we obtain $x{\circ}z \;\bar{R}\; y{\circ}z$.

2. Let $R^2 = H^2$. For any $\lambda \in x{\circ}x$, since $(\lambda,y) \in R^2$, there is μ such that $(\lambda,\mu) \in R$, $(\mu,y) \in R$ whence $(y,\mu) \in R$, so $\mu \in y{\circ}y$. Therefore, $x{\circ}x \;\bar{R}\; y{\circ}y$ for every $(x,y) \in H^2$. Then R is regular on both sides.

The second statement follows from Theorem 12 and from [437, Theorem 29]. ■

In this paragraph, the analysis of Rosenberg hypergroup, associated with union, intersection, product of relations is continued in depth, obtaining several results among which also the mutual associativity plays a part.

30. Proposition. *Let R be a relation on H. If H_R is a hypergroup, then $\forall n \in \mathbb{N}^*$, H_{R^n} is a hypergroup.*

Proof. It is immediate that $\mathbb{D}(R) = \mathbb{R}(R) = H$ and $R \subset R^2$ imply $\mathbb{D}(R^n) = \mathbb{R}(R^n) = H$. Moreover, for every $1 \leq s \leq t$, we have $R^s \subset R^t$, in particular $R^n \subset R^{2n}$, for every $n \in \mathbb{N}^*$. It remains to prove 4) in Theorem 8. Let x be an outer element for R^n, that is there exists $h \in H : (h, x) \notin R^{2n}$, whence $(h, x) \notin R^2$, that is also x is an outer element for R.

Suppose $(a, x) \in R^{2n}$. Then $\exists u_1 \in H : (a, u_1) \in R^{2n-2}$ and $(u_1, x) \in R^2$. Since x is an outer element for R, clearly $(u_1, x) \in R$, so $(a, x) \in R^{2n-1}$. Continuing in the same manner, we obtain $(a, x) \in R \subset R^n$. ∎

Let us denote by $C_t(R)$ the transitive closure of a relation R.

31. Theorem. *Let R and S be two relations on H, such that $R \subset S \subset S^2 \subset C_t(R)$. If H_R is a hypergroup, then also H_S is a hypergroup.*

Proof. Since H_R is a hypergroup, we have $\mathbb{D}(R) = \mathbb{R}(R) = H$, whence $\mathbb{D}(S) = \mathbb{R}(S) = H$. Now, let us consider an outer element x for S, that is $\exists h \in H : (h, x) \notin S^2$. Hence, x is an outer element also for R.

We show $(a, x) \in C_t(R) \Longrightarrow (a, x) \in R$. Indeed, let $(a, x) \in C_t(R)$. Denote by ℓ the least integer such that $(a, x) \in R^\ell$. Then there exist $a = u_0, u_1, ..., u_\ell = x$ such that $(u_i, u_{i+1}) \in R$ for all $i \in \{0, 1, ..., \ell - 1\}$. If $\ell \geq 2$, then $(u_{\ell-2}, x) \in R^2$ and x outer for R would yield $(u_{\ell-2}, x) \in R$ in contradiction to the minimality of ℓ. Thus $\ell = 1$ and $(a, x) \in R$.

Consider $(b, x) \in S^2$. Then $(b, x) \in C_t(R)$ and so $(b, x) \in R \subseteq S$ proving 4) for S. ∎

32. Corollary. *Let R be a relation on H, such that $C_t(R) = H \times H$ and H_R is a hypergroup. Then for each relation S on H, such that $R \subset S \subset S^2$, H_S is also a hypergroup.*

33. Corollary. *Let R and S be relations on H, such that H_R is a hypergroup and let $k \geq 1$, and $s \geq 1$. Then*

1. *if $S \subset S^2 \subset C_t(R)$, then also $H_{R^s \cup S^k}$ is a hypergroup;*

2. *if $T \subseteq C_t(R)$ is reflexive then also $H_{R^s \cup T}$ is a hypergroup.*

Proof. 1. Since H_R is a hypergroup, $R \subset R^2$. Hence the assumptions imply

$$R \subset R^s \cup S^k \subset (R^s \cup S^k)^2 \subset C_t(R).$$

Apply the theorem.

2. In 1) set $S = T$ and $k = 1$. ∎

34. Corollary. *Let R and S be relations on H, such that $H_{R \cap S}$ is a hypergroup and $R \subset R^2 \subset C_t(R \cap S)$. Then also H_R is a hypergroup.*

Proof. Apply the previous theorem to $R' = R \cap S$ and $S' = R$. ∎

35. Corollary. *Let R and S be two relations on H, such that $R \subset S \subset S^2 \subset C_t(R)$. If H_R is a hypergroup then for all positive k_1 and k_2, also $H_{R^{k_1} S^{k_2}}$ and $H_{S^{k_2} R^{k_1}}$ are hypergroups.*

Proof. From $R \subset R^2$,

$$R \subset R^{k_1+k_2} \subset R^{k_1} S^{k_2} \subset R^{k_1+k_2+k_1} S^{k_2} \subset R^{k_1} S^{k_2} R^{k_1} S^{k_2} =$$
$$= (R^{k_1} S^{k_2})^2 \subseteq C_t(R).$$

Theorem 31 applied to R and $S' = R^{k_1} S^{k_2}$ yields that $H_{R^{k_1} S^{k_2}}$ is a hypergroup. The proof that $H_{S^{k_2} R^{k_2}}$ is a hypergroup is similar. ∎

36. Corollary. *Let R and S be two reflexive relations on H, such that $S \subset C_t(R)$. If $H_{R \cup S}$ is a hypergroup, then for all positive k_1 and k_2, also $H_{R^{k_1} S^{k_2}}$ and $H_{S^{k_2} R^{k_1}}$ are hypergroups.*

Proof. Set $R' = R \cup S$ and $S' = R^{k_1}S^{k_2}$. Since both R and S are reflexive, $R' \subset S' \subset S'^2 \subset C_t(R)$. Then apply Theorem 31 to R' and S' to obtain that $H_{R^{k_1}S^{k_2}}$ is a hypergroup.

By symmetry, also $H_{S^{k_2}R^{k_1}}$ is a hypergroup. ■

Now let us mention some results about mutually asociative H_R hypergroups.

First, recall the definition of mutually associative partial hypergroupoids:

37. Definition. We say that two partial hypergroupoids $< H, \circ_1 >$ and $< H, \circ_2 >$ are *mutually associative* (m.a.) if $\forall (x, y, z) \in H^3$ we have

$$(*) \qquad (x \circ_1 y) \circ_2 z = x \circ_1 (y \circ_2 z), \ (x \circ_2 y) \circ_1 z = x \circ_2 (y \circ_1 z).$$

For a relation R on H and $X \subset H$ set

$$R(X) = \{y \mid (x, y) \in R \text{ for some } x \in X\}.$$

If $X = \{x_1, ..., x_n\}$ we write $R(x_1, ..., x_n)$ for $R(X)$.

38. Proposition. *Let R and S be relations on H with full domain. Then $H_R =< H; \circ_R >$ and $H_S =< H; \circ_S >$ are mutually associative if and only if for all $(x, y, z) \in H^3$*

$$(**) \qquad RS(x, y) \cup S(z) = R(x) \cup SR(y, z).$$

Proof. We have: if $(c \in \mathbb{D}(S) \Longrightarrow a \circ_R b \neq \emptyset)$, then $(a \circ_R b) \circ_S c =$ $= \{t \in H \mid (a, t) \in RS \text{ or } (b, t) \in RS \text{ or } (c, t) \in S\}$; if $(a \in \mathbb{D}(S)$ $\Longrightarrow b \circ_S c \neq \emptyset)$, then $a \circ_R (b \circ_S c) = \{t \in H \mid (b, t) \in SR$ or $(c, t) \in SR$ or $(a, t) \in R\}$. Hence $(**)$ is the first equality of $(*)$ for H_R and H_S. Since both H_R and H_S are commutative, the second equality of $(*)$ coincides with the first one. ■

39. Proposition. *Let R and S be two relations on H such that H_R and H_S are mutually associative hypergroups. Then also $H_{R \cup S}$ is a hypergroup.*

Proof. $\mathbb{D}(R \cup S) = H = \mathbb{R}(R \cup S)$ because H_R is a hypergroup and so $\mathbb{D}(R) = H = \mathbb{R}(R)$. Next $R \subset R^2$ and $S \subset S^2$ and therefore $R \cup S \subset R^2 \cup S^2 \subset (R \cup S)^2$. To prove 4) in Theorem 8 let x be an outer element for $R \cup S$. Then $(h, x) \notin (R \cup S)^2 = R^2 \cup RS \cup SR \cup S^2$ for some h; in particular $(h, x) \notin R^2$ and x is outer for R.

Similarly, x is outer for S. Now consider $(a, x) \in (R \cup S)^2$. If $(a, x) \in R^2$ then $(a, x) \in R$ because H_R is a hypergroup and x is outer for R. By symmetry the same holds for $(a, x) \in S^2$. Again by symmetry it suffices to consider $(a, x) \in RS$. Setting $x = a$ and $y = z = h$ we obtain

$$RS(a, h) \cup S(h) = R(a) \cup SR(h).$$

Here $x \in RS(a)$ but $x \notin SR(h)$ due to $(x, h) \notin (R \cup S)^2$. Thus $x \in R(a)$ proving the required $(a, x) \in R$. ■

40. Proposition. *Let R and S be relations on H, such that $R \subset RS$ and $SR \cap \{(x, x) \mid x \in H\} = \emptyset$. If H_R is a hypergroup and H_R, H_S are mutually associative, then also H_{RS} is a hypergroup.*

Proof. Since $R \subset RS$ and H_R is a hypergroup, it results

$$\mathbb{D}(RS) = \mathbb{R}(RS) = H.$$

Moreover, from $R \subset RS$, it results $RS \subset (RS)^2$.

Now, let us consider x an outer element for RS, so x is also an outer element for R. If $(a, x) \in (RS)^2$, then $\exists b \in H$, such that $(a, b) \in RS \ni (b, x)$. Then $b \in (a \circ_R b) \circ_S b = a \circ_R (b \circ_S b)$ and since $(b, b) \notin SR$, it results $(a, b) \in R$.

Similarly, we have $x \in (b \circ_R x) \circ_S x = b \circ_R (x \circ_S x)$ and since $(x, x) \notin SR$, it results $(b, x) \in R$.

Therefore $(a, x) \in R^2$ and since x is an outer element for R, it results $(a, x) \in R \subset R \circ S$.

Then H_{RS} is a hypergroup. ■

41. Proposition. *Let R and S be relations on H, such that $R \subset RS$ and $\mathbb{D}(SR) \neq H$. If H_R is a hypergroup and H_R, H_S are mutually associative, then H_{RS} is a hypergroup.*

Proof. As in the proof of the above proposition, we have

$$\mathbb{D}(RS) = \mathbb{R}(RS) = H \text{ and } RS \subset (RS)^2.$$

Let x be an outer element for RS. If $(a, x) \in (RS)^2$, then $\exists b \in H$, such that $(a, b) \in RS \ni (b, x)$. Let $h \in H - \mathbb{D}(SR)$. We have $b \in (a \circ_R h) \circ_S h = a \circ_R (h \circ_S h)$ and since $(h, b) \notin SR$, it results $(a, b) \in R$.

Similarly, $x \in (b \circ_1 h) \circ_2 h = b \circ_1 (h \circ_2 h)$ and since $(h, x) \notin SR$ it results $(b, x) \in R$. Then $(a, x) \in R^2$, so $(a, x) \in R \subset RS$. Then H_{RS} is a hypergroup. ■

§4. Relation β in semihypergroups

Recall that with each binary relation R on a set H, a partial hypergroupoid $H_R = < H; \circ >$ is associated as follows:

$\forall (x, z) \in H^2$, $x \circ_R x = \{y \in H \mid (x, y) \in R\}$, $x \circ_R z = x \circ_R x \cup z \circ_R z$.

x is an outer element for R if $\exists h \in H : (h, x) \notin R^2$.

Recall Theorem 8 of this chapter:

H_R is a hypergroup if and only if:

1. $H = \mathbb{D}(R)$;
2. $H = \mathbb{D}(R)$;
3. $R \subset R^2$;
4. *if x is an outer element for R, then $\forall a \in H$,* $(a, x) \in R^2 \Longrightarrow (a, x) \in R$.

For a relation T on H set $H_T = < H; \circ_T >$ and for two relations R and S on H, let $RS = \{(x, y) \mid (x, u) \in R,\ (u, y) \in S \text{ for some } u\}$.

42. Proposition. *Let R and S be two relations on H. Then for all $a, b, c \in H$, we have:*

(i) $(a \circ_R a) \circ_R (a \circ_R a) = \bigcup_{t \in a \circ_R a} t \circ_R t$;

(ii) $a \circ_R a \circ_R a = a \circ_R a \cup (a \circ_R a) \circ_R (a \circ_R a)$;

(iii) *if* $R \subset R^2$, *then* $(a, x) \in R^2 \Longleftrightarrow x \in a \circ_R a \circ_R a$;

(iv) *if* $a \circ_S a \neq \emptyset \Longrightarrow a \circ_R a \neq \emptyset$, *then* $(a \circ_R a) \circ_S a = a \circ_S a \cup a \circ_{RS} a$;

(v) $(a \circ_R a) \circ_S (a \circ_R a) = a \circ_{RS} a$;

(vi) $a \circ_{R \cup S} a = a \circ_R a \cup a \circ_S a$; $a \circ_{R \cap S} a = a \circ_R a \cap a \circ_S a$; $a \circ_{R \cup S} a \circ_{R \cap S} a = a \circ_R a \circ_R a \cup a \circ_S a \circ_S a \cup a \circ_{RS} a \cup a \circ_{SR} a$;

(vii) *if* $c \in \mathbb{D}(S) \Longrightarrow a \circ_R b \neq \emptyset$, *then* $(a \circ_R b) \circ_S c =$
$= \{t \in H \mid (a, t) \in RS$ *or* $(b, t) \in RS$ *or* $(c, t) \in S\}$;
if $a \in \mathbb{D}(S) \Longrightarrow b \circ_S c \neq \emptyset$, *then* $a \circ_R (b \circ_S c) =$
$= \{t \in H \mid (b, t) \in SR$ *or* $(c, t) \in SR$ *or* $(a, t) \in R\}$.

Proof. A straightforward verification. ∎

43. Corollary. *If* $R \subset R^2$, *then* x *is an outer element for* R *if and only if* $\exists a \in H$, *such that* $x \notin a \circ_R a \circ_R a$.

44. Remark. If $R \subset R^2$ then there are no outer elements for R if and only if $\forall a \in H$, we have $a \circ_R a \circ_R a = H$.

45. Proposition. *The following two conditions are equivalent for a relation* R *on* H, *such that* $R \subset R^2$:

(i) $\forall (a, c) \in H^2$, *we have* $(R^2 - R)(a) \subset R^2(c)$;

(ii) *if* x *is an outer element for* R, *then* $(a, x) \in R^2 \Longrightarrow (a, x) \in R$.

46. Remarks.

1. If R is a relation on H, such that $R \subset R^2$, then R is transitive if and only if for all $a \in H$, we have $a \circ_R a \circ_R a = a \circ_R a$.

2. If R is a relation on H, then $R \subset R^2$ if and only if for all $a \in H$, we have $a \circ_R a \circ_R a = (a \circ_R a) \circ_R (a \circ_R a)$.

3. If R is a symmetric nontransitive relation on H, such that $R \subset R^2$, then H_R is a hypergroup if and only if $\forall x \in H$, we have $x \circ_R x \circ_R x = H$.

Proof. "$\Longrightarrow$" It results by (iii) of Theorem 12 and (iii) of Proposition 42.

"$\Longleftarrow$" The conditions of Theorem 8 are verified. ∎

Now, let $\langle H, \circ \rangle$ be a semihypergroup. Set

$$P(H) = \left\{ \prod_{i=1}^{n} a_i \mid n \in \mathbb{N}^*;\ \forall i \in \{1, 2, ..., n\},\ a_i \in H \right\}.$$

We have:

$$\forall x \in H, x \circ_\beta x = \{y \in H \mid x\beta y\} =$$
$$= \{y \in H \mid \exists P_0 \in P(H) : \{x, y\} \subset P_0\} = \bigcup_{P_0 \in P(H); x \in P_0} P_0.$$

Denote

$$\bigcup_{P_0 \in P(H); x \in P_0} P_0 = \mathcal{C}_1(x) \text{ and}$$
$$\forall n \in \mathbb{N}^*, \bigcup \{P_0 \in P(H) \mid P_0 \cap \mathcal{C}_n(x) \neq \emptyset\} = \mathcal{C}_{n+1}(x)$$

47. Theorem. *Let H be a semihypergroup. Then the relation β is transitive if and only if $\mathcal{C}(x) = \mathcal{C}_1(x)$, for all $x \in H$, where by $\mathcal{C}(x)$ we have denoted the complete closure of x.*

Proof. By Remark 46, 1, it results that β is transitive if and only if

$$\forall x \in H,\ x \circ_\beta x \circ_\beta x = x \circ_\beta x.$$

We have:

$$x \circ_\beta x \circ_\beta x = \bigcup_{a \in x \circ_\beta x} a \circ_\beta a \cup x \circ_\beta x =$$
$$= \{t \in H \mid t \in \mathcal{C}_1(a), a \in \mathcal{C}_1(x)\} \cup \mathcal{C}_1(x) =$$
$$= \{t \in H \mid t \in \bigcup_{P_0 \in P(H); a \in P_0} P_0,\ a \in \mathcal{C}_1(x)\} \cup \mathcal{C}_1(x) =$$
$$= \cup\{P_0 \in P(H) \mid P_0 \cap \mathcal{C}_1(x) \neq \emptyset\} \cup \mathcal{C}_1(x) = \mathcal{C}_2(x) \cup \mathcal{C}_1(x).$$

Therefore, β is transitive if and only if $\forall x \in H, \mathcal{C}_2(x) \cup \mathcal{C}(x) = \mathcal{C}_1(x)$, that is $\forall x \in H$, $\mathcal{C}_2(x) \subset \mathcal{C}_1(x)$. Then $\forall n \in \mathbb{N}^*$, $\mathcal{C}_{n+1}(x) \subset \mathcal{C}_n(x)$.

Indeed, if we suppose $\mathcal{C}_k(x) \subset \mathcal{C}_{k-1}(x)$, where $k \in \mathbb{N}^*$, then $\mathcal{C}_{k+1}(x) = \cup\{P_0 \in P(H) \mid P_0 \cap \mathcal{C}_k(x) \neq \emptyset\} \subset \cup\{P_0 \in P(H) \mid P_0 \cap \mathcal{C}_{k-1}(x) \neq \emptyset\} = \mathcal{C}_k(x)$. Since $\mathcal{C}(x) = \bigcup_{i \in \mathbb{N}^*} \mathcal{C}_i(x)$, it results that β is transitive if and only if $\forall x \in H$, $\mathcal{C}(x) = \mathcal{C}_1(x)$.

48. Proposition. *Let* $\mathbb{H} = \langle H, \cdot \rangle$ *be a semihypergroup such that the relation* β *is not transitive. Then* H_β *is a hypergroup if and only if* $\beta^2 = H \times H$.

Proof. It results by Theorem 8.

49. Remarks. Let $\langle H, \cdot \rangle$ be a hypergroup.

1. If $(x, y) \in H^2$, such that $x \in x \cdot y$ (or $(x \in y \cdot x)$ then $x \cdot y \subset x \circ_\beta y$ (respectively, $y \cdot x \subset x \circ_\beta y$).

2. $\forall x \in H$, if $x \in x \cdot x$, then $x \cdot x \subset x \circ_\beta x$.

Chapter 4

Lattices

Introduced by Ch.S. Pierce and E. Schröder and independently by R. Dedekind, and afterwards developed by G. Birkhoff, V. Glivenko, K. Menger, J. von Neumann, O. Ore and others, Lattice Theory is a highly topical field, with many applications in mathematics.

Distributive lattices represent the starting point in Lattice Theory; their study is required by more and more frequent situations when distributivity is imposed by applications.

A weaker condition of distributivity is the modularity, introduced by R. Dedekind.

Modularity and distributivity are characterized in this chaper, using hyperstructures, particularly join spaces.

§1. Distributive lattices and join spaces

The following hyperoperation was associated with an arbitrary lattice $(L, \vee, \wedge)$, by J.C. Varlet:

$$\forall (a,b) \in L^2, \; a \circ b = \{x \in L \mid a \wedge b \leq x \leq a \vee b\}.$$

The study of this hyperoperation will be continued in §2.

The importance of the hyperstructure $(L, \circ)$ consists in the fact that it is frequently used in *machine learning applications.*

The following proposition can be easily verified:

1. Proposition. *The following properties hold:*

1. $\forall (a,b) \in L^2,\ \{a,b\} \subset a{\circ}b$;
2. $\forall (a,b) \in L^2,\ a{\circ}b = b{\circ}a$;
3. $\forall (a,b) \in L^2,\ a/b \neq \emptyset$ *since* $a \in a/b = \{x \in L \mid x \wedge b \leq a \leq x \vee b\}$;
4. $\forall a \in L,\ a/a = L$;
5. $\forall (a,b) \in L^2,\ a/b \ni b$ *if and only if* $a = b$;
6. *if* a *has the unique complement* b, *then* $a/b = \{a\}$ *and* $b/a = \{b\}$;
7. $x \in a/b \cap b/a$ *if and only if* $a \wedge x = b \wedge x$ *and* $a \vee x = b \vee x$.

J.C. Varlet [397] obtained the following result:

2. Theorem. *For a lattice* L, *the following are equivalent:*

(1) L *is distributive;*

(2) $(L, \circ)$ *is a join space.*

Proof. (1)$\Longrightarrow$(2). First of all, we shall verify the associativity of the hyperoperation "$\circ$". Let a, b, c be arbitrary in L. The least and greatest elements of $a{\circ}(b{\circ}c)$ are $a \wedge b \wedge c$ and $a \vee b \vee c$ respectively, hence $a{\circ}(b{\circ}c) \subseteq [a \wedge b \wedge c, a \vee b \vee c]$.

Let us consider an arbitrary element x of $[a \wedge b \wedge c, a \vee b \vee c]$. If $y = (x \wedge (b \vee c)) \vee (b \wedge c)$, then $b \wedge c \leq y \leq b \vee c$, that is $y \in b{\circ}c$. Moreover, $a \wedge y \leq c \leq a \vee y$. Indeed, using distributivity, we have:

$$a \wedge ((x \wedge (b \vee c)) \vee (b \wedge c)) = (a \wedge x \wedge (b \vee c)) \vee (a \wedge b \wedge c) \leq x$$

and

$$a \vee ((x \wedge (b \vee c)) \vee (b \wedge c)) = (a \vee (b \wedge c) \vee x) \wedge (a \vee (b \wedge c) \vee (b \vee c)) \geq x.$$

Hence

$$x \in a\circ(b\circ c) \text{ and } a\circ(b\circ c) = [a \wedge b \wedge c, a \vee b \vee c].$$

Similarly, we have $(a\circ b)\circ c = [a \wedge b \wedge c, a \vee b \vee c]$, whence it follows the associativity.

Now, let us assume that $a/b \cap c/d \neq \emptyset$, that is there exists $x \in L$ such that $a \in b\circ x$ and $c \in d\circ x$. We have to prove that there exists $y \in L$ such that $y \in a\circ d \cap b\circ c$; which is equivalent to

$$(a \wedge d) \vee (b \wedge c) \leq y \leq (a \vee d) \wedge (b \vee c).$$

From $b \wedge x \leq a \leq b \vee x$ and $d \wedge x \leq c \leq d \vee x$, we deduce

$$a \wedge d \leq (b \vee x) \wedge d = (b \wedge d) \vee (x \wedge d) \leq (b \wedge d) \vee c \leq b \vee c.$$

Since $a\wedge d \leq b\vee c$ and $b\wedge c \leq b\vee c$, we have $(a\wedge d)\vee(b\wedge c) \leq b\vee c$.

Similarly, $(b \wedge c) \vee (a \wedge d) \leq a \vee d$, therefore $(a \wedge d) \vee (b \wedge c) \leq$ $\leq (a \vee d) \wedge (b \vee c)$, so $a\circ d \cap b\circ c \neq \emptyset$.

(2)$\Longrightarrow$(1). First, let us notice that $a/b \cap b/d \neq \emptyset$ implies $a\circ d \cap b\circ b \neq \emptyset$ and since $b\circ b = \{b\}$, it follows $b \in a\circ d$.

Therefore $a/b \cap b/a \neq \emptyset$ implies $b \in a\circ a = \{a\}$, whence $a = b$.

Let us suppose L is not distributive. Then L contains a five–element sublattice $\{a, b, c, d, e\}$, with $a \vee c = b \vee c = e$, $a \wedge c =$ $= b \wedge c = d$ and either $a > b$ or a, b, c mutually non–comparable. In both cases, a/b contains a and c, but not d.

We have $c \in a/b \cap b/a$ and yet $a \neq b$, contradiction.

Therefore, L is a distributive lattice. ■

§2. Lattice ordered join space

Ordered hypergroupoids and hypergroups have been studied by M. Konstantinidou and S. Serafimidis. In the following, an important example of *lattice–ordered join space* is considered, which is that one presented in §1. This topic has been explored by Ath. Kehagias and M. Konstantinidou and we present here some of their results.

Let $(L, \vee, \wedge)$ be a distributive lattice. Denote by "$\leq$" the associated order.

3. Notation. The class of intervals of elements of L is denoted by $I(L)$, that is:

$$I(L) = \{[a, b] \mid (a, b) \in L^2,\ a \leq b\}.$$

We consider on the distributive lattice $(L, \vee, \wedge)$ the following hyperoperation

$$\forall\,(a, b) \in L^2,\ a{\circ}b = \{x \in L \mid a \wedge b \leq x \leq a \vee b\} = [a \wedge b, a \vee b].$$

4. Proposition. *We have:*

$$I(L) = \{a{\circ}b \mid (a{\circ}b) \in L^2\}.$$

Proof. If $[a, b] \in I(L)$, then, by definition, we have $a \leq b$ so $a{\circ}b = [a \wedge b, a \vee b] = [a, b]$. On the other hand, any $a{\circ}b$ is an interval, by definition. ■

The following properties of intervals $[a, b]$ (where $(a, b) \in L^2$, $a \leq b$), are useful to prove again that if $(L, \leq)$ is a distributive lattice, then $(L, \circ)$ is a hypergroup.

5. Proposition. *Let* $(a, b, x, y) \in L^4$, *such that* $x < y$ *and* $a \leq b$. *We have:*

(i) $a{\circ}[x, y] = [a \wedge x, a \vee y]$;

(ii) $[a, b]{\circ}[x, y] = [a \wedge x, b \vee y]$.

Proof. i) We have $a{\circ}[x, y] = \bigcup_{x \leq z \leq y} a{\circ}z$. If $u \in a{\circ}[x, y]$, then there is $z_u \in [x, y]$, such that $a \wedge z_u \leq u \leq a \vee z_u$.

Since $x \leq z_u$ and $z_u \leq y$ it follows $a \wedge x \leq a \wedge z_u$ and $a \vee z_u \leq a \vee y$.

Therefore, $a \wedge x \leq u \leq a \vee y$, whence $u \in [a \wedge x, a \vee y]$. Hence $a \circ [x, y] \subseteq [a \wedge x, a \vee y]$.

On the other hand, if $v \in [a \wedge x, a \vee y]$ and we set $z_v = (v \vee x) \wedge y$, then, by distributivity, we also have $z_v = (v \wedge y) \vee x$. Then $x \leq \leq (v \wedge y) \vee x = z_v = (v \vee x) \wedge y \leq y$, that is $z_v \in [x, y]$. We also have

$$z_v \wedge a = [(v \vee x) \wedge y] \wedge a = (v \vee x) \wedge (y \wedge a) = (v \wedge y \wedge a) \vee (x \wedge y \wedge a).$$

From $v \wedge y \wedge a \leq v$ and $x \wedge y \wedge a = x \wedge a \leq v$, it follows $z_v \wedge a \leq v$. Similarly, we can verify that $v \subseteq z_v \vee a$. So, $z_v \wedge a \leq v \leq z_v \vee a$, whence $v \in a \circ z_v$. Hence, $z_v \in [x, y]$ and $v \in a \circ z_v$, which implies that $v \in a \circ [x, y]$. Thus, $[x \wedge a, y \vee a] \subseteq a \circ [x, y]$. We can conclude that $[x \wedge a, y \vee a] = a \circ [x, y]$.

ii) First of all, we shall verify that $[a, b] \circ [x, y] \subseteq [a \wedge x, b \vee y]$. If $u \in [a, b] \circ [x, y] = \bigcup_{a \leq z \leq b} z \circ [x, y] = \bigcup_{a \leq z \leq b} [z \wedge x, z \vee y]$, then there is $z_1 \in [a, b]$, such that $z_1 \wedge x \leq u \leq z_1 \vee y$. On the other hand, $a \wedge x \leq z_1 \wedge x$ and $z_1 \vee y \leq b \vee y$, whence $a \wedge x \leq z_1 \wedge x \leq u \leq z_1 \vee y \leq \leq b \vee y$, so $u \in [a \wedge x, b \vee y]$. Therefore $[a, b] \circ [x, y] \subseteq [a \wedge x, b \vee y]$. Conversely, let $v \in [a \wedge x, b \vee y]$, that is $a \wedge x \leq v \leq b \vee y$. Set $z_1 = (v \vee x) \wedge y = (v \wedge y) \vee x$ and $z_2 = (v \vee a) \wedge b = (v \wedge b) \vee a$. It easily results that $z_1 \in [x, y]$ and $z_2 \in [a, b]$. We have

$$\begin{aligned} z_1 \wedge z_2 = [(v \vee x) \wedge y] \wedge [(v \vee a) \wedge b] = [v \vee (a \wedge x)] \wedge [b \wedge y] = \\ = [v \wedge (b \wedge y)] \vee [a \wedge x] \leq v. \end{aligned}$$

Similarly, we verify that $v \leq z_1 \vee z_2$, so $v \in z_1 \circ z_2 \subseteq [x, y] \circ [a, b]$. Therefore, $[a, b] \circ [x, y] = [a \wedge x, b \vee y]$. ■

6. Proposition. *For any $(a, b, c) \in L^3$, the following properties hold:*

(i) $(a \circ b) \circ c = a \circ (b \circ c)$;

(ii) $a \circ L = L$.

Moreover, $\nexists u \in L$ *such that* $\forall\, x \in L$, *we have* $|u{\circ}x| = 1$.

Proof. i) $(a{\circ}b){\circ}c = [a{\wedge}b, a{\vee}b]{\circ}c = [a{\wedge}b{\wedge}c, a{\vee}b{\vee}c]$, by the previous proposition. Similarly, $a{\circ}(b{\circ}c) = a{\circ}[b\wedge c, b\vee c] = [a\wedge b\wedge c, a\vee b\vee c]$.

ii) For any $a \in L$, we have $a{\circ}L = \bigcup\limits_{x\in L} a{\circ}x \supseteq \bigcup\limits_{x\in L} x = L$. On the other hand, we have $a{\circ}L \subseteq L$, so $a{\circ}L = L$. Finally, notice that for any $a \in L$ and $x \in L$, $x \neq a$, we have $\{a, x\} \subset a{\circ}x$, therefore $|a{\circ}x| \geq 2$. ■

7. Corollary. $(L, \circ)$ *is a hypergroup.*

8. Proposition. *For any* $(a, b) \in L^2$, *we have that* $(a{\circ}b, \circ)$ *is a subhypergroup of* L.

Proof. Let $(a, b) \in L^2$. We shall verify that for any x and y in (a, b), we have:

$$1)\ x{\circ}y \subseteq a{\circ}b \quad \text{and} \quad 2)\ x{\circ}(a{\circ}b) = a{\circ}b.$$

1) We have $a{\wedge}b \leq x{\wedge}y \leq x{\vee}y \leq a{\vee}b$, that means $x\circ y \subseteq a{\circ}b$.

2) We have $x \circ (a{\circ}b) = [x \wedge a \wedge b, x \vee a \vee b] = [a \wedge b, a \vee b]$, since $a \wedge b \leq x \leq a \vee b$.

Therefore, $x \circ (a{\circ}b) = a{\circ}b$. ■

9. Proposition. *Let* $(a, b, c) \in L^3$. *We have:*

(i) $a{\circ}(b \vee c) = (a{\circ}b) \vee (a{\circ}c)$;

(ii) $a{\circ}(b \wedge c) = (a{\circ}b) \wedge (a{\circ}c)$.

Proof. i) Let $u \in a{\circ}(b \vee c)$ and set $x = (a \vee b) \wedge u$, $y = (a \vee c) \wedge u$. From $a{\circ}(b \vee c) = [a \wedge (b \vee c), a \vee b \vee c]$ it follows $u \leq a \vee b \vee c$. We also have $x \vee y = [(a \vee b) \wedge u] \vee [(a \vee c) \wedge u] = (a \vee b \vee c) \wedge u = u$.

On the other hand, from $x = (a \vee b) \wedge u$ it follows $x \leq a \vee b$. From $a \wedge (b \vee c) \leq u$, we obtain $a \wedge b \leq u$; so $a \wedge b \leq (a \vee b) \wedge u = x$. Thus, $x \in a{\circ}b$. Similarly, we can verify that $y \in a{\circ}c$. Therefore,

$\forall u \in a\circ(b \vee c)$, $\exists x \in a\circ b$, $\exists y \in a\circ c$ such that $u = x \vee y$. Hence $a\circ(b \vee c) \subseteq (a\circ b) \vee (a\circ c)$. Now, consider $v \in (a\circ b) \vee (a\circ c)$. Then there is $x \in a\circ b$ and $y \in a\circ c$, such that $v = x \vee y$. So, $a \wedge (b \vee c) = = (a \wedge b) \vee (a \wedge c) \leq x \vee y = v$. Similarly, $v \leq a \vee (b \vee c)$. Hence $v \in a\circ(b \vee c)$, that means $(a\circ b) \vee (a\circ c) \subseteq a\circ(b \vee c)$. Therefore, $a\circ(b \vee c) = (a\circ b) \vee (a\circ c)$.

ii) It follows by duality. ∎

10. Definition. The structure $(L, \leq, *)$ is called a *strictly lattice–ordered hypergroup* (respectively, *join space*) if and only if

(i) $(L, \leq)$ is a lattice;

(ii) $(L, *)$ is a hypergroup (respectively a join space);

(iii) $\forall\,(x, y) \in L^2$, $x\circ y$ is an interval;

(iv) $\forall\,(a, x, y) \in L^3$, we have:

$$a * (x \vee y) = (a * x) \vee (a * y) \text{ and}$$
$$a * (x \wedge y) = (a * x) \wedge (a * y).$$

11. Remark. The structure $(L, \leq, \circ)$ is a strictly lattice–ordered join space, according to Theorem 2.

12. Remark. In the hypergroup $(L, \circ)$, we have:

$$\forall\,(a, b) \in L^2,\ a/b = \{x \in L \mid a \in x\circ b\} = \{x \in L \mid x \wedge b \leq a \leq x \vee b\}.$$

From here it follows $\forall\, a \in L,\ a \in a/b$.

13. Proposition. *For any $(a, b, c, d) \in L^4$, the following conditions are equivalent:*

1) *$a \wedge d \leq b \vee c$ and $b \wedge c \leq a \vee d$;*

2) *$a\circ d \cap b\circ c \neq \emptyset$.*

Proof. 1)$\Longrightarrow$2). From $a \wedge d \leq b \vee c$ and $b \wedge c \leq a \vee d$, we obtain $a \wedge d \leq (a \wedge d) \vee (b \wedge c) \leq a \vee d$ and $b \wedge c \leq (a \wedge d) \vee (b \wedge c) \leq b \vee c$. In a similar way, it follows:

$$a \wedge d \leq (a \vee d) \wedge (b \vee c) \leq a \vee d \text{ and}$$
$$b \wedge c \leq (a \vee d) \wedge (b \vee c) \leq b \vee c.$$

Moreover, we have

$$(a \wedge d) \vee (b \wedge c) \leq (a \vee d) \wedge (b \vee c).$$

Set $u = (a \wedge d) \vee (b \wedge c)$ and $v = (a \vee d) \wedge (b \vee c)$. By the previous inequalities we obtain: $[u, v] \subseteq (a \circ d) \cap (b \circ c)$, so 2) holds.

2)$\Longrightarrow$1) Let $p \in a \circ d \cap b \circ c$. Then $a \wedge d \leq p \leq b \vee c$ and $b \wedge c \leq p \leq a \vee d$, whence we obtain 1). ■

§3. Modular lattices and join spaces

In the following, the hypergroupoids attached to semi–lattices and lattices are studied. Moreover, characterizations for modular lattices are presented. Results on this direction have been obtained by St. Comer, J. Mittas, M. Konstantinidou and afterwards by G. Călugăreanu and V. Leoreanu. In the following, we mention some of them.

Let $(L, \leq, \vee)$ be a semi–lattice and let us consider the following hyperoperation on L, introduced by Nakano [298]:

$$\forall (x, y) \in L^2,\ x \oplus y = \{z \in T \mid z \vee x = z \vee y = x \vee y\}.$$

We notice that

$$\forall (x, y) \in L^2,\ x \vee y \in x \oplus y.$$

$< L, \oplus >$ is called the *attached hypergroupoid to the semi–lattice* $(L, \leq, \vee)$. Notice that $< L, \oplus >$ is a quasi–hypergroup and if L has a zero, then 0 is a scalar identity of $< L, \oplus >$.

14. Theorem. (Comer) *If* $< L, \leq >$ *is a modular lattice with zero, then* $(L, \oplus)$ *is a canonical hypergroup.*

We shall prove this theorem by a different way (see Lemma 24 – Prop. 35).

15. Proposition. *For any* $(x, y, z, w) \in L^4$, *we have:*

$$(x \oplus y) \cap (z \oplus \omega) \neq \emptyset \Longrightarrow (x \oplus z) \cap (y \oplus w) \neq \emptyset.$$

Proof. Let $t \in (x \oplus y) \cap (z \oplus w)$. Then

$$\begin{aligned} t \in x \oplus y \quad & \Longrightarrow y \in x \oplus t \Longrightarrow y \in x \oplus (z \oplus w) = (x \oplus z) \oplus w \Longrightarrow \\ & \Longrightarrow \exists s \in x \oplus z,\ y \in s \oplus w \Longrightarrow s \in y \oplus w. \end{aligned}$$

So, $(x \oplus z) \cap (y \oplus w) \neq \emptyset$. ■

16. Corollary. *For any* $(x, y) \in L^2$, *we have*

$$(x \oplus x) \cap (y \oplus y) \neq \emptyset.$$

17. Proposition. *If the hypergroupoid* $< L, \oplus >$ *associated with a semi–lattice is a hypergroup, then it is a join space.*

Proof. It follows by the above proposition and the equality:
$\forall (x, y) \in L^2,\ x/y = x \oplus y$. ■

Let us suppose in the following that $< L, \oplus >$ is a hypergroup.

18. Proposition. *For any* $n \in \mathbb{N}^*$, $\forall i \in \{1, 2, ..., n\}$, $x_i \in L$, *we have*

$$\bigcap_{i=1}^{n} x_i \oplus x_i \neq \emptyset$$

Proof. We prove it by induction on n.

For $n = 2$ we have just verified the thesis, so we suppose $\bigcap_{i=1}^{n-1} x_i \oplus x_i \neq \emptyset$.

Let $z \in \bigcap_{i=1}^{n-1} x_i \oplus x_i$ and $w \in x_n \oplus x_n$. We have $z \oplus z \cap w \oplus w \neq \emptyset$, whence there is $u \in z \oplus z \cap w \oplus w$. We have $u \leq z$ and $u \leq w$, hence for any $i \in \{1, 2, ..., n\}$, $u \leq x_i$, whence $u \in \bigcap_{i=1}^{n} x_i \oplus x_i$. ■

Let us consider now

$$I = \bigcap_{x \in L} x \oplus x$$

and suppose that $I \neq \emptyset$. If $z \in I$, then we have $z \in x \oplus x$, that means $z \leq x$, for any $x \in L$. So, L has a minimum, such that $z = z \oplus z$. It follows:

19. Proposition. *If $I \neq \emptyset$, then L has minimum, which we denote by 0 (it is a scalar) and the attached hypergroup $(L, \oplus)$ is a canonical one and conversely, if $(L, \oplus)$ is a canonical hypergroup, then $I \neq \emptyset$.*

20. Remark. For any $x \in L$, we have that $h_x = x \oplus x$ is an invertible subhypergroup of the hypergroup $(L, \oplus)$.

21. Proposition. *Let h be a subhypergroup of $< L, \oplus >$. Then $h = \bigcup_{x \in h} (x \oplus x)$.*

Proof. Let $z \in h$. It follows $z \oplus z \subseteq \bigcup_{x \in h} x \oplus x$ and since $z \in z \oplus z$ it follows that $z \in \bigcup_{x \in h} x \oplus x$, hence $h \subset \bigcup_{x \in h} x \oplus x \subset h$, so we have the equality. ■

22. Remark. For any $(x, y) \in L^2$, we have:

(i) $h_{xy} = (x \oplus x) \cap (y \oplus y)$ is a subhypergroup of L.

(ii) if there is $\inf(x, y) = x \wedge y$, then
$x \oplus x \cap y \oplus y = (x \wedge y) \oplus (x \wedge y) = h_{x \wedge y}$.

23. Proposition. *For any* $(x, y) \in L^2$, *one has*

$$(x \oplus x) \cup (y \oplus y) \subseteq (x \oplus x) \oplus (y \oplus y) = (x \vee y) \oplus (x \vee y) = h_{x \vee y}.$$

Proof. Let us consider the following equivalence relation on L, denote by $\operatorname{Mod} a$, where $a \in L$:

$$x \equiv y (\operatorname{mod} a) \Longleftrightarrow a \vee x = a \vee y.$$

The equivalence class of x is

$$C_a(x) = \{y \in T \mid a \vee x = a \vee y\}$$

First of all, let us prove that

$$\forall (x, y) \in T^2, \ C_a(x) \oplus C_a(y) = (a \vee x) \oplus (a \vee y).$$

Indeed, if $z \in C_a(x) \oplus C_a(y)$, then $\exists x' \in C_a(x)$, $\exists y' \in C_a(y)$ such that $z \in x' \oplus y'$, so $z \vee x' = z \vee y' = x' \vee y'$, whence $z \vee (a \vee x') =$ $= z \vee (a \vee y') = (a \vee x') \vee (a \vee y')$, hence $z \in (a \vee x) \oplus (a \vee y)$. Then, $C_a(x) \oplus C_a(y) \subseteq (a \vee x) \oplus (a \vee y)$. Obviously, we have

$$(a \vee x) \oplus (a \vee y) \subseteq C_a(x) \oplus C_a(y).$$

Then $C_a(x) \oplus C_a(y) = (a \vee x) \oplus (a \vee y)$. On the other hand, $\forall x \in L$, we have $C_a(x) = x \oplus (a \oplus a)$. Indeed, if $z \in C_a(x)$, then $a \vee z = a \vee x$, hence $z \oplus (a \oplus a) = x \oplus (a \oplus a)$, whence $z \in x \oplus (a \oplus a)$. Conversely, if $z \in x \oplus (a \oplus a)$, then $z \oplus (a \oplus a) \subseteq x \oplus (a \oplus a)$ so $z \oplus (a \oplus a) = x \oplus (a \oplus a)$, hence $a \vee z = a \vee x$, that means $z \in C_a(x)$.

We have $(x \vee y) \oplus (x \vee y) = C_x(y) \oplus C_x(y) = [y \oplus (x \oplus x)] \oplus$ $\oplus [y \oplus (x \oplus x)] = (x \oplus x) \oplus (y \oplus y)$. If $z \in (x \oplus x \cup (y \oplus y)$, then $z \leq x \vee y$, that means $z \in (x \vee y) \oplus (x \vee y)$, so $(x \oplus x) \cup (y \oplus y) \subseteq (x \oplus x) \oplus (y \oplus y)$. ■

Now, let L be a lattice and we define the hyperoperation on L, as above: for each $(a,b) \in L^2$, $a \oplus b = \{x \in L \mid a \vee x = b \vee x = a \vee b\}$.

24. Lemma. *For* $(a,b,c) \in L^3$, *if* $S = \{y \in L \mid a \vee b \vee y = a \vee c \vee y = b \vee c \vee y = a \vee b \vee c\}$ *then* $(a \oplus b) \oplus c \subseteq S$.

Proof. Let $y \in (a \oplus b) \oplus c$. Then $\exists x \in L : a \vee x = b \vee x = a \vee b$ and $x \vee c = x \vee y = y \vee c$ and hence $(a \vee b) \vee y = (a \vee x) \vee y = a \vee (x \vee y) = a \vee (x \vee c) = \begin{cases} (a \vee x) \vee c = (a \vee b) \vee c \\ a \vee (c \vee y) = (a \vee c) \vee y \end{cases}$ respectively $(a \vee b) \vee y = (b \vee x) \vee y = b \vee (x \vee y) = b \vee (c \vee y) = (b \vee c) \vee y$. Therefore $y \in S$. ■

25. Corollary. *If* $y \in S$, *then* $b \vee c \leq b \vee y \vee a$; $b \vee c \leq c \vee y \vee a$ *and* $y \vee a \leq y \vee b \vee c$; $y \vee a \leq a \vee b \vee c$.

26. Lemma. *If* L *is a modular lattice, then* $S \subseteq a \oplus (b \oplus c)$.

Proof. For an arbitrary $y \in S$ set $z = (y \vee a) \wedge (b \vee c)$. We verify $z \in b \oplus c$ and $y \in a \oplus z$.

Indeed, $b \vee z = b \vee [(y \vee a) \wedge (b \vee c)] \overset{\text{mod}}{=\!=} (b \vee y \vee a) \wedge (b \vee c) \overset{*}{=\!=} b \vee c$ and, similarly, $c \vee z = b \vee c$. On the other hand,

$$y \vee z = y \vee [(b \vee c) \wedge (y \vee a)] \overset{\text{mod}}{=\!=} (b \vee c) \wedge (y \vee a) \overset{*}{=\!=} y \vee a$$

and, similarly, $a \vee z = y \vee a$ (the $*$-equalities hold, according to the above consequence). Hence $y \in a \oplus z \subseteq a \oplus (b \oplus c)$. ■

27. Corollary. *In a modular lattice, we have:* $\forall (a,b,c) \in L^3$, $(a \oplus b) \oplus c = a \oplus (b \oplus c)$.

Proof. We already have $(a \oplus b) \oplus c \subseteq S \subseteq a \oplus (b \oplus c)$. The subset S is invariant to permutations of $\{a,b,c\}$ so we also obtain $(b \oplus c) \oplus a \subseteq S$. By the commutativity, we have $a \oplus (b \oplus c) \subseteq S$ and so $a \oplus (b \oplus c) = S$. Analogously, $(a \oplus b) \oplus c = S$. ■

28. Corollary. *If* L *is a modular lattice, then* $< L, \oplus >$ *is a semihypergroup.*

29. Remark. In an arbitrary lattice, the hyperoperation "$\oplus$" is not generally associative.

Indeed, in the 5-elements non-modular lattice N_5 ($N_5=\{0,a,b,c,1\}$, where $0<b<a<1$, $0<c<1$ and $a\|c$, $b\|c$, where "$\|$" means that the corresponding elements are not comparable) one verifies that $(a\oplus b)\oplus c=\{1\}\neq\{1;c\}=a\oplus(b\oplus c)$.

Moreover, the following interesting characterization holds

30. Theorem. *The hyperoperation "$\oplus$" is associative if and only if the lattice L is modular.*

Proof. If L is not modular, using a well–known characterization, L contains a 5-elements sublattice isomorphic to the above one: $\{m,a,b,c,M\}$, where $m<b<a<M$, $m<c<M$ and $a\|c$, $b\|c$.

But then $c\in a\oplus(b\oplus c)$ and $c\notin(a\oplus b)\oplus c$, hence "$\oplus$" is not associative.

Indeed, $c\in\{y\in L\mid a\vee y=M\vee y=a\vee M=M\}=a\oplus M\subseteq$ $\subseteq a\oplus(b\oplus c)$, because $M\in\{x\in L\mid b\vee x=c\vee x=b\vee c=M\}=b\oplus c$. Finally, $a\oplus b=\{x\in L\mid a\vee x=b\vee x=a\vee b=a\}=\{x\in L\mid x\leq a=$ $=x\vee b\}$ so that $a\oplus b\cap\{x\in L\mid x\leq b\}=\emptyset$. On the other hand, $c\in(a\oplus b)\oplus c=\{y\in L\mid x\vee y=c\vee y=x\vee c$, where $x\in a\oplus b\}$. So, if $x\in a\oplus b$, then $x\leq c$, whence $x\leq\inf(a;c)=m$ and so $x\leq b$, contradiction with the above void intersection. ■

31. Lemma. *$<L,\oplus>$ is a quasi–hypergroup.*

Proof. Indeed, $\forall(a,b)\in L^2$; $\exists x=a\vee b: a\in b\oplus x=x\oplus b$, because $a\vee x=b\vee x=a\vee b$. Hence $\forall b\in L: b\oplus L=L\oplus b=L$. ■

32. Corollary. *For a modular lattice L, $<L,\oplus>$ is a hypergroup.*

33. Remark. Each element in L is a partial identity in $<L,\oplus>$.

Indeed, $x\in x\oplus x$ holds for each $x\in L$.

34. Remark. For an arbitrary lattice, $\forall(a,b,c)\in L^3$:$a\in b\oplus c$ $\Longrightarrow b\in a\oplus c$; $c\in a\oplus b$.

35. Proposition. *If* L *is a modular lattice with zero, then* $< L, \oplus >$ *is a canonical hypergroup.*

Proof. Indeed, 0 is the unique scalar identity (that is $\forall a \in L$, $\{a\} = 0 \oplus a$ and for any identity we have $e = e \oplus 0 = 0 \oplus e = 0$; each element has a unique inverse: itself (indeed, $0 \in a \oplus a$ and $0 \in a \oplus b \Longrightarrow a = b$) and the reversibility follows from the previous remark. ■

Moreover,

36. Theorem. *Let* L *be a modular lattice. The following conditions are equivalent:*

(i) $< L, \oplus >$ *is a regular hypergroup;*

(ii) $< L, \oplus >$ *is a regular reversible hypergroup;*

(iii) $< L, \oplus >$ *is a canonical hypergroup;*

(iv) L *has a zero.*

Proof. According to the proof of the above theorem it remains only to remark that if m is the identity then $\forall a \in L$ we have $a \in m \oplus a = \{x \in L \mid m \vee a = m \vee x = x \vee a\}$ and so $\forall a \in L$, $m \leq a$. ■

37. Theorem. *For a modular lattice* L, $< L, \oplus >$ *is a join space.*

Proof. We only have to verify that $a/b \cap c/d \neq \emptyset$ implies $(a \oplus d) \cap$ $\cap (b \oplus c) \neq \emptyset$ where $a/b = \{x \in L \mid a \in x \oplus b\}$. But $a/b = a \oplus b$ and so we have to verify that if $x \in (a \oplus b) \cap (c \oplus d)$ (that is $a \vee x = b \vee x = a \vee b$ and $c \vee x = d \vee x = c \vee d$)) then there is an element $y \in a \oplus d \cap b \oplus c$. Set $y = (a \vee d) \wedge (b \vee c)$. We have $a \vee y = a \vee [(a \vee d) \wedge (b \vee c)] \overset{\text{mod}}{=} (a \vee d) \wedge (a \vee b \vee c) = (a \vee d) \wedge (a \vee x \vee c) =$ $= (a \vee d) \wedge (a \vee c \vee d) = a \vee d$ and $d \vee y = d \vee [(a \vee d) \wedge (b \vee c)] \overset{\text{mod}}{=}$ $(a \vee d) \wedge (b \vee c \vee d) = (a \vee d) \wedge (b \vee x \vee d) = (a \vee d) \wedge (a \vee b \vee d) = a \vee d$ so that $y \in a \oplus d$ and, similarly, we have $y \in b \oplus c$. ■

Notice that Theorem 37 can also be obtained from Proposition 17. From Theorem 30 and theorem 37 it follows

38. Corollary. *The lattice* $(L, \vee, \wedge)$ *is modular if and only if* $< L, \oplus >$ *is a join space.*

39. Remark. The hypergroup $< L, \oplus >$ is not complete.

We can also consider the dual hyperoperation, that is $\forall\,(a, b) \in L^2$, $a \circledast b = \{x \in L \mid a \wedge x = b \wedge x = a \wedge b\}$. By duality, the following results are verified:

40. Theorem. *Let L be a modular lattice. The following conditions are equivalent:*

(i) $< L,\ \circledast >$ *is a regular hypergroup,*

(ii) $< L,\ \circledast >$ *is a regular reversible hypergroup,*

(iii) $< L,\ \circledast >$ *is a canonical hypergroup,*

(iv) L *has a greatest element.*

41. Theorem. *For a modular lattice* L, $< L,\ \circledast >$ *is a join space.*

42. Theorem. *For a lattice L, the following conditions are equivalent:*

(i) L *is modular;*

(ii) $< L, \oplus >$ *is a hypergroup;*

(iii) $< L,\ \circledast >$ *is a hypergroup.*

43. Proposition. *Let L be a modular lattice. A subset I of L is an (invertible) subhypergroup of* $< L, \oplus >$ *if and only if I is an ideal of L.*

Proof. If I is a subhypergroup of $< L, \oplus >$, then for every $(a, b) \in I^2$ we have $a \vee b \in a \oplus b \subseteq I$. Moreover, if $a \in I$ and $x \leq a$; $x \in L$ then $x \in a \oplus a \subseteq I$ and so, I is an ideal of L.

Conversely, let I be an ideal of L. For $(a,b) \in I^2$, if $t \in a \oplus b$, then $t \leq a \vee b$ and so $t \in I$. For every $(a,b) \in I^2$, there is an element $x = a \vee b \in I$ such that $a \in b \oplus x$. Hence, I is a subhypergroup of $< L, \oplus >$.

We finally remark that if I is a subhypergroup of $< L, \oplus >$ then it is invertible.

Dually, it follows the following

44. Proposition. *Let L be a modular lattice. A subset I of L is an (invertible) subhypergroup of $< L, \circledast >$ if and only if I is a filter of L.*

Moreover, we have

45. Proposition. *If L is a modular lattice, the only ultraclosed subhypergroup of $< L, \circledast >$ (resp. $< L, \circledast >$) is $< L, \oplus >$ (resp. $< L, \circledast >$).*

Proof. Suppose that I is a ultraclosed subhypergroup of L. If $I \neq L$, set $a \notin I$ and $t \in I$. Then $a \vee t \in I$ and so

$$a \vee t \in (a \oplus I) \cap (a \oplus (L - I)).$$

Hence $(a \oplus I) \cap (a \oplus (L - I)) = \emptyset$ holds for every $a \in L$ only if $I = L$.

46. Corollary. *If L is a modular lattice, then*

$$\omega_{<L,\oplus>} = \omega_{<L,\circledast>} = L.$$

Now, we shall mention some important (for which follows) properties of the join space $< L, \oplus >$ associated with a modular lattice $(L, \vee, \wedge)$.

One can verify these properties, using the equivalence relation

$$(1) \qquad x \in a \oplus a \Longleftrightarrow x \leq a.$$

47. Proposition. *For a modular lattice L, the associated join space $< L, \oplus >$ satisfies the following properties:*

(i) $\forall a \in L,\ a \in a \oplus a$; *$a \oplus a$ is a subhypergroup of* $< L, \oplus >$;

(ii) $\forall (a,b) \in L^2, \quad \bigcap_{\{a,b\} \subseteq x \oplus x} x \oplus x = (a \vee b) \oplus (a \vee b)$;

(iii) $\forall (a,b) \in L^2,\ a \oplus a \cap b \oplus b \subseteq (a \wedge b) \oplus (a \wedge b)$ *and* $a \wedge b \in a \oplus a \cap b \oplus b$;

(iv) $\forall (a,b) \in L^2,\ \{a,b\} \subseteq a \oplus b \Longrightarrow a = b$;

(v) $\forall a \in L,\ a \oplus a \oplus a = a \oplus a$;

(vi) $\forall (a,b) \in L^2,\ a \oplus b = [a \oplus (a \vee b)] \cap [b \oplus (a \vee b)]$;

(vii) *if* $a < b,\ a \oplus b = \{b\} \cup \{x \in L \mid x < b,\ x \| a,\ \nexists y \in L,\ a < y < b,\ x < y\}$, *where we denote by* $x \| a$ *two incomparable elements of* L.

In the following, we shall characterize the join space associated with modular lattices.

We notice that in a join space $< L, \oplus >$, associated with a modular lattice the following condition holds:

$$(\alpha) \qquad \begin{gathered} \forall (a,b) \in L^2,\ \exists x \in L,\ \exists t \in L,\ \{a,b\} \subseteq x \oplus x, \\ \bigcap_{\{a,b\} \subseteq x \oplus x} x \oplus x = t \oplus t,\ a \oplus b = a \oplus t \cap b \oplus t. \end{gathered}$$

Moreover, if $a \in t \oplus t - \{t\}$, then

$$a \oplus t = \{t\} \cup \left\{ \begin{matrix} u \in L \mid u \in t \oplus t - \{t\},\ u \notin a \oplus a,\ a \notin u \oplus u,\ \nexists y \in L, \\ a \in y \oplus y - \{y\},\ u \in y \oplus y - \{y\},\ y \in t \oplus t - \{t\} \end{matrix} \right\}.$$

The condition (α) is equivalent to the set of conditions (ii), (vi) and (vii), written using only the hyperoperation "$\oplus$" (not the order "$\leq$").

48. Theorem. *A join space* $< H, \circ >$ *is associated with a lattice* $(H, \vee, \wedge)$ *if and only if it satisfies* (α) *and the following conditions:*

(1) $\forall (a,b) \in H^2,\ a/b = a \oplus b$;

(2) $\forall a \in H,\ a \oplus a \oplus a = a \oplus a;$

(3) $\forall (a,b) \in H^2,\ \exists s \in a \oplus a \cap b \oplus b,\ a \oplus a \cap b \oplus b \subseteq s o s;$

(4) $\forall (a,b) \in H^2,\ \{a,b\} \subseteq a \oplus b \Longleftrightarrow a = b.$

Proof. From Proposition 47, it follows that the above conditions are necessary. For the sufficiency, we define a binary relation on H as follows:

$$a \leq b \Longleftrightarrow a \in b \oplus b \overset{(1)}{\Longleftrightarrow} b \in a \oplus b.$$

This is an order on H according to (1) [reflexivity: $\forall a \in H,\ a \in a \oplus a$], [transitivity: $a \in b \oplus b, b \in c \oplus c \Longrightarrow a \in c \oplus c \oplus c \oplus c = c \oplus c \oplus c = c \oplus c$] and again (1) [antisymmetry].

In order to obtain a lattice structure, for arbitrary elements $a, b \in H$, we consider t, where

$$\bigcap_{\{a,b\} \subseteq x \oplus x} x \oplus x = t \oplus t$$

and verify that $t = \sup(a,b)$. Indeed, $\{a,b\} \in t \oplus t$ so that $a \leq t$, $b \leq t$; moreover, if $a \leq s, b \leq s$, then

$$t \in t \oplus t = \bigcap_{\{a,b\} \subseteq x \oplus x} x \oplus x \subseteq s \oplus s$$

because $\{a,b\} \in s \oplus s$ so $t \leq s$. The antisymmetry proves that t is unique. On the other hand, for an arbitrary element $(a,b) \in H^2$, we consider s, such that

$$s \in a \oplus a \cap b \oplus b \text{ and } a \oplus a \cap b \oplus b \subseteq s \oplus s$$

and verify that $s = \inf(a,b)$. Obviously, we have $s \leq a$, $s \leq b$ and if $u \leq a, u \leq b$ then $u \in a \oplus a \cap b \oplus b \subseteq s \oplus s$, whence $u \leq s$. The element s is unique because $\{s_1, s_2\} \subseteq a \oplus a \cap b \oplus b \subseteq s_1 \oplus s_1 \cap s_2 \oplus s_2$ implies $s_1 \in s_2 \oplus s_2$ and $s_2 \in s_1 \oplus s_1$ and so $s_1 = s_2$. Hence, $(H, \sup, \inf)$ is a lattice. Its modularity is easily checked.

In what follows, we use the standard notations

$$\sup(a,b) = a \vee b,\ \inf(a,b) = a \wedge b.$$

Now, we verify the inclusion $a \oplus b \subseteq \{x \in H | a \vee x = b \vee x = a \vee b\}$: let $x \in a \oplus b$; from $\{a,b\} \subseteq t \oplus t$ where $t = a \vee b$ it follows $x \in a \oplus b \subseteq t \oplus t \oplus t \oplus t = t \oplus t$ and so $x \in (a \vee b) \oplus (a \vee b)$ that is $x \leq a \vee b$. Hence $a \vee x \leq a \vee b$ and $b \vee x \leq a \vee b$. But $\{b,x\} \subseteq (b \vee x) \oplus (b \vee x)$ and so $b \oplus x \subseteq (b \vee x) \oplus (b \vee x) \oplus (b \vee x) \oplus (b \vee x) = (b \vee x) \oplus (b \vee x)$. Using (1), we have $x \in a \oplus b = a/b$ and so $a \in b \oplus x \subseteq (b \vee x) \oplus (b \vee x)$ whence $a \leq b \vee x$ so $a \vee b \leq b \vee x$. We obtain $a \vee b = b \vee x$. Similarly, we have $a \vee b = a \vee x$ and hence $x \in \{z \in H | a \vee z = b \vee z = a \vee b\}$.

Conversely, let $x \in H$ be such that $a \vee x = b \vee x = a \vee b$.

It follows $x \leq a \vee x = b \vee x = a \vee b$.

We distinguish the following cases:

Case 1: if $a = b$ then $x \leq a$ and so $x \in a \oplus a = a \oplus b$.

Case 2: if $b < a$ then $x \leq a = a \vee b$. If $x = a$ nothing is to be proved. If $x < a$ then $b \leq x$ is not possible (otherwise $a = b \vee x = x$) nor $b \geq x$ (otherwise $a = b \vee x = b$) and so $b \| x$. Moreover, there is no element $y \in H$ such that $b < y < a$, $x < y$ (otherwise $a = b \vee x \leq y < a$). Therefore we obtain $x \in a \oplus b$, using (α).

Case 3: Similarly, if $a < b$, one verifies that $x \in a \oplus b$.

Case 4: if $a \| b$ then $a < a \vee b, b < a \vee b$. We first check that $x \in a \oplus (a \vee b)$. This is clear for $x = a \vee b$ so in what follows we suppose $x \neq a \vee b$. Now $x \leq a \Longrightarrow a \vee x \leq a \Longleftrightarrow a \vee b \leq a \Longleftrightarrow a \vee b = a$ and analogously $a \leq x \Longrightarrow x = a \vee b$, both contradictions, so that $x \| a$ and $x < a \vee b$. As above $x \in$ $\in \{a \vee b\} \cup \{u \in H | u < a \vee b, u \| a, \nexists y \in H, a < y < a \vee b, u < y\} =$ $= a \oplus (a \vee b)$.

Similarly, $x \in b \oplus (a \vee b)$ and so $x \in a \oplus (a \vee b) \cap b \oplus (a \vee b) =$ $= a \oplus t \cap b \oplus t$. Hence, by the condition (α), it follows $x \in a \oplus b$ and this completes our proof. ■

49. Lemma. *Let $(L, \vee, \wedge)$ be a lattice and $f : L \to L$ be a bijective map. The following conditions are equivalent:*

a) $\forall (a,b) \in L^2,\ f(a \vee b) = f(a) \wedge f(b)$;

b) $\forall (a,b) \in L^2,\ f(a \oplus b) = f(a) \circledast f(b)$.

Proof. (a)$\Longrightarrow$(b). Clearly, $f(a \oplus b) = \{f(x) \mid x \in L;\ x \vee a = x \vee b = a \vee b\}$ and so $f(x) \in f(a) \circledast f(b)$, by (a).

Conversely, if $t \in f(a) \circledast f(b)$, there is an element $x \in L$, such that $t = f(x)$, since f is onto, and so $f(x \vee a) = f(x \vee b) = f(a \vee b)$, again by (a). Since f is an one-to-one map, it follows $x \in a \oplus b$ and $t \in f(a \oplus b)$.

(b)$\Longrightarrow$(a). For every $x \in a \oplus b$, it follows $f(x) \in f(a) \circledast f(b)$ and so $f(x) \wedge f(a) = f(x) \wedge f(b) = f(a) \wedge f(b) \leq f(x)$. Set $x = a \vee b$ we obtain $f(a) \wedge f(b) \leq f(a \vee b)$. Conversely, observe that $f(x) \in f(a) \circledast f(a)$ holds for each $x \in a \oplus a$ (and each $a \in L$). Hence $f(a) = f(x) \wedge f(a)$ whence $f(a) \leq f(x)$. Again, setting $x = a \vee b$, we have $f(a \vee b) \leq f(a)$. Similarly, $f(a \vee b) \leq f(b)$ and so $f(a \vee b) \leq f(a) \wedge f(b)$.

Dually, it follows the following

50. Lemma. *Let $(L, \vee, \wedge)$ be a lattice and $f : L \longrightarrow L$ be a bijective map. The following conditions are equivalent:*

(a$'$) $f(a \wedge b) = f(a) \vee f(b);\ \forall (a,b) \in L^2$;

(b$'$) $f(a \circledast b) = f(a) \oplus f(b);\ \forall (a,b) \in L^2$.

51. Remark. If $(L, \vee, \wedge)$ is a Boole lattice and $f : L \longrightarrow L$ is defined by $f(a) = a'$, $\forall a \in L$, then all the above conditions are fulfilled.

52. Remark. The condition (a) characterizes the hypergroup isomorphisms $f : < L, \oplus > \longrightarrow < L, \circledast >$.

§4. Direct limit and inverse limit of join spaces associated with lattices

In this paragraph, we prove that the direct limit (inverse limit) of a direct (respectively, inverse) family of join spaces associated with modular lattices is also a join space associated with a modular lattice.

We have utilised the notions of direct limit and inverse limit done by Grätzer in [447].

If $(H, \vee, \wedge)$ is a modular lattice, then we can associate (as in §3) a join space structure on H as follows:

$$\forall (x, y) \in H^2,\ x \circ y = \{z \in H \mid x \vee y = x \vee z = y \vee z\}$$

Let us denote by JSL the class of join spaces associated with modular lattices, as above.

In the following, we shall utilise the following result proved in the previous paragraph.

53. Theorem. *A join space $< H, \circ >$ belongs to the class* JSL *iff it satisfies the following conditions:*

1) $\forall (a, b) \in H^2,\ a/b = a \circ b;$

2) $\forall (a, b) \in H^2,\ a \circ a \circ a = a \circ a;$

3) $\forall (a, b) \in H^2,\ \exists s \in a \circ a \cap b \circ b,\ a \circ a \cap b \circ b \subseteq s \circ s;$

4) $\forall (a, b) \in H^2,\ \{a, b\} \subseteq a \circ b \Longleftrightarrow a = b;$

5) $\forall (a, b) \in H^2,\ \exists x \in H,\ \exists t \in H,\ \{a, b\} \subseteq x \circ x,$
$\bigcap_{\{a,b\} \subseteq x \circ x} x \circ x = t \circ t,\ a \circ b = a \circ t \cap b \circ t;$

6) $\forall a \in b \circ b - \{b\}$, *we have*

$$a \circ b = \{b\} \cup \{u \in H \mid u \in b \circ b - \{b\}, u \notin a \circ a, a \notin u \circ u;$$
$$\nexists y \in H,\ a \in y \circ y - \{y\}, u \in y \circ y - \{y\}, y \in b \circ b - \{b\}\}.$$

1. Direct limit of a direct family of join spaces associated with modular lattices

54. Definition. A family $\{(H_i, \otimes_i\}_{i \in I}$ of join spaces is called *a direct family* if:

1) $(I, \leq)$ is a directed partially ordered set;

2) $\forall (i,j) \in I^2$, we have $i \neq j \Longleftrightarrow H_i \cap H_j = \emptyset$;

3) $\forall (i,j) \in I^2$, $i \leq j$, there is a homomorphism $\varphi_{ij} : H_i \to H_j$ such that if $i \leq j \leq k$, then $\varphi_{jk} \circ \varphi_{ij} = \varphi_{ik}$ and $\forall i \in I$, φ_{ii} is the identity mapping.

Let us define on $H = \bigcup_{i \in I} H_i$, the following equivalence relation: $x \sim y$ iff the following implication is satisfied: $(x,y) \in H_i \times H_j \Longrightarrow \exists k \in I,\ k \geq i,\ k \geq j$, such that $\varphi_{ik}(x) = \varphi_{jk}(y)$.

If $x_i \in H_i$ and $i \leq j$, we denote $\varphi_{ij}(x_i)$ by x_j and we consider $\overline{H} = \{\bar{x} \mid x \in H\}$ the set of equivalence classes.

$\overline{H}$ is a hypergroup with respect to the following hyperoperation:

$$\bar{x} \circ \bar{y} = \{\bar{z} \mid \exists i \in I,\ \exists x_i \in \bar{x} \cap H_i,\ \exists y_i \in \bar{y} \cap H_i,\\ \exists z_i \in \bar{z} \cap H_i : z_i \in x_i \otimes_i y_i\}$$

and it is called the *direct limit* of the direct family $\{(H_i, \otimes_i)\}_{i \in I}$.

55. Proposition. *If $\{(H_i, \otimes_i)\}_{i \in I}$ is a direct family of semihypergroups, such that $\forall i \in I$, $\exists k \in I$, $i \leq k$, for which $(H_k, \otimes_k)$ is a join space, then $(\overline{H}, *)$ is a join space.*

For each $i \in I$, we shall associate the join space $(H_i, \circ_i)$ with the modular lattice $(H_i, \vee_i, \wedge_i)$.

So, $\forall (x_i, y_i) \in H_i^2$, we have:

$$x_i \circ_i y_i = \{z_i \in H_i \mid x_i \vee_i y_i = x_i \vee_i z_i = y_i \vee_i z_i\}.$$

56. Theorem. *The direct limit of a direct family of semihypergroups, $\{(H_i, \circ_i)\}_{i \in I}$, such that $\forall i \in I$, $\exists k \in I$, $k \geq i : (H_k, \circ_k) \in$ JSL, is a join space $(\overline{H}, \circ)$ which belongs to* JSL.

Note. To simplify the notations, we shall denote $H_i \in$ JSL instead of $(H_i, \circ_i) \in$ JSL.

Proof. We shall verify the conditions of the Theorem 53.

1) $\forall(\bar{a},\bar{b}) \in \overline{H}^2$, we have

$$\bar{a}/\bar{b} = \{\bar{c} \in \overline{H} \mid \bar{a} \in \bar{b} \circ \bar{c}\} = \{\bar{c} \mid \exists i \in I : a_i \in b_i \circ_i c_i\} =$$
$$= \{\bar{c} \mid \exists k \in I,\ k \geq i : H_k \in \text{JSL},\ c_k \in a_k/b_k = a_k \circ_k b_k\} =$$
$$= \{\bar{c} \mid \bar{c} \in \bar{a} \circ \bar{b}\} = \bar{a} \circ \bar{b},$$

2) $\forall\bar{a} \in \overline{H}$, we have

$$\bar{a} \circ \bar{a} \circ \bar{a} = \bigcup_{\bar{t} \in \bar{a} \circ \bar{a}} \bar{t} \circ \bar{a} = \bigcup_{\bar{t} \in \bar{a} \circ \bar{a}} \{\bar{c} \in \overline{H} \mid \exists i \in I : c_i \in t_i \circ_i a_i\} =$$
$$= \{\bar{c} \in \overline{H} \mid \exists i \in I : c_i \in t_i \circ_i a_i,\ \exists j \in I : t_j \in a_j \circ_j a_j\} =$$
$$= \{\bar{c} \in \overline{H} \mid \exists k \in I,\ k \geq i,\ k \geq j : H_k \in \text{JSL},$$
$$c_k \in t_k \circ_k a_k,\ t_k \in a_k \circ_k a_k\} =$$
$$= \{\bar{c} \in \overline{H} \mid \exists k \in I; c_k \in a_k \circ_k a_k \circ_k a_k = a_k \circ_k a_k\} = \bar{a} \circ \bar{a};$$

3) $\forall(\bar{a},\bar{b}) \in \overline{H}^2$, we shall prove that $\exists\bar{s} \in \overline{H}$, such that $\bar{s} \in \bar{a} \circ \bar{a} \cap \bar{b} \circ \bar{b}$ and $\bar{a} \circ \bar{a} \cap \bar{b} \circ \bar{b} \subset \bar{s} \circ \bar{s}$.

Indeed, if $\bar{t} \in \bar{a} \circ \bar{a} \cap \bar{b} \circ \bar{b}$, then $\exists i \in I : t_i \in a_i \circ_i a_i$ and $\exists j \in I : t_j \in b_j \circ_j b_j$, hence $\exists k \in I$, $k \geq i$, $k \geq j : H_k \in$ JSL and $t_k \in a_k \circ_k a_k \cap b_k \circ_k b_k \subset s_k \circ_k s_k$, where $s_k \in a_k \circ_k a_k \cap b_k \circ_k b_k$. Therefore $\bar{t} \in \bar{s} \circ \bar{s}$, whence $\bar{a} \circ \bar{a} \cap \bar{b} \circ \bar{b} \subset \bar{s} \circ \bar{s}$ and $\bar{s} \in \bar{a} \circ \bar{a} \cap \bar{b} \circ \bar{b}$.

4) Notice that $\forall\bar{a} \in \overline{H}$, $\bar{a} \in \bar{a} \circ \bar{a}$, since $\exists k \in I$, such that $H_k \in$ JSL, so $\forall a_k \in H_k$, $a_k \in a_k \circ_k a_k$.

On the other hand, if $(\bar{a},\bar{b}) \in \overline{H}^2$ such that $\{\bar{a},\bar{b}\} \subset \bar{a} \circ \bar{b}$, then $\exists i \in I : a_i \in a_i \circ_i b_i$ and $\exists j \in I : b_j \in a_j \circ_j b_j$, whence $\exists k \in I$, $k \geq i$, $k \geq j : H_k \in$ JSL and $\{a_k, b_k\} \subset a_k \circ_k b_k$, hence $a_k = b_k$. So, $\bar{a} = \bar{b}$.

5) We shall prove that $\forall(\bar{a},\bar{b}) \in \overline{H}^2$, $\exists(\bar{x},\bar{t}) \in \overline{H}^2 : \{\bar{a},\bar{b}\} \subset \bar{x} \circ \bar{x}$, $\bigcap_{\{\bar{a},\bar{b}\} \subset \bar{x} \circ \bar{x}} \bar{x} \circ \bar{x} = \bar{t} \circ \bar{t}$ and $\bar{a} \circ \bar{b} = \bar{a} \circ \bar{t} \cap \bar{b} \circ \bar{t}$. Indeed, since $\exists k \in I$, $H_k \in$ JSL it follows $\exists x_k \in H_k : \{a_k, b_k\} \subset x_k \circ_k x_k$, whence $\{\bar{a},\bar{b}\} \subset \bar{x} \circ \bar{x}$. Moreover, $\exists t_k \in H_k : \bigcap_{\{a_k,b_k\} \subset x_k \circ_k x_k} x_k \circ_k x_k = t_k \circ_k t_k$, that means

$\{a_k, b_k\} \subset t_k \circ_k t_k \subset x_k \circ_k x_k$, for any $x_k : \{a_k, b_k\} \subset x_k \circ_k x_k$, whence $\{\bar{a}, \bar{b}\} \subset \bar{t} \circ \bar{t} \subset \bar{x} \circ \bar{x}$, for any $\{\bar{a}, \bar{b}\} \subset \bar{x} \circ \bar{x}$. Hence, $\bigcap_{\{\bar{a},\bar{b}\} \subset \bar{x} \circ \bar{x}} \bar{x} \circ \bar{x} = \bar{t} \circ \bar{t}$.

On the other hand, $a_k \circ_k b_k = a_k \circ_k t_k \cap b_k \circ_k t_k$, since $H_k \in$ JSL.

Let $\bar{c} \in \bar{a} \circ \bar{b}$. It follows $\exists i \in I : c_i \in a_i \circ_i b_i$. Let $j \in I$, $j \geq i$, such that $H_j \in$ JSL. We have $c_j \in a_j \circ_j b_j = a_j \circ_j t_j \cap b_j \circ_j t_j$. So, $\bar{c} \in \bar{a} \circ \bar{t} \cap \bar{b} \circ \bar{t}$.

Conversely, if $\bar{u} \in \bar{a} \circ \bar{t} \cap \bar{b} \circ \bar{t}$, then $\exists i \in I : u_i \in a_i \circ_i t_i$ and $\exists j \in I : u_j \in b_j \circ_j t_j$.

By hypothesis, it follows $\exists k \in I$, $k \geq i$, $k \geq j$ such that $H_k \in$ JSL. One obtains $u_k \in a_k \circ_k t_k \cap b_k \circ_k t_k = a_k \circ_k b_k$, hence $\bar{u} \in \bar{a} \circ \bar{b}$.

Then, $\bar{a} \circ \bar{b} = \bar{a} \circ \bar{t} \cap \bar{b} \circ \bar{t}$.

6) Let $\bar{b} \in \overline{H}$ and $\bar{a} \in \bar{b} \circ \bar{b} - \{\bar{b}\}$. We denote by $\overline{A}$ the set

$$\{\bar{u} \in \overline{H} \mid \bar{u} \in \bar{b} \circ \bar{b} - \{\bar{b}\}, \bar{u} \notin \bar{a} \circ \bar{a}, \bar{a} \notin \bar{u} \circ \bar{u},$$
$$\nexists \bar{y} \in \overline{H} : \bar{a} \in \bar{y} \circ \bar{y} - \{\bar{y}\}, \bar{u} \in \bar{y} \circ \bar{y} - \{\bar{y}\}, \bar{y} \in \bar{b} \circ \bar{b} - \{\bar{b}\}\}.$$

We shall prove that $\bar{a} \circ \bar{b} = \{\bar{b}\} \cup \overline{A}$.

For any $i \in I$, we denote by A_i the set:

$$\{u_i \in H_i \mid u_i \in b_i \circ_i b_i - \{b_i\},\ u_i \notin a_i \circ_i a_i,\ a_i \notin u_i \circ_i u_i,$$
$$\nexists y_i \in H_i : a_i \in y_i \circ_i y_i - \{y_i\}, u_i \in y_i \circ_i y_i - \{y_i\}, y_i \in b_i \circ_i b_i - \{b_i\}\}.$$

Let $\bar{u} \in \bar{a} \circ \bar{b}$, where $\bar{a} \in \bar{b} \circ \bar{b} - \{\bar{b}\}$; so, $\exists i_1 \in I : u_{i_1} \in a_{i_1} \circ_{i_1} b_{i_1}$. Since $\bar{a} \in \bar{b} \circ \bar{b} - \{\bar{b}\}$, it follows that $\exists i_2 \in I : a_{i_2} \in b_{i_2} \circ_{i_2} b_{i_2}$ and $\forall i \in I$, $a_i \neq b_i$.

By hypothesis, there is $i \in I$, $i \geq i_1$, $i \geq i_2$, such that $H_i \in$ JSL; hence $a_i \in b_i \circ_i b_i - \{b_i\}$ and $u_i \in a_i \circ_i b_i = \{b_i\} \cup A_i$.

Case 1°. If $u_i = b_i$, then $\bar{u} = \bar{b}$. In the following, we suppose that $\hat{u} \neq \hat{b}$.

Case 2°. If $u_i \in A_i$, then $u_i \in b_i \circ_i b_i - \{b_i\}$, hence $\bar{u} \in \bar{b} \circ \bar{b}$.

According to the above assumption, we have $\bar{u} \in \bar{b} \circ \bar{b} - \{\bar{b}\}$.

Suppose $\bar{u} \in \bar{a} \circ \bar{a}$. It follows $\exists j \in I : u_j \in a_j \circ_j a_j$. There is $k \in I$, $k \geq i$, $k \geq j$, such that $H_k \in$ JSL. We have $u_k \in a_k \circ_k a_k$, $u_k \in a_k \circ_k b_k$, whence $u_k = b_k$ or $u_k \in A_k$. Since $\bar{u} \neq \bar{b}$, it follows $u_k \in A_k$, so $u_k \notin a_k \circ_k a_k$, contradiction. Therefore, $\bar{u} \notin \bar{a} \circ \bar{a}$.

In a similar way, we can verify that $\bar{a} \notin \bar{u} \circ \bar{u}$.

Suppose now that $\exists \bar{y} \in \overline{H} : \bar{a} \in \bar{y} \circ \bar{y} - \{\bar{y}\}$, $\bar{u} \in \bar{y} \circ \bar{y} - \{\bar{y}\}$ and $\bar{y} \in \bar{b} \circ \bar{b} - \{\bar{b}\}$.

Since $\bar{a} \in \bar{y} \circ \bar{y}$ it follows $\exists p \in I : a_p \in y_p \circ_p y_p$ and since $\bar{a} \neq \bar{y}$, it follows $\forall i \in I$, $a_i \neq y_i$. Similarly, we have $\exists r \in I : u_r \in y_r \circ_r y_r$, $\exists \ell \in I : y_\ell \in b_\ell \circ_\ell b_\ell$, and $\forall i \in I$, $u_i \neq y_i \neq b_i$.

Let $s \in I$, $s \geq p$, $s \geq r$, $s \geq \ell$, $s \geq i$ and such that $H_s \in$ JSL. We obtain

$$\exists y_s \in H_s : a_s \in y_s \circ_s y_s - \{y_s\},$$
$$u_s \in y_s \circ_s y_s - \{y_s\} \text{ and } y_s \in b_s \circ_s b_s - \{b_s\}.$$

On the other hand, since $u_i \in a_i \circ_i b_i$ it follows $u_s \in a_s \circ_s b_s$ and since $H_s \in$ JSL and $\bar{u} \neq \bar{b}$, it follows $u_s \in A_s$, whence one obtains that: $\nexists y_s \in H_s : a_s \in y_s \circ_s y_s - \{y_s\}$, $u_s \in y_s \circ_s y_s - \{y_s\}$ and $y_s \in b_s \circ_s b_s - \{b_s\}$, contradiction.

Therefore the last assumption is false, so

$$\nexists \bar{y} \in \overline{H} : \bar{a} \in \bar{y} \circ \bar{y} - \{\bar{y}\}, \bar{u} \in \bar{y} \circ \bar{y} - \{\bar{y}\} \text{ and } \bar{y} \in \bar{b} \circ \bar{b} - \{\bar{b}\}.$$

Then, we can conclude that $\bar{a} \circ \bar{b} \subseteq \{\bar{b}\} \cup \overline{A}$. Conversely, we have $\bar{b} \in \bar{a} \circ \bar{b}$, since $\bar{a} \in \bar{b} \circ \bar{b} = \bar{b}/\bar{b}$.

Let $\bar{u} \in \overline{A}$. Then $\exists j \in I : u_j \in b_j \circ_j b_j$ and $\forall i \in I : u_i \neq b_i$, $u_i \notin a_i \circ_i a_i$, $a_i \notin u_i \circ_i u_i$.

Moreover, since $\nexists \bar{y} \in \overline{H} : \bar{a} \in \bar{y} \circ \bar{y} - \{\bar{y}\}$, $\bar{u} \in \bar{y} \circ \bar{y} - \{\bar{y}\}$, $\bar{y} \in \bar{b} \circ \bar{b} - \{\bar{b}\}$, it follows that $\forall i \in I$, $\nexists y_i \in H_i$ such that $a_i \in y_i \circ_i y_i$, $u_i \in y_i \circ_i y_i$, $y_i \in b_i \circ_i b_i$ and $a_i \not\sim y_i$, $u_i \not\sim y_i$, $y_i \not\sim b_i$.

Since $\bar{a} \in \bar{b} \circ \bar{b} - \{\bar{b}\}$, that is $\exists k \in I : a_k \in b_k \circ_k b_k$ and $\forall i \in I$, $a_i \neq b_i$ and since $\bar{u} \in \overline{A}$, it follows $\exists j \in I : u_j \in b_j \circ_j b_j$, $\forall i \in I : u_i \neq b_i$, $u_i \notin a_i \circ_i a_i$, $a_i \notin u_i \circ_i u_i$ and $\forall i \in I$, $\nexists y_i \in H_i$, $a_i \in y_i \circ_i y_i$, $u_i \in y_i \circ_i y_i$, $y_i \in b_i \circ_i b_i$ and $a_\imath \not\sim y_i$, $u_i \not\sim y_i$, $y_i \not\sim b_i$.

Let $s \in I$, $s \geq k$, $s \geq j$ and such that $H_s \in$ JSL. We have $a_s \in b_s \circ_s b_s - \{b_s\}$ and $u_s \in \{v_s \in H_s \mid v_s \in b_s \circ_s b_s - \{b_s\},$

$v_s \notin a_s \circ_s a_s$, $a_s \notin v_s \circ_s v_s$, $\not\exists y_s \in H_s : a_s \in y_s \circ_s y_s - \{y_s\}$, $v_s \in y_s \circ_s y_s - \{y_s\}$, $y_s \in b_s \circ_s b_s - \{b_s\}\} = A_s$, whence it follows $u_s \in a_s \circ_s b_s$, hence $\bar{u} \in \bar{a} \circ \bar{b}$. Therefore, $\bar{a} \circ \bar{b} = \{\bar{b}\} \cup \overline{A}$ and we can conclude that $\overline{H} \in$ JSL. ∎

57. Remark.

1° In the lattice $(\overline{H}, \vee, \wedge)$ we have $\bar{a} \leq \bar{b} \Longleftrightarrow \bar{a} \in \bar{b} \circ \bar{b} \Longleftrightarrow \exists i \in I :$ $a_i \in b_i \circ_i b_i$. If $\exists (i,j) \in I^2$, such that $H_i \in$ JSL, $H_j \in$ JSL and $a_i \in b_i \circ_i b_i$ (that means $a_i \leq b_i$) and $b_j \in a_j \circ_j a_j$ (that means $b_j \leq a_j$), then $\exists k \in I$, $k \geq i$, $k \geq j$, such that $H_k \in$ JSL and $a_k \leq b_k \leq a_k$, whence $a_k = b_k$, hence $\bar{a} = \bar{b}$.

2° For any $(\bar{a}, \bar{b}) \in \overline{H}^2$, $\sup(\bar{a}, \bar{b}) = \bar{t}$ (where $\bar{t}$ satisfies the condition 5)) and $\inf(\bar{a}, \bar{b}) = \bar{s}$ (where $\bar{s}$ satisfies the condition 3)). (This follows by the proof of Theorem 53.)

2. Inverse limit of an inverse family of join spaces associated with modular lattice

First, let us recall the notion of inverse limit of an inverse family of join spaces.

58. Definition. A family of join spaces $\{(H_i, \otimes_i)\}_{i \in I}$ is called an *inverse family* if:

1) $(I, \leq)$ is a directed partially ordered set;

2) $\forall (i,j) \in I^2$, we have $H_i \cap H_j = \emptyset \Longleftrightarrow i \neq j$;

3) $\forall (i,j) \in I^2, i \leq j$, there is a homomorphism of join spaces $\psi_{ij} : H_i \to H_j$, such that: if $i \geq j \geq k$, $\psi_{jk} \circ \psi_{ij} = \psi_{ik}$ and $\forall i \in I$, ψ_{ii} is the identity mapping.

Let $\left(H = \prod_{i \in I} H_i, \otimes\right)$ be the direct product of the family $\{(H_i, \otimes_i)\}_{i \in I}$ and $\widetilde{H} = \{x \in H \mid \psi_{ij}(x_i) = x_j, \forall i \geq j\}$, where $x = (x_i)_{i \in I}$.

If $\widetilde{H} \neq \emptyset$, then we define on $\widetilde{H}$ the hyperoperation: $\widetilde{x}\Box\widetilde{y} = \widetilde{x} \otimes \widetilde{y} \cap \widetilde{H}$.

If I has a maximum s, then $\widetilde{H} \neq \emptyset$ and for each $(\widetilde{x}, \widetilde{y}) \in \widetilde{H}^2$, $\widetilde{x}\Box\widetilde{y} \neq \emptyset$. Indeed, if $z \in \widetilde{x} \otimes \widetilde{y}$, then $z_s \in x_s \circ_s y_s$, whence $\forall i \in I$, $\psi_{si}(z_s) \in x_i \circ_i y_i$, hence $\widetilde{z} = (\psi_{si}(z_s))_{i\in I} \in \widetilde{H}$ and $\widetilde{z} \in \widetilde{x}\Box\widetilde{y}$.

If $\widetilde{H} \neq \emptyset$, then $(\widetilde{H}, \Box)$ is called the *inverse limit* of the inverse family $\{(H_i, \otimes_i)\}_{i\in I}$.

59. Theorem. *Let $\{(H_i, \circ_i)\}_{i\in I}$ be an inverse family of join spaces, such that $\forall i \in I$, $H_i \in$ JSL. Moreover, let us supose that I has a maximum. Then $(\widetilde{H}, \Box)$ is a join space and moreover $\widetilde{H} \in$ JSL.*

Proof. We shall verify the conditions of Theorem 53.

1) $\forall(\widetilde{x}, \widetilde{y}) \in \widetilde{H}^2$, $\widetilde{x}/\widetilde{y} = \{\widetilde{z} \in \widetilde{H} \mid \widetilde{x} \in \widetilde{y}\Box\widetilde{z}\} = \{\widetilde{z} \in \widetilde{H} \mid \forall i \in I : x_i \in y_i \circ_i z_i\} = \{\widetilde{z} \in \widetilde{H} \mid \forall i \in : z_i \in x_i/y_i = x_i \circ_i y_i\} = \widetilde{x} \otimes \widetilde{y} \cap \widetilde{H} = \widetilde{x}\Box\widetilde{y}$, whence $\widetilde{x} = (x_i)_{i\in I}$, $\widetilde{y} = (y_i)_{i\in I}$.

2) $\forall \widetilde{x} \in \widetilde{H}$, $\widetilde{x} = (x_i)_{i\in I}$, we have

$$\begin{aligned}
\widetilde{x}\Box\widetilde{x}\Box\widetilde{x} &= \bigcup_{\widetilde{t}\in\widetilde{x}\Box\widetilde{x}} \widetilde{t}\Box\widetilde{x} = \bigcup_{\widetilde{t}\in\widetilde{x}\Box\widetilde{x}} \{\widetilde{u} \in \widetilde{H} \mid \forall i \in I : u_i \in t_i \circ_i x_i\} = \\
&= \{\widetilde{u} \in \widetilde{H} \mid \forall i \in I : u_i \in t_i \circ_i x_i,\ t_i \in x_i \circ_i x_i\} \subseteq \\
&\subseteq \{\widetilde{u} \in \widetilde{H} \mid \forall i \in I : u_i \in x_i \circ_i x_i \circ_i x_i = x_i \circ_i x_i\} = \\
&= \widetilde{x} \otimes \widetilde{x} \cap \widetilde{H} = \widetilde{x}\Box\widetilde{x}.
\end{aligned}$$

Conversely, $\forall \widetilde{x} \in \widetilde{H}$, $\widetilde{x} \in \widetilde{x}\Box\widetilde{x}$, because $\forall i \in I$, $x_i \in x_i \circ_i x_i$. So, $\widetilde{x}\Box\widetilde{x} \subset \widetilde{x}\Box\widetilde{x}\Box\widetilde{x}$, hence we obtain the equality.

3) For any $\widetilde{a} = (a_i)_{i\in I}$, $\widetilde{b} = (b_i)_{i\in I}$ of $\widetilde{H}$, we shall prove that there is $\widetilde{z} \in \widetilde{a}\Box\widetilde{a} \cap \widetilde{b}\Box\widetilde{b}$, such that $\widetilde{a}\Box\widetilde{a} \cap \widetilde{b}\Box\widetilde{b} \subseteq \widetilde{z}\Box\widetilde{z}$.

Let $s = \max I$. Since $H_s \in$ JSL, there is

$$z_s \in H_s : z_s \in a_s \circ_s a_s \cap b_s \circ_s b_s \subset z_s \circ_s z_s.$$

Hence $\forall i \in I$ we have

$$\psi_{si}(z_s) \in \psi_{si}(a_s) \circ_i \psi_{si}(a_s) \cap \psi_{si}(b_s) \circ_i \psi_{si}(b_s) = a_i \circ_i a_i \cap b_i \circ_i b_i.$$

Therefore, $\exists \widetilde{z} = (\psi_{si}(z_s))_{i\in I} \in \widetilde{H}$, such that $\widetilde{z} \in \widetilde{a}\Box\widetilde{a} \cap \widetilde{b}\Box\widetilde{b}$.

Let $\widetilde{t} \in \widetilde{a}\Box\widetilde{a} \cap \widetilde{b}\Box\widetilde{b}$. Then $t_s \in a_s \circ_s a_s \cap b_s \circ_s b_s \subset z_s \circ_s z_s$, whence $\forall i \in I$, $t_i \in \psi_{si}(z_s) \circ_i \psi_{si}(z_s)$, so $\widetilde{t} \in \widetilde{z}\Box\widetilde{z}$. Therefore, $\widetilde{a}\Box\widetilde{a} \cap \widetilde{b}\Box\widetilde{b} \subseteq \widetilde{z}\Box\widetilde{z}$.

4) For any $\widetilde{x} = (x_i)_{i\in I}$, $\widetilde{y} = (y_i)_{i\in I}$ we have $\{\widetilde{x}, \widetilde{y}\} \subset \widetilde{x}\Box\widetilde{y}$ iff $\widetilde{x} = \widetilde{y}$.

Indeed, if $\{\widetilde{x}, \widetilde{y}\} \subset \widetilde{x}\Box\widetilde{y}$, we have $\forall i \in I$, $\{x_i, y_i\} \subset x_i \circ_i y_i$, whence $x_i = y_i$, since $H_i \in$ JSL. Hence $\widetilde{x} = \widetilde{y}$. Conversely, we have $\forall \widetilde{x} \in \widetilde{H}$, $\widetilde{x} \in \widetilde{x}\Box\widetilde{x}$, because $\forall i \in I$, $x_i \in x_i \circ_i x_i$, where $\widetilde{x} = (x_i)_{i\in I}$.

5) We shall prove now that $\forall \widetilde{a} = (a_i)_{i\in I} \in \widetilde{H}$, $\forall \widetilde{b} = (b_i)_{i\in I} \in \widetilde{H}$, $\exists \widetilde{x} \in \widetilde{H}$, $\exists \widetilde{t} \in \widetilde{H} : \{\widetilde{a}, \widetilde{b}\} \subset \widetilde{x}\Box\widetilde{x}$, $\bigcap_{\{\widetilde{a},\widetilde{b}\}\subset\widetilde{x}\Box\widetilde{x}} \widetilde{x}\Box\widetilde{x} = \widetilde{t}\Box\widetilde{t}$ and $\widetilde{a}\Box\widetilde{b} = \widetilde{a}\Box\widetilde{t} \cap \widetilde{b}\Box\widetilde{t}$.

Indeed, if $s = \max I$, then $\exists x_s \in H_s : \{a_s, b_s\} \subset x_s \circ_s x_s$, $\exists t_s \in I : \bigcap_{\{a_s,b_s\}\subset x_s\circ_s x_s} x_s \circ_s x_s = t_s \circ_s t_s$ and $a_s \circ_s b_s = a_s \circ_s t_s \cap b_s \circ_s t_s$, since $H_s \in$ JSL. It follows $\forall i \in I$,

$$\{\psi_{si}(a_s) = a_i,\ \psi_{si}(b_s) = b_i\} \subset \psi_{si}(x_s) \circ_i \psi_{si}(x_s),$$

whence $\exists \widetilde{x} = (\psi_{si}(x_s))_{i\in I} \in \widetilde{H} : \{\widetilde{a}, \widetilde{b}\} \subset \widetilde{x}\Box\widetilde{x}$.

On the other hand, let $\widetilde{t} = (\psi_{si}(t_s))_{i\in I} \in \widetilde{H}$ and $\widetilde{v} \in \widetilde{t}\Box\widetilde{t}$. It follows

$$v_s \in t_s \circ_s t_s = \bigcap_{\{a_s,b_s\}\subset x_s\circ_s x_s} x_s \circ_s x_s,$$

whence $\forall i \in I$,

$$v_i \in \bigcap_{\{a_i,b_i\}\subset\psi_{si}(x_s)\circ_i\psi_{si}(x_s)} \psi_{si}(s_s) \circ_i \psi_{si}(x_s),$$

hence $\widetilde{v} \in \bigcap_{\{\widetilde{a},\widetilde{b}\}\subset\widetilde{x}\Box\widetilde{x}} \widetilde{x}\Box\widetilde{x}$.

Conversely, since $\{a_s, b_s\} \subset t_s \circ_s t_s$, it follows $\forall i \in I$,

$$\{a_i b_i\} \subset \psi_{si}(t_s) \circ_i \psi_{si}(t_s),$$

so $\{\widetilde{a},\widetilde{b}\} \subset \widetilde{t}\Box\widetilde{t}$. Therefore, $\bigcap\limits_{\{\widetilde{a},\widetilde{b}\}\subset\widetilde{x}\Box\widetilde{x}} \widetilde{x}\Box\widetilde{x} = \widetilde{t}\Box\widetilde{t}$.

Finally, we have to verify the equality:

$$\widetilde{a}\Box\widetilde{b} = \widetilde{a}\Box\widetilde{t} \cap \widetilde{b}\Box\widetilde{t}.$$

We have $\widetilde{u} \in \widetilde{a}\Box\widetilde{b}$ iff $\forall i \in I$, $u_i \in a_i \circ_i b_i$, that means $\forall i \in I$, $u_i \in a_i \circ_i t_i \cap b_i \circ_i t_i$ (since $\forall i \in I$, $H_i \in$ JSL), that is

$$\widetilde{u} \in \widetilde{a} \otimes \widetilde{t} \cap \widetilde{b} \otimes \widetilde{t} \cap \widetilde{H} = \widetilde{a}\Box\widetilde{t} \cap \widetilde{b}\Box\widetilde{t}.$$

Therefore, $\widetilde{a}\Box\widetilde{b} = \widetilde{a}\Box\widetilde{t} \cap \widetilde{b}\Box\widetilde{t}$.

6) Let $(\widetilde{a},\widetilde{b}) \in \widetilde{H}^2 : \widetilde{a} \in \widetilde{b}\Box\widetilde{b} - \{\widetilde{b}\}$ and let us denote

$$\widetilde{A} = \{\widetilde{u} \in \widetilde{H} \mid \widetilde{u} \in \widetilde{b}\Box\widetilde{b} - \{\widetilde{b}\}, \widetilde{u} \notin \widetilde{a}\Box\widetilde{a}, \widetilde{a} \notin \widetilde{u}\Box\widetilde{u},$$
$$\nexists\widetilde{y} \in \widetilde{H} : \widetilde{a} \in \widetilde{y}\Box\widetilde{y} - \{\widetilde{y}\}, \widetilde{u} \in \widetilde{y}\Box\widetilde{y} - \{\widetilde{y}\}, \widetilde{y} \in \widetilde{b}\Box\widetilde{b} - \{\widetilde{b}\}\}.$$

We shall verify that $\widetilde{a}\Box\widetilde{b} = \{\widetilde{b}\} \cup \widetilde{A}$. Since $\widetilde{a} \in \widetilde{b}\Box\widetilde{b} - \{\widetilde{b}\}$ it follows $\forall i \in I, a_i \in b_i \circ_i b_i$ and $\exists i_0 \in I : a_{i_0} \neq b_{i_0}$. It follows $a_s \neq b_s$, since otherwise from $a_s = b_s$ one obtains $\forall i \in I$, $a_i = \psi_{si}(a_s) = \psi_{si}(b_s) = b_i$, which is false. So, $a_s \in b_s \circ_s b_s - \{b_s\}$ and since $H_s \in$ JSL it follows $a_s \circ_s b_s = \{b_s\} \cup A_s$.

Let $\widetilde{u} \in \widetilde{a}\Box\widetilde{b}$, that is $\forall i \in I$, $u_i \in a_i \circ_i b_i$. Then $u_s \in \{b_s\} \cup A_s$.

Case 1°. If $u_s = b_s$, then $\forall i \in I$, $u_i = \psi_{si}(u_s) = \psi_{si}(b_s) = b_i$, whence $\widetilde{u} = \widetilde{b}$.

Case 2°. If $u_s \in A_s$, then we have $u_s \in b_s \circ_s b_s - \{b_s\}$, $u_s \notin a_s \circ_s a_s$, $a_s \notin u_s \circ_s u_s$, $\nexists y_s \in H_s : a_s \in y_s \circ_s y_s - \{y_s\}$, $u_s \in y_s \circ_s y_s - \{y_s\}$ and $y_s \in b_s \circ_s b_s - \{b_s\}$.

It follows that $\forall i \in I$, $u_i = \psi_{si}(u_s) \in b_i \circ_i b_i$ and since $\widetilde{u} \in \widetilde{H}$, we obtain $\widetilde{u} \in \widetilde{b}\Box\widetilde{b} - \{\widetilde{b}\}$, because $u_s \neq b_s$.

Now, suppose that $\widetilde{u} \in \widetilde{a}\Box\widetilde{a}$, that is $\forall i \in I$, $u_i \in a_i \circ_i a_i$, contradiction with $u_s \notin a_s \circ_s a_s$. So, $\widetilde{u} \notin \widetilde{a}\Box\widetilde{a}$ and similarly we have $\widetilde{a} \notin \widetilde{u}\Box\widetilde{u}$. We suppose now that $\exists\widetilde{y} \in \widetilde{H} : \widetilde{a} \in \widetilde{y}\Box\widetilde{y} - \{\widetilde{y}\}$, $\widetilde{u} \in \widetilde{y}\Box\widetilde{y} - \{\widetilde{y}\}$ and $\widetilde{y} \in \widetilde{b}\Box\widetilde{b} - \{\widetilde{b}\}$. So, $\forall i \in I$, $\{a_i, u_i\} \subset y_i \circ_i y_i$,

$y_i \in b_i \circ_i b_i$ and $\exists(i_1, i_2, i_3) \in I^3$, such that $a_{i_1} \neq y_{i_1}$,$u_{i_2} = y_{i_2}$ and $y_{i_3} = b_{i_3}$.

From $a_{i_1} \neq y_{i_1}$ we obtain $a_s \neq y_s$ and similarly we have $u_s \neq \neq y_s \neq b_s$. Hence, $\exists y_s \in H_s : a_s \in y_s \circ_s y_s - \{y_s\}$, $u_s \in y_s \circ_s y_s - \{y_s\}$ and $y_s \in b_s \circ_s b_s - \{b_s\}$, which is false. Therefore, we can conclude that $\widetilde{u} \in \widetilde{A}$. Then $\widetilde{a}\Box\widetilde{b} \subseteq \{\widetilde{b}\} \cup \widetilde{A}$.

Conversely, we have $\widetilde{b} \in \widetilde{a}\Box\widetilde{b}$ since $\widetilde{a} \in \widetilde{b}\Box\widetilde{b} = \widetilde{b}/\widetilde{b}$.

Let $\widetilde{v} \in \widetilde{A}$. It follows that $v_s \in b_s \circ_s b_s = \{b_s\}$, otherwise if $v_s = b_s$, then $\forall i \in I$, $v_i = \psi_{si}(v_s) = \psi_{si}(b_s) = b_i$, whence $\widetilde{v} = \widetilde{b}$, which is false.

Moreover, $v_s \notin a_s \circ_s a_s$, otherwise we obtain $\forall i \in I$,

$$v_i = \psi_{si}(v_s) \in \psi_{si}(a_s) \circ_i \psi_{si}(a_s) = a_i \circ_i a_i,$$

whence $\widetilde{v} \in \widetilde{a}\Box\widetilde{a}$, which is false. Similarly, we have $a_s \notin v_s \circ_s v_s$.

Furthermore, $\nexists y_s \in H_s$, such that $a_s \in y_s \circ_s y_s - \{y_s\}$, $u_s \in y_s \circ_s y_s - \{y_s\}$, $y_s \in b_s \circ_s b_s - \{b_s\}$.

Indeed, if we suppose the contrary, it follows $\forall i \in I$, $a_i = = \psi_{si}(a_s) \in \psi_{si}(y_s) \circ_i \psi_{si}(y_s) = y_i \circ_i y_i$ whence $\widetilde{a} \in \widetilde{y}\Box\widetilde{y}$ and on the other hand we have $\widetilde{a} \neq \widetilde{y}$. Similarly, $u_s \in y_s \circ_s y_s - \{y_s\}$, $y_s \in b_s \circ_s b_s - \{b_s\}$ imply $\widetilde{u} \in \widetilde{y}\Box\widetilde{y} - \{\widetilde{y}\}$, $\widetilde{y} \in \widetilde{b}\Box\widetilde{b} - \{\widetilde{b}\}$, so we obtain a contradiction.

Therefore, $\widetilde{v} \in \widetilde{A}$ implies $v_s \in A_s$.

On the other hand, from $\widetilde{a} \in \widetilde{b}\Box\widetilde{b} - \{\widetilde{b}\}$ it follows $a_s \in b_s \circ_s b_s - \{b_s\}$. Since $H_s \in$ JSL, we obtain: $a_s \circ_s b_s = \{b_s\} \cup A_s$, whence $v_s \in a_s \circ_s b_s$. It results that $\forall i \in I$, $v_i = \psi_{si}(v_s) \in \psi_{si}(a_s) \circ_i \psi_{si}(b_s) = = a_i \circ_i b_i$, hence $\widetilde{v} = \widetilde{a}\Box\widetilde{b}$.

Therefore $\widetilde{a}\Box\widetilde{b}=\{\widetilde{b}\} \cup \widetilde{A}$ and we can conclude that $\widetilde{H} \in$ JSL. ∎

60. Remark.

1°) In the lattice $(\widetilde{H}, \vee, \wedge)$ we have: $\widetilde{a} \leq \widetilde{b} \iff \widetilde{a} \in \widetilde{b}\Box\widetilde{b} \iff \forall i \in I$, $a_i \in b_i \circ_i b_i \iff \forall i \in I$, $a_i \leq b_i$, where $\widetilde{a} = (a_i)_{i \in I}$ and $\widetilde{b} = (b_i)_{i \in I}$.

2°) For any $(\widetilde{a}, \widetilde{b}) \in \widetilde{H}^2$, $\sup(\widetilde{a}, \widetilde{b}) = \widetilde{t}$, which satisfies the condition 5) and $\inf(\widetilde{a}, \widetilde{b}) = \widetilde{z}$, which satisfies the condition 3). (This follows by the proof of Theorem 53.)

§5. Hyperlattices and join spaces

The *hyperlattices* have been introduced by M. Konstantinidou and J. Mittas. In the following, a connection between hyperlattices and join spaces is established.

61. Definition. Let H be a set, V a hyperoperation on H and $\wedge$ an operation. We say that $(H, \vee, \wedge)$ is a *hyperlattice* if the following conditions are satisfied, for all $(a, b, c) \in H^3$:

1. $a \in a \vee a$ and $a \wedge a = a$;
2. $a \vee b = b \vee a$ and $a \wedge b = b \wedge a$;
3. $(a \vee b) \vee c = a \vee (b \vee c)$ and $(a \wedge b) \wedge c = a \wedge (b \wedge c)$;
4. $a \in [a \vee (a \wedge b)] \wedge [a \wedge (a \vee b)]$;
5. $a \in a \vee b \Longrightarrow b = a \wedge b$.

I) Let X and Z be sets and $s : X \to \mathcal{P}^*(Z)$ a function.
A.R. Ashrafi [10] defined on X the following hyperoperation:

$$\forall (a, b) \in X^2,\ a \overset{s}{*} b = \{x \in X \mid s(x) \subseteq s(a) \cup s(b)\}.$$

We present here some of Ashrafi's results about this subject:

62. Proposition. *If $s(X)$ is a $\vee$-subsemilattice of $P^\star(Z)$ then $(X, \overset{s}{*})$ is a commutative hypergroup.*

Proof. Let $y \in (a \overset{s}{*} b) \overset{s}{*} c$. Then there exists $x \in X$ such that $s(x) \subseteq s(a) \cup s(b)$ and $s(y) \subseteq s(x) \cup s(c)$. Therefore, $s(y) \subseteq (s(a) \cup \cup s(b)) \cup s(c) = s(a) \cup (s(b) \cup s(c))$. Since $s(X)$ is a $\vee$-subsemilattice of $P^\star(Z)$, there exists $t \in X$ such that $s(b) \cup s(c) = s(t)$ and so $s(y) \subseteq s(a) \cup s(t)$. Thus, $y \in a \overset{s}{*} (b \overset{s}{*} c)$, that is $(a \overset{s}{*} b) \overset{s}{*} c \subseteq a \overset{s}{*} (b \overset{s}{*} c)$. Similarly, we have $a \overset{s}{*} (b \overset{s}{*} c) \subseteq (a \overset{s}{*} b) \overset{s}{*} c$. Therefore, the associative law holds. ∎

63. Corollary. *If $s(X)$ is a $\vee$-subsemilattice, then we have*

$$a_1 \overset{s}{*} a_2 \overset{s}{*} \cdots \overset{s}{*} a_n = \{x \in X \mid s(x) \subseteq s(a_1) \cup \cdots \cup s(a_n)\}.$$

Proof. Let $U = a_1 \overset{s}{*} a_2 \overset{s}{*} \cdots \overset{s}{*} a_n$ and $V = \{x \in X \mid s(x) \subseteq \subseteq s(a_1) \cup \cdots \cup s(a_n)\}$. We must prove $U = V$. We have $U \subseteq V$. Now, let $y \in V$. Then $s(y) \subseteq s(a_1) \cup \cdots \cup s(a_n)$. Since $s(X)$ is a $\vee$-subsemilattice of $P^\star(Z)$, hence there exists an element $x \in X$ such that $s(x) = s(a_1) \cup \cdots \cup s(a_{n-1})$. By induction, we have $x \in a_1 \overset{s}{*} a_2 \overset{s}{*} \cdots \overset{s}{*} a_{n-1}$ and $y \in x \overset{s}{*} a_n$. Therefore, $y \in U$ and so $V \subseteq U$, hence $U = V$. ∎

64. Proposition. *If* $s(X)$ *is a partition of* Z *then* $(X, \overset{s}{*})$ *is a commutative hypergroup.*

Proof. It is enough to verify the associative law. Let $(a, b, c) \in X^3$,

$$\begin{aligned}(a \overset{s}{*} b) \overset{s}{*} c &= \{x \in X \mid s(x) \subseteq s(a) \cup s(b)\} \overset{s}{*} c \\ &= \bigcup_{s(x)\subseteq s(a)\cup s(b)} x \overset{s}{*} c\end{aligned}$$

Denote $T = \{x \in X \mid s(x) \subseteq s(a) \cup s(b) \cup s(c)\}$. Now we check that $T = (a \overset{s}{*} b) \overset{s}{*} c$. It is easy to see that $(a \overset{s}{*} b) \overset{s}{*} c \subseteq T$. Let $y \in T$. Then $s(y) \subseteq s(a) \cup s(b) \cup s(c)$ and so $s(y) = (s(y) \cap s(a)) \cup (s(y) \cap \cap s(b)) \cup (s(y) \cap s(c))$. Since $\{s(x) \mid x \in X\}$ is a partition of Z, we shall consider the following cases:

Case 1) $s(y) = s(a)$ *or* $[s(y) \neq s(a)$ *and* $s(y) = s(b)]$. In this case we choose $x = y$ and we have,

$$y \in x \overset{s}{*} c \text{ and } s(x) = s(y) \subseteq s(a) \cup s(b).$$

Therefore, $y \in (a \overset{s}{*} b) \overset{s}{*} c$.

Case 2) $s(y) \neq s(a)$, $s(y) \neq s(b)$ *and* $s(y) = s(c)$. In this case we choose $x = a$ and we have,

$$y \in x \overset{s}{*} c \text{ and } s(x) = s(a) \subseteq s(a) \cup s(b).$$

Thus, $y \in (a \overset{s}{*} b) \overset{s}{*} c$.

Case 3) $s(y) \neq s(a)$, $s(y) \neq s(b)$ *and* $s(y) \neq s(c)$. It follows $s(y)=\emptyset$, which is absurd. Similarly, $T = a \overset{s}{*} (b \overset{s}{*} c)$ and so $(a \overset{s}{*} b) \overset{s}{*} c = = a \overset{s}{*} (b \overset{s}{*} c)$. ∎

65. Corollary. *If $s(X)$ is a partition of Z, then we have*

$$a_1 \overset{s}{*} a_2 \overset{s}{*} \cdots \overset{s}{*} a_n = \{x \in X \mid s(x) \subseteq s(a_1) \cup \cdots \cup s(a_n)\}.$$

Proof. Let $a_1 \overset{s}{*} \cdots \overset{s}{*} a_{n-1} = \{x \mid s(x) \subseteq s(a_1) \cup \cdots s(a_{n-1})\}$. Then we have,

$$\begin{aligned} a_1 \overset{s}{*} \cdots \overset{s}{*} a_n &= \{x \mid s(x) \subseteq s(a_1) \cup \cdots s(a_{n-1})\} \overset{s}{*} a_n \\ &= \bigcup_{s(x) \subseteq s(a_1) \cup \cdots \cup s(a_{n-1})} x \overset{s}{*} a_n \\ &= \bigcup_{s(x) \subseteq s(a_1) \cup \cdots \cup s(a_{n-1})} \{x \in X \mid s(x) \subseteq s(g) \cup s(a_n)\} \end{aligned}$$

Denote

$$T = a_1 \overset{s}{*} \cdots \overset{s}{*} a_n \text{ and } S = \{x \in X \mid s(x) \subseteq s(a_1) \cup \cdots \cup s(a_n)\}.$$

It is obvious that $T \subseteq S$, so it is enough to verify that $S \subseteq T$. Suppose $\bar{x} \in S$, then $s(\bar{x}) \subseteq s(a_1) \cup \cdots \cup s(a_n)$, and we have

$$\begin{aligned} s(\bar{x}) &= s(\bar{x}) \cap (s(a_1) \cup \cdots \cup s(a_n)) \\ &= [s(\bar{x}) \cap (s(a_1) \cup \cdots \cup s(a_{n-1})] \cup [s(\bar{x}) \cap s(a_n)] \end{aligned}$$

If $s(\bar{x}) = s(a_n)$ then we choose $x = a_1$ and we have $s(x) \subseteq s(a_1) \cup \cdots \cup s(a_{n-1})$, so $s(\bar{x}) \subseteq s(x) \cup s(\bar{x}) \subseteq s(a_1) \cup \cdots \cup s(a_n)$. Now, we assume that $s(\bar{x}) \cap s(a_n) = \emptyset$, therefore $s(\bar{x}) = s(\bar{x}) \cap (s(a_1) \cup \cdots \cup s(a_{n-1}))$. Choose $x = \bar{x}$. We have $s(\bar{x}) \subseteq s(x) \cup s(a_n)$, so $s(x) = s(\bar{x}) \subseteq s(a_1) \cup \cdots \cup s(a_{n-1})$. Hence $T = S$. ∎

66. Proposition. *Let $s(X)$ be a $\vee$-subsemilattice of $P^\star(Z)$. If $(X, \overset{s}{*})$ is a join space, then $s(a) \cap s(b) \neq \emptyset$, for all $(a, b) \in X^2$.*

Proof. Suppose there is $(a,b) \in X^2$, such that $s(a) \cap s(b) = \emptyset$. By hypothesis, there exists $t \in X$ such that $s(t) = s(a) \cup s(b)$. We have $t \in a/b \cap b/a$, but $a \overset{s}{*} a \cap b \overset{s}{*} b = \emptyset$. Therefore, $(X, \overset{s}{*})$ is not a join space, which is a contradiction. ■

67. Lemma. *If $s(X)$ is a sublattice of $P^*(Z)$, then $(X, \overset{s}{*})$ is a join space.*

Proof. By Proposition 62, $(X, \overset{s}{*})$ is a commutative hypergroup. Now we suppose that $(a,b,c,d) \in X^4$. Set $s(a) \cup s(d) = s(u)$, $s(b) \cup s(c) = s(v)$ and $s(u) \cap s(v) = s(w)$. It follows $w \in a \overset{s}{*} d \cap b \overset{s}{*} c$. Hence $(X, \overset{s}{*})$ is a join space. ■

68. Proposition. *If $s(X)$ is a partition of Z then $(X, \overset{s}{*})$ is a join space.*

Proof. Suppose $s(X)$ is a partition of Z and $(a,b,c,d) \in X^4$, such that $a/b \cap c/d \neq \emptyset$. If $s(a) = s(b)$, then $a \in a \overset{s}{*} d \cap b \overset{s}{*} c$ and if $s(c) = s(d)$, then $c \in a \overset{s}{*} d \cap b \overset{s}{*} c$. Therefore, we can assume $s(a) \neq s(b)$ and $s(c) \neq s(d)$. Now, since $s(X)$ is a partition of Z, it follows $a/b = s^{-1}(s(a))$ and $c/d = s^{-1}(s(c))$. By assumption, we have $s^{-1}(s(a)) \cap s^{-1}(s(c)) \neq \emptyset$ and so $s(a) = s(c)$, that is, $a \in a \overset{s}{*} d \cap b \overset{s}{*} c$. Therefore, $(X, \overset{s}{*})$ is a join space. ■

In the following, we shall consider the set of all subhypergroups of the hypergroup $(X, \overset{s}{*})$ and define a hyperlattice structure on this set.

Let $(X, \overset{s}{*})$ be a hypergroup and $\mathcal{L}(X)$ the set of all sub–hypergroups of X.

Let $X_A = \{g \in X \mid s(g) \subseteq A\}$. If $A \in \mathcal{P}^*(Z)$ then we suppose $X_A \neq \emptyset$.

69. Proposition. *Let Z be a finite set and $s : X \longrightarrow P^*(Z)$ be a function such that $(X, \overset{s}{*})$ is a hypergroup. Also, we assume that*

$$a_1 \overset{s}{*} a_2 \overset{s}{*} \cdots \overset{s}{*} a_n = \{g \in X \mid s(g) \subseteq s(a_1) \cup \cdots \cup s(a_n)\},$$

and H is a subhypergroup of X. Then there exists a set T such that $H = X_T$.

Proof. Let H be a subhypergroup of X and $T = \bigcup_{b \in H} s(b)$. We claim that $H = X_T$. Indeed, suppose $x \in H$. Then $s(x) \subseteq \bigcup_{b \in H} s(b) = T$ and so $x \in X_T$, that is, $H \subseteq X_T$. Now we assume that $x \in X_T$. Then $s(x) \subseteq T = \bigcup_{b \in H} s(b)$. We choose the elements $b_1, b_2, \cdots, b_r$ of H such that $s(x) \subseteq s(b_1) \cup \cdots \cup s(b_r)$. We have

$$x \in \{g \in X \mid s(g) \subseteq s(b_1) \cup \cdots \cup s(b_r)\} = b_1 \overset{s}{*} b_2 \overset{s}{*} \cdots \overset{s}{*} b_r$$

and H is a subhypergroup of X, hence $x \in H$. Therefore, $H = X_T$. ■

We have $X_{A \cap B} = X_A \cap X_B$, for all $A, B \in P^\star(Z)$. On the other hand, we have $X_{A \cup B} = X_A \cup X_B$. Let us consider the following example.

70. Example. Let $X = P(Z)$ and s be the identity function on $P^\star(Z)$ with $s(\emptyset) = Z$, $|Z| \geq 3$ and let a, b, c be distinct elements of Z. Set $R = \{a, b\}$ and $S = \{c\}$. Then $X_R = P^\star(R), X_S = P^\star(S)$ and $X_{R \cup S} = P^\star(R \cup S)$. We have $X_{R \cup S} \neq X_R \cup X_S$. ■

By Corollary 63, Corollary 65 and Proposition 69, if $s(X)$ is a $\vee$-subsemilattice or if it is a partition of Z, then $\mathcal{L}(X) = \{X_T \mid T \in P^\star(Z)$ and $X_T \neq \emptyset\}$. In this case, we define a hyperoperation $\vee$ and an operation $\wedge$ on $\mathcal{L}(X)$ such that $(\mathcal{L}(X), \vee, \wedge)$ is a hyperlattice. We assume that

$$X_A \wedge X_B = X_{A \cap B} \text{ and } X_A \vee X_B = \{X_T \mid A \cup B \subseteq T\}.$$

In the following lemmas we investigate the conditions of a hyperlattice.

71. Lemma. *$X_A \in X_A \vee X_A$, $X_A \wedge X_A = X_A$, $X_A \vee X_B = X_B \vee X_A$ and $X_A \wedge X_B = X_{A \wedge B} = X_B \wedge X_A$.*

Proof. Immediate. ■

72. Lemma. *$(X_A \vee X_B) \vee X_C = X_A \vee (X_B \vee X_C)$ and $(X_A \wedge X_B) \wedge$ $\wedge X_C = X_A \wedge (X_B \wedge X_C)$.*

Proof. The associativity of $\wedge$ is immediate. We verify the associativity of $\vee$. Let $A, B, C \in P^\star(Z)$. Then

$$\begin{aligned}(X_A \vee X_B) \vee X_C &= \{X_T \mid A \cup B \subseteq T\} \vee X_C = \\ &= \bigcup_{A \cup B \subseteq T} X_T \vee X_C = \\ &= \bigcup_{A \cup B \subseteq T} \{X_U \mid T \cup C \subseteq U\} = \\ &= \{X_V \mid A \cup B \cup C \leq V\}\end{aligned}$$

Similarly, we have $X_A \vee (X_B \vee X_C) = \{X_U \mid A \cup B \cup C \subseteq U\}$, hence also $\vee$ is associative. ∎

73. Lemma. *$X_A \in [X_A \vee (X_A \wedge X_B)] \cap [(X_A \wedge (X_A \vee X_B)]$, for all $A, B \in P^\star(Z)$.*

Proof. Let A and B be arbitrary elements of $P^\star(Z)$. Then we have,

$$\begin{aligned}X_A \vee (X_A \wedge X_B) &= X_A \vee X_{A \wedge B} = \\ &= \{X_T \mid A \cup (A \wedge B) \subseteq T\} = \\ &= \{X_T \mid A \subseteq T\}\end{aligned}$$

Therefore, $X_A \in X_A \vee (X_A \wedge X_B)$. On the other hand, $X_A =$ $= X_{A \wedge (A \vee B)} = X_A \wedge X_{A \cup B} \in X_A \wedge (X_A \vee X_B)$, as required. ∎

74. Lemma. *If $X_A \in X_A \vee X_B$, then $X_B = X_A \wedge X_B$.*

Proof. Let $X_A \in X_A \vee X_B$. Then there exists $T \in P^\star(Z)$ such that $X_A = X_T$ and $A \cup B \subseteq T$. Thus, $B = B \cap T$ and so $X_B =$ $= X_{B \cap T} = X_B \wedge X_T = X_A \wedge X_B$. Therefore, $X_B = X_A \wedge X_B$ and the lemma is proved. ∎

We summarize the above lemmas in the following theorem:

75. Theorem. *Let $s : X \longrightarrow P^\star(Z)$ be a function such that $(X, \overset{s}{*})$ is a hypergroup. Also, we assume that for all positive integer n and the elements $a_1, \cdots, a_n$ of X, we have*

$$a_1 \overset{s}{*} a_2 \overset{s}{*} \cdots \overset{s}{*} a_n = \{g \in X \mid s(g) \subseteq s(a_1) \cup \cdots \cup s(a_n)\}.$$

Then $(\mathcal{L}(X), \vee, \wedge)$ is a hyperlattice.

We now investigate the distributivity of $\mathcal{L}(X)$ and show that this hyperlattice is not distributive, in general. In fact, we can consider the following example.

76. Example. There exists a function $s : X \longrightarrow P^\star(Z)$ such that $(X, \overset{s}{*})$ is a hypergroup which satisfies the conditions of Theorem 75, but $\mathcal{L}(X)$ is not distributive. Indeed, let us assume that H is a finite group, $\Pi_e(H) = \{\text{ord}(x) \mid x \in H\}$ and $s : P(H) \longrightarrow P^\star(\Pi_e(H))$ defined by $s(A) = \{\text{ord}(x) \mid x \in A\}$ and $s(\emptyset) = \Pi_e(H)$. It is easy to see that the function s is onto, so by Theorem 75 $\mathcal{L}(P^\star(H))$ is a hyperlattice. Suppose, $H = Z_4 = \{e, a, a^2, a^3\}$, the cyclic group of order four, and $X = P^\star(H)$. Then $\Pi_e(Z_4) = \{1, 2, 4\}$. Set, $A = \{1, 2\}, B = \{1\}, C = \{2, 4\}$ and $D = \{2\}$. It is clear that $X_A \wedge (X_B \vee X_C) = X_A \wedge X_{\Pi_e(X)} = X_A$ and $(X_A \wedge X_B) \vee (X_A \wedge X_C) = = X_B \vee X_D = \{X_A, X_{\Pi_e(X)}\}$. This shows that $X_A \wedge (X_B \vee X_C) \neq \neq (X_A \wedge X_B) \vee (X_A \wedge X_C)$. Therefore, $\mathcal{L}(P(Z_4))$ is a hyperlattice which is not distributive. ■

II) We present here some results on a special type of hyperlattices, called *P-hyperlattices*, introduced and studied by M. Konstantinidou.

Let us recall what a hyperlattice is.

77. Definition. Let $H \neq \emptyset$ and $\vee : H \times H \rightarrow \mathcal{P}^*(H)$, $\wedge : H \times H \rightarrow H$ be such that $\forall\, (a, b, c) \in H^3$, we have:

(i) $a \in a \vee a, \ a = a \wedge a;$

(ii) $a \vee b = b \vee a, \ a \wedge b = b \wedge a;$

(iii) $(a \vee b) \vee c = a \vee (b \vee c)$, $(a \vee b) \wedge c = a \wedge (b \vee c)$;

(iv) $a \in [a \wedge (a \vee b)] \cap [a \vee (a \wedge b)]$;

(v) $b \leq a \Longleftrightarrow a \in a \vee b$.

Then the hyperstructure $(H, \vee, \wedge)$ is called *hyperlattice*.

Notice that in a hyperlattice $(H, \vee, \wedge)$ the following properties hold, and they can be proved easily:

1) $\forall (a, b) \in H^2$, $a \vee b \subseteq a \vee (a \vee b)$ (if H is a lattice, we have the equality, whence $a \leq a \vee b$).

2) if $a \leq b$ are elements of H and x is arbitrary in H, then $b \vee x \subseteq (a \vee x) \vee (b \vee x)$ (if H is a lattice, we have the equality, whence $a \leq b \Longrightarrow a \vee x \leq b \vee x$).

3) if $a \leq c$ and $b \leq d$ are elements of H, then $c \vee d \subseteq (a \vee b) \vee (c \vee d)$ (if H is a lattice, then we have the equality, whence ($a \leq c$ and $b \leq d$) $\Longrightarrow a \vee b \leq c \vee d$).

4) if $(a, b, c, d) \in H^4$, then $a \vee b \subseteq (a \vee b) \vee (a \wedge c) \vee (b \wedge d)$ (if H is a lattice, then we have the equality, whence $(a \wedge c) \vee (b \wedge d) \leq a \vee b$).

5) $\forall (a, b) \in H^2$, we have $a \wedge b \in (a \wedge b) \vee (a \vee b)$ (if H is a lattice, then we have the equality, whence $a \wedge b \leq a \vee b$).

Let $(L, \vee, \wedge)$ be a lattice and $P \subseteq L$, $P \neq \emptyset$. We define the following hyperoperation on L:

$$\forall (a, b) \in L^2, \ a \overset{P}{\bigvee} b = a \vee b \vee P = \{a \vee b \vee q \mid q \in P\}.$$

78. Remark. Let $a \in L$. We have $a \in a \overset{P}{\bigvee} a$ if and only if $\exists q \in P$ such that $q \leq a$.

Proof. "$\Longleftarrow$" Let $q \in P$ such that $q \leq a$. Then $a \overset{P}{\bigvee} a = a \vee P \ni$ $\ni a \vee q = a$.

"$\Longrightarrow$" If $a \in a \overset{P}{\bigvee} a = a \vee P$, then there is $q \in P$ such that $a = a \vee q$, whence $q \leq a$. ■

Notation. Let I^L be the set

$$\{P \subseteq L \mid \forall x \in L,\ \exists q \in P : q \leq x\}.$$

79. Theorem. *The hyperstructure* $(L, \overset{P}{\bigvee}, \wedge)$ *is a hyperlattice if and only if* $P \in I^L$.

Proof. "$\Longleftarrow$" Let $P \in I^L$. For any $(a, b, c) \in L^3$, we have:

(i) $a \in a\overset{P}{\bigvee}a$ (by the previous remark);

(ii) $a\overset{P}{\bigvee}a = b\overset{P}{\bigvee}a = a \vee b \vee P$;

(iii) $(a\overset{P}{\bigvee}b)\overset{P}{\bigvee}c = a\overset{P}{\bigvee}(b\overset{P}{\bigvee}c) = \{a \vee b \vee c \vee q \vee r \mid (q, r) \in P^2\} =$
$= a \vee b \vee c \vee P \vee P$;

(iv) Let $a \in L$ and $q \in P$ be such that $q \leq a$. We have
$a = a \wedge (a \vee b) = a \wedge (a \vee b \vee q) \in a \wedge (a \vee b \vee P) = a \wedge (a\overset{P}{\bigvee}b)$ and
$a = a \vee (a \wedge b) = a \vee (a \wedge b) \vee q \in a \vee (a \wedge b) \vee P = a\overset{P}{\bigvee}(a \wedge b)$.

(v) If $b \leq a$, then $a = a \vee q \in a \vee P = a \vee b \vee P = a\overset{P}{\bigvee}b$. Conversely, if $a \in a\overset{P}{\bigvee}b = a \vee b \vee P$ then $\exists t \in P$ such that $a = a \vee b \vee t$, that is $b \leq a$.

Therefore, $(L, \overset{P}{\bigvee}, \wedge)$ is a hyperlattice.

"$\Longrightarrow$" Let $(L, \overset{P}{\bigvee}, \wedge)$ be a hyperlattice. Then $\forall a \in L$, we have $a \in a\overset{P}{\bigvee}a$, therefore $P \in I^L$, according to the previous remark. ∎

80. Corollary. $(L, \overset{P}{\bigvee}, \wedge)$ *is a hyperlattice if and only if* $\forall a \in L$, $a \in a\overset{P}{\bigvee}a$.

81. Proposition. *Let* $\mathcal{J}$ *be an ideal of a lattice* L. *Then* $\mathcal{J} \in I^L$.

Proof. Since $\mathcal{J}$ is an ideal of L, it follows that $L \wedge \mathcal{J} \subseteq \mathcal{J}$, whence $\forall\,(a, q) \in L \times \mathcal{J}$, $\exists \bar{q} \in J$ such that $a \wedge q = \bar{q}$, so $\bar{q} \leq a$. Hence $\mathcal{J} \in I^L$. ∎

82. Remark. The converse of the previous proposition is not true. Indeed, if $L = \{o, a, b\}$ and $o \leq a \leq b$, then $\mathcal{J} = \{o, b\} \in I^L$ because $o \in \mathcal{J}$, but $\mathcal{J}$ is not an ideal of L.

83. Definition. The hyperlattice $(L, \overset{P}{\bigvee}, \wedge)$ is called *P-hyperlattice.*

84. Remark. $\forall\,(a, b) \in L^2$ we have $a \vee b \in a\overset{P}{\bigvee}b$, so if a P-hyperlattice degenerates into a lattice, this coincides with the supporting lattice.

85. Remark. There are hyperlattices, where there does not exist the supremum for all pairs of their elements. Indeed, let $H = \{a, b, c, d, x, y, z\}$, where $a < b < d < x < y < z$, $a < c < d$ and $b \| c$.

If we consider the following hyperlattice on $H : \forall\,(\alpha, \beta) \in H^2$, $\alpha \leq \beta$, we have $\alpha \vee \beta = \{\gamma \in H \mid \beta \leq \gamma\}$ and $b \vee c = H - \{a, b, c\}$. Then the $\sup(b, c)$ does not exist. Hyperlattices of this kind cannot be P-hyperlattices.

Chapter 5

Fuzzy sets and rough sets

Fuzzy Sets and Hyperstructures introduced by Zadeh, in 1965, and by Marty, in 1934, respectively, are now used in the world both on the theoretical point of view and for their many applications. The Rough Sets considered for the first time by Shafer in 1976, have been reintroduced in the international scientific circle by Pawlak, in 1991 especially in connection with Artificial Intelligence. The relations between Rough Sets and Fuzzy Sets have been already considered by Dubois and Prade [137]*, those between Fuzzy Sets and Hyperstructures by Corsini, Corsini–Leoreanu, Corsini–Tofan, Ameri–Zahedi and others, those between Rough Sets and Hyperstructures by Davvaz. More recently, M. Konstantinidou and A. Kehagias have obtained interesting results on hyperstructures and fuzzy subsets.*

§1. Join spaces associated with fuzzy subsets

The first connection between fuzzy subsets and join spaces has been established by P. Corsini. Afterwards, P. Corsini and V. Leoreanu have obtained more results concerning this connection. We present some of Corsini and Leoreanu results here.

Let $\mu : H \rightarrow I$ be a function from a nonempty set H to the closed interval $I = [0,1]$ that is $< H; \mu >$ is a fuzzy subset. Let us define on H the hyperoperation: for all $(x,y) \in H^2$ such that $\mu(x) \leq \mu(y)$,

$$y \circ x = x \circ y = \{x \in H \mid \mu(x) \leq \mu(z) \leq \mu(y)\}$$

1. Theorem. *The hypergroupoid* $< H; \circ >$ *is a join space.*

Proof. It is clear that $\circ$ is associative and reproducible, that is $< H; \circ >$ is a (clearly commutative) hypergroup. It remains to prove that $< H; \circ >$ satisfies the condition $a/b \cap c/d \neq \emptyset \Longrightarrow a \circ d \cap b \circ c \neq \emptyset$. Let us suppose $x \in a/b \cap c/d$, that is

$$\mu(a) \in [\mu(x), \mu(b)], \ \mu(c) \in [\mu(x), \mu(d)].$$

We distinguish four cases:

1. $\mu(x) \leq \mu(b), \ \mu(x) \geq \mu(d)$.
   Then we have: $\mu(x) \leq \mu(a) \leq \mu(b)$, $\mu(d) \leq \mu(c) \leq \mu(x)$, from which: $\mu(d) \leq \mu(c) \leq \mu(x) \leq \mu(a) \leq \mu(b)$, whence $[\mu(c), \mu(x)] \subset [\mu(a), \mu(d)] \cap [\mu(c), \mu(b)]$, therefore $a \circ d \cap b \circ c \neq \emptyset$.

2. $\mu(x) \geq \mu(b), \mu(x) \leq \mu(d)$.
   Then we have: $\mu(b) \leq \mu(a) \leq \mu(x)$, $\mu(x) \leq \mu(c) \leq \mu(d)$, whence $\mu(b) \leq \mu(a) \leq \mu(x) \leq \mu(c) \leq \mu(d)$, from which $[\mu(a), \mu(x)] \subset [\mu(b), \mu(c)] \cap [\mu(a), \mu(d)]$, therefore $a \circ d \cap b \circ c \neq \emptyset$.

3. $\mu(x) < \mu(b), \mu(d) \geq \mu(x)$. Then $\mu(a) \leq \mu(b), \ \mu(c) \leq \mu(d)$.
   We can distinguish two cases:

   (i) $\mu(b) \leq \mu(d)$;
   we have: $\mu(a) \leq \mu(b) \leq \mu(d)$, therefore $a \circ d \cap b \circ c \neq \emptyset$.

   (ii) $\mu(b) \geq \mu(d)$;
   then we have: $\mu(c) \leq \mu(d) \leq \mu(b)$, whence $b \circ c \cap a \circ d \neq \emptyset$.

4. $\mu(x) \geq \mu(b), \mu(x) \geq \mu(d)$.
   From which $\mu(b) \leq \mu(a), \mu(d) \leq \mu(c)$.

We can distinguish two cases:

(i) $\mu(a) \leq \mu(c)$; then $\mu(b) \leq \mu(a) \leq \mu(c)$, therefore $b \circ c \cap a \circ d \neq \emptyset$.

(ii) $\mu(a) \geq \mu(c)$; then $\mu(d) \leq \mu(c) \leq \mu(a)$, therefore $a \circ d \cap b \circ c \neq \emptyset$. ∎

2. Theorem.

1) *We have* $\forall\, n \in \mathbb{N}^*$, $\forall\,(z_1, z_2, ..., z_n) \in H^n$

$$\prod_{i=1}^{n} z_i = \left\{ u \mid \bigwedge_{i=1}^{n} \mu(z_i) \leq \mu(u) \leq \bigvee_{i=1}^{n} \mu(z_i) \right\}.$$

Moreover, if R is the equivalence relation defined on H:

$$xRy \Longleftrightarrow \mu(x) = \mu(y),$$

we have

2) $R \subset \beta_2$;

3) *H/R is a hypergroup with respect to the hyperoperation* $\bar{x} \,\bar{\circ}\, \bar{y} = \{\bar{z} \mid z \in x \circ y\}$.

Proof. 1) It follows inductively from the definition. Indeed we have

$$\prod_{i=1}^{n} z_i = \prod_{i=1}^{n-1} z_i \circ z_n = \bigcup_{v \in \prod_{i=1}^{n-1} z_i} v \circ z_n =$$

$$= \bigcup_{v \in \prod_{i=1}^{n-1} z_i} \{\lambda \mid \mu(v) \wedge \mu(z_n) \leq \mu(\lambda) \leq \mu(v) \vee \mu(z_n)\}.$$

Let us suppose by induction

$$\prod_{i=1}^{n-1} z_i = \left\{ \delta \mid \bigwedge_{i=1}^{n-1} \mu(z_i) \leq \mu(\delta) \leq \bigvee_{i=1}^{n-1} \mu(z_i) \right\}.$$

Then we obtain

$$\prod_{i=1}^{n} z_i = \bigcup_{\bigwedge_{i=1}^{n-1}\mu(z_i)\leq\mu(v)\leq\bigvee_{n-1}^{n-1}\mu(z_i)} \{\mu(v)\wedge\mu(z_n)\leq\mu(\lambda)\leq\mu(v)\vee\mu(z_n)\} =$$

$$= \left\{\lambda \mid \bigwedge_{i=1}^{n}\mu(z_i)\leq\mu(\lambda)\leq \overset{n}{\underset{i=1}{\mu}}(z_i)\right\}.$$

2) Since $< H;\circ >$ is a join space, it is a hypergroup, whence $\forall\,(a,b)\in H^2$, there is $q\in H$ such that $a\in b{\circ}q$. So, if aRa', it follows $a'\in\mu^{-1}\mu(a)\subset b{\circ}q$ then $R(a)\subset b{\circ}q$ and therefore $R\subset\beta_2$.

3) R is regular. Indeed, aRa', bRb' implies $a{\circ}b = a'{\circ}b'$. Then by Theorem 29 [437], $< H/R;\bar{\circ} >$ is a hypergroup. ■

In the following, we shall give a necessary and sufficient condition for the isomorphism of two join spaces, associated with fuzzy subsets, on the same universe.

We shall find the number of isomorphism classes of such join spaces, in the case of a finite universe.

To different fuzzy subsets μ_A, μ_B, isomorphic join spaces can correspond, for instance, if $\mu_{\bar{A}}$ is the complement of μ_A, then $< H;\circ_A >$ and $< H;\circ_{\bar{A}} >$ are isomorphic.

First, we shall find the number of isomorphism classes of join spaces associated with fuzzy subsets on a universe H such that $|H| = n < \aleph_0$.

Let us set $H = I(n) = \{1,2,...,n\}$, let $\widetilde{\mathcal{P}}$ be the set of fuzzy subsets on H and let $\mu_A\in\widetilde{P}(H)$.

Let us define on H the equivalence relation

$$u\sim_A v \text{ if and only if } \mu_A(u)=\mu_A(v).$$

Let us set $H' = H/{\sim_A}$, $|H'| = s$, and let us order H' such that

II) $\forall\,(h,k)\in H^2$, $\bar{h}<\bar{k}$ if and only if $\mu_A(h)<\mu_A(k)$.

Let $H' = \{\bar{h}_1, \bar{h}_2, ..., \bar{h}_s\}$ and let $\lambda(\mu_A)$ be the ordered partition of n into s parts defined as follows:

III) $\forall (\bar{h}_1, ..., \bar{h}_s) \in H'^s$, $\lambda(\mu_A) = (a_1, a_2, ..., a_s)$ if and only if $\forall i$, $a_i = \left|\mu_A^{-1}(\mu_A(h_i))\right|$ and $\forall (i,j) \in I(s) \times I(s)$ such that $i \neq j$, $i < j \Longrightarrow \bar{h}_i < \bar{h}_j$.

Clearly, we have $\sum_{i=1}^{s} a_i = n$, and $\forall i$, $a_i \geq 1$.

IV) Let $(a_1, a_2, ..., a_s)$ be an ordered partition of n into s parts. Let us set $\Psi(a_1, a_2, ..., a_s) = (b_1, ..., b_s)$ where $\forall i : 1 \leq i \leq s$, $b_i = a_{s-i+1}$.

We shall prove the following

3. Theorem. *If μ_A, μ_B are fuzzy subsets on a finite universe H, then the join spaces $< H; \circ_A >$ and $< H; \circ_B >$ are isomorphic if and only if either $\lambda(\mu_A) = \lambda(\mu_B)$ or $\lambda(\mu_B) = \Psi(\lambda(\mu_A))$.*

Proof. We shall prove before the implication "$\Longleftrightarrow$".

Let us suppose $\lambda(\mu_A) = \lambda(\mu_B) = (a_1, ..., a_s)$. We can set $H = \bigcup_{j=1}^{s} H = \bigcup_{j=1}^{s} H'_j$, where $\forall j \in I(s)$, $H_j = \mu_A^{-1}(\mu_A(h_j))$ and $H'_j = \mu_B^{-1}(\mu_B(h_j))$.

Let $H_j = \{x_{1,j}, x_{2,j}, ..., x_{a_j,j}\}$, $H'_j = \{x'_{1,j}, x_{2,j}, ..., x'_{a_j,j}\}$.

Let us order H in the following manner

$$\forall (j,j') \in I(s) \times I(s),\ \forall (h', h') \in I(a_j) \times I(a_j)$$
$$x_{h,j} < x_{h',j} \Longleftrightarrow h < h'.$$

If $j \neq j'$, $\forall (h, h')$,

$$x_{h,j} < x_{h',j'} \Longleftrightarrow j < j'.$$

Moreover, $\forall (i,j) \in I(s) \times I(s)$ we shall denote $i \vee j = \max\{i,j\}$, $i \wedge j = \min\{i,j\}$. Then $\forall (i,j) \in I(s) \times I(s)$, $\forall (h,k) \in I(a_i) \times I(a_j)$, by I) we have

$$x_{h,i} \circ_A x_{k,j} = \bigcup_{i \wedge j \leq r \leq i \vee j} H_r,$$

$$x_{h,i} \circ_B x_{k,j} = \bigcup_{i \wedge j \leq r \leq i \vee j} H'_r.$$

Therefore, if $f :< H; \circ_A > \to < H; \circ_B >$ is the function defined as follows: $\forall (u,t) \in I(a_t) \times I(s)$, $f(x_{u,t}) = x'_{u,t}$; then we have

$$f(x_{h,i} \circ_A x_{k,j}) = x'_{h,i} \circ_B x'_{k,j} = f(x_{h,i}) \circ_B f(x_{k,j}),$$

where $< H; \circ_A >$ and $< H' \circ_B >$ are isomorphic hypergroups.

Let us suppose now $\lambda(\mu_B) = \Psi(\lambda(\mu_A))$.

Let us set $H = \bigcup_{1 \leq j \leq s} H_j$ where $\forall j \in I(s)$, $H_j = \{x_{1,j}, ..., x_{a_j,j}\}$, and $H'_j = \bigcup_{1 \leq j' \leq s} H'_{j'}$, where $j' = \Psi(j) = s - j + 1$, and $H'_{j'} = \{x'_{1,j'}, ..., x'_{a'_{j'},j'}\}$ with $a'_{j'} = a_j$.

Let us define $\forall (h,j) \in I(a_j) \times I(s)$, $f(x_{h,j}) = x'_{h,j'}$. We have

$$f(x_{h,i} \circ_A x_{k,j}) = f\left(\bigcup_{i \wedge j \leq r \leq i \vee j} H_r\right) = \bigcup_{\Psi(i \vee j) \leq \Psi(r) \leq \Psi(i \wedge j)} H'_{\Psi(r)} =$$

$$= \bigcup_{i' \wedge j' \leq r' \leq i' \vee j'} H'_r = x'_{h,i'} \circ_B x'_{k,j'} = f(x_{h,i}) \circ_B f(x_{k,j}),$$

whence f is an isomorphism.

Let us prove now the implication "$\Longrightarrow$".

Let $p : H \to I(s)$ be the function defined as follows: $\forall x \in H$, $p(x) = j$ where j is that unique element of $I(s)$, such that $\exists k \in I(a_j)$ so that $x = x_{kj} \in H_j$. Analogously we define $p' : H \to I(s')$ where $s' = |H/{\sim_B}|$.

Let $f :< H : \circ_A > \longrightarrow < H; \circ_B >$ be the isomorphism of these two join spaces. Then $\forall (x,y) \in H^2$, we have

$$f(x \circ_A y) = f\left(\bigcup_{p(x) \wedge p(y) \leq j \leq p(x) \vee p(y)} H_j\right) = \bigcup_{p(x) \wedge p(y) \leq j \leq p(x) \vee p(y)} f(H_j).$$

But we also have, if we set $\mu_B^{-1}\mu_B(x'_{k,r}) = H'_{r'}$

$$f(x \circ_A y) = f(x) \circ_B f(y) = \bigcup_{p'(f(x))\wedge p'(f(y))\leq r\leq p'(f(x))\vee p'(f(y))} H'_r.$$

For $\forall\,(u,v) \in N\times N$, we shall denote $I(u,v)$ the set

$$\{z \in N \mid u \wedge v \leq z \leq u \vee v\}.$$

Let us remark now

1) $r_1 \neq r_2 \Longrightarrow H'_{r_1} \cap H'_{r_2} = \emptyset$,

2) if $\{x,y\} \subset H$, we have $x \circ_A y = H = x \circ_A x = y \circ_A y$, whence $f(x) \circ_B f(y) = f(x \circ_A y) = f(x \circ_A x) = f(H_j) = = f(x) \circ_B f(x)$,

3) by 1) and 2), there is only one $t = p'(f(x))$ such that $f(H_j) = = f(x) \circ_B f(y) = H'_t$.

We shall set $t = \varphi(j)$, whence $f(H_j) = H'_{\varphi(j)}$.

So we have $\varphi : I(p(x),p(y)) \rightarrow I(p'(f(x)),p'(f(y)))$ and we have clearly $\forall\, x \in H$, $\varphi(p(x)) = p'(f(x))$.

We shall prove that φ is a bijection.

φ is clearly an one-to-one function.

Indeed, if there were j_1, j_2 such that $j_1 \neq j_2$ and $\varphi(j_1) = \varphi(j_2)$, it follows $f(H_{j_1}) = f(H_{j_2})$ from which $\forall\, k : k \in I(a_{j_1})$, $h \in I(a_{j_2})$ exists such that $f(x_{k,j_1}){=}f(x_{k,j_2})$ which is absurd since $H_{j_1}{\cap}H_{j_2}{=}\emptyset$, and f is one-to-one function.

φ is also onto.

Indeed, since

$$f(x_{k,i} \circ_A x_{h,j}) = f(x_{k,i}) \circ_B f(x_{h,j}) = \bigcup_{r\in I(p'(f(x_{k,i}))p'(f(x_{h,j})))} H'_r$$

and $p'(f(x_{k,i})) = \varphi(i)$, $p'(f(x_{h,j})) = \varphi(j)$.

We have that $\forall\, r{\in} I(\varphi(i),\varphi(j))$, $\exists t{\in} I(i,j)$ such that $\varphi(t){=}r$.

Therefore, φ is a bijection from the interval $I(i,j)$ to the interval $(I(\varphi(i),\varphi(j))$. Particularly $\varphi : I(1,s) \rightarrow (I(\varphi(1),\varphi(s))$. On the other hand, as f is a bijection, it follows $\forall y \in H$, $\exists!(k,j) \in$ $\in I(a_j) \times I(s)$ such that $f(x_{k,j}) = y$, whence

$$f(H) = H = \bigcup_{j \in I(s)} f(H_j) = \bigcup_{j \in I(s)} H'_{\varphi(j)} = \bigcup_{r \in I(\varphi(1), \varphi(s))} H'_s.$$

Moreover, clearly φ is a function from $I(s)$ to $I(s')$ and since $s' = |H/{\sim_B}|$ and f is an isomorphism, we have

$$|f(H)/{\sim_B}| = |H/{\sim_B}| = |p'(H)| = |p'(f(H))| = |\varphi(p(H))|,$$

whence $s' = |H/{\sim_B}| = |\varphi(p(H))| = |p(H)| = s$.

Moreover, $\forall j \in I(s)$, we have

$$\forall j \in I(s),\ a_j = |H_j| = |f(H_j)| = \left|H_{\varphi(j)}\right| = a'_{\varphi(j)}.$$

On the other side ,$\forall k \in I(a_1)$, $\forall h \in I(a_s)$, we have

$$f(H) = f(x_{k,1} \circ_A x_{h,s}) = \bigcup_{t \in I(1,s)} H'_t = H.$$

Therefore, the interval $I(\varphi(1), \varphi(s))$ coincides with the interval $I(1,s)=I(s)$. It follows $\{\varphi(1), \varphi(s))\}=\{1,s\}$. Hence $I(2, s-1)=$ $= I(1,s)-\{1,s\} = I(\varphi(1), \varphi(s)) - \{\varphi(1), \varphi(s)\} = \varphi(I(s)) -$ $-\{\varphi(1), \varphi(s)\} = \varphi(I(2, s-1)) = I(\varphi(2), \varphi(s-1)))$ from $I(2, s-1) =$ $= I(\varphi(2), \varphi(s-1))$.

One obtains analogously $\{\varphi(2), \varphi(s-1)\} = \{2, s-1\}$.

In general, we have

(ε) $\forall k,\ \varphi(k) \in \{k, s-k+1\}$.

Let V be the set of the permutations of $I(s,)$ which satisfy (ε).

$\eta)$ We shall prove now that either φ is the identity function I of $I(s)$ or it is the permutation Ψ of $I(s)$ defined:

$$\forall k \in I(s),\ \Psi(k) = s - k + 1.$$

If $s \leq 3$ we have $V = \{I_{I(s)}, \Psi\}$. If $s > 3$ and one supposes $\varphi(1) = 1$, $\varphi(2) = \Psi(2) = s - 2 + 1 = s - 1$, it follows $\varphi(I(1,2)) = I(\varphi(1), \varphi(2)) = I(1, s-1)$, whence $2 = |(I(1,2)| = |\varphi(I(1,2))| \neq |I(1, s-1)| \geq 3$, absurd.

Analogously if $\varphi(1) = s$, then $\varphi(2) = 2$.

Therefore, either $\varphi_{I(2)} = I_{I(2)}$ or $\varphi_{I(2)} = \Psi_{I(2)}$.

Let k be in $I(s)$ and let us suppose

$$\varphi_{I(k)} = I_{I(k)},\ \varphi(k+1) = \Psi(k+1).$$

Then we have $k+1 = |I(k+1)| = |\varphi(I(k+1))| = |I(\varphi(1), \varphi(k))| + |I(\varphi(k), \Psi(k+1))| - 1 = k + |I(k, s-k)| - 1 = k + s - 2k$, from which $s = 2k+1$, hence $\varphi(k+1) = s - (k+1) + 1 = 2k + 1 - k - 1 + 1 = k + 1$, from which $\varphi_{I(k+1)} = I_{I(k+1)}$.

Analogously, if one supposes

$$\varphi_{I(k)} = \Psi_{I(k)},\ \varphi(k+1) = k+1,$$

then we have $k+1 = |\varphi(I(k+1))| = |\varphi(I_k)| + |\varphi(I(k, s-k))| - 1 = k + s - sk + 1 - 1$ whence $s - 2k = 1$ that is $s = 2k+1$. Then $\Psi(k+1) = s - (k+1) + 1 = 2k + 1 - k - 1 + 1 = k + 1 = I(k+1)$ from which $\varphi_{I(k+1)} = \Psi_{I(k+1)}$.

Therefore, by induction, η) is proved, hence the implication $\Longrightarrow$ follows and consequently the theorem is proved. ∎

4. Theorem. *Let H be $I(n)$ and let $J_\mu(n)$ be the set of isomorphism classes of the join spaces $< H; \circ_A >$ associated with the fuzzy subsets μ_A on the universe H. Then*

$$\begin{array}{lll} \text{if} & n = 2k+1, & |J_\mu(n)| = 2^{k-1}(2^k + 1) \\ \text{if} & n = 2k, & |J_\mu(n)| = 2^{k-1}(2^{k-1} + 1). \end{array}$$

To calculate $|J_\mu(n)|$ by Theorem 3, it is enough to remember that if $(p.o.)(n)$ is the set of the ordered partitions of n, we have $|(p.o.)| = 2^{n-1}$ (see [448]), and to find those $p \in (p.o.)(n)$ such that $\Psi(p) = p$.

An ordered partition $(a_1, a_2, ..., a_s)$ of the integer n is called *symmetrical* if $\Psi(a_1, ..., a_s) = (a_1, ..., a_s)$. Let us denote $(s.o.p.)(n)$, the set of the symmetrical ordered partitions of n. To calculate the number $|(s.o.p.)(n)|$, we shall distinguish the case n is odd from that it is even.

Let us suppose $n = 2k + 1$, and $X \in (s.o.p.)(n)$. Then either $X = (2k + 1)$ or it is of the type $(i_1, i_2, ..., i_s, 2t + 1, i_s, i_{s-1}, ..., i_1)$ where $t \in \{0, 1, ..., k - 1\}$, $s \in \{k - t, k - t - 1, ..., 1\}$, and we have $2\sum_{r=1}^{s} i_r + 2t + 1 = 2k + 1$, whence $\sum_{r=1}^{s} i_r = k - t$.

For any t, we have $|(s.o.p.)(k - t)| = 2^{k-t-1}$ (see [448]).

It follows $|(s.o.p.)(n)| = \sum_{t=0}^{k-1} 2^{k-t-1} + 1$.

Let us set $k - t - 1 = v$, then

$$\sum_{t=0}^{k-1} 2^{k-t-1} = \sum_{v=k-1}^{0} 2^v = 2^{(k-1)+1} - 1$$

from which $|(s.o.p.)(2k + 1)| = 2^k$.

Let us suppose now $n=2k$. Then if $X \in (s.o.p.)(n)$, either $X \in (2k)$ or it is of the type: $(i_1, ..., i_s, 2t, i_s, ..., i_1)$, where $t \in \{0, 1, ..., k - 1\}$, $s \in \{k - t, k - t - 1, ..., 1\}$ and $\sum_{j=1}^{s} i_j = k - t$. We have $|(s.o.p.)(k - t)| = 2^{k-t-1}$, from which

$$|(s.o.p.)(2k)| = \sum_{t=0}^{k-1} 2^{k-t-1} + 1 = 2^k.$$

Now we can conclude.

If $n = 2k + 1$, then

$$|J_\mu(n)| = 2^k + (2^{n-1} - 2^k)\frac{1}{2} = 2^k + 2^{2k-1} - 2^{k-1} = 2^{k-1}(2^k - 1).$$

If $n = 2k$, then

$$|J_\mu(n)| = 2^k + (2^{k-1} - 2^k)\frac{1}{2} = 2^{k-1}(2 + 2^{k-1} - 1) = 2^{k-1}(2^{k-1} + 1).$$

■

Now, it is interesting to study how the isomorphism problem of two join spaces associated with fuzzy subsets on a finite universe, can be generalized for the case of an arbitrary universe.

Before see it, let us make some notations.

Let μ_A be a fuzzy subset on an arbitrary universe H and let us define the equivalence relation

$$u \sim_A v \quad \text{if and only if} \quad \mu_A(u) = \mu_A(v).$$

Now, if μ_A and μ_B are two fuzzy subsets on H, let us set $H/{\sim_A} = \{H_i \mid i \in I\}$ and $H/{\sim_B} = \{H'_{i'} \mid i' \in I'\}$, where

$$\begin{array}{llll} \forall\, i \in I, & H_i = \{x_{k,i} \mid k \in K_i\}; & |K_i| = |H_i| = a_i & \text{and} \\ \forall\, i' \in I', & H'_{i'} = \{x'_{k',i'} \mid k' \in K'_{i'}\}; & |K'_{i'}| = |H'_{i'}| = a'_{i'}. & \end{array}$$

We order I (I', respectively) such that: $i < j \iff (\forall\,(x,y) \in H_i{\times}H_j,\ \mu_A(x) < \mu_A(y))$ ($i' < j' \iff (\forall\,(x',y') \in H'_{i'}{\times}H'_{j'},\ \mu_B(x') < \mu_B(y'))$, respectively) and we order also $H/{\sim_A}$ ($H/{\sim_B}$) such that: $H_i < H_j \iff i < j$ ($H'_{i'} < H'_{j'} \iff i' < j'$, respectively).

We have the following

5. Theorem. *If μ_A, μ_B are fuzzy subsets on a universe H, then the join spaces $< H; \circ_A >$ and $< H; \circ_B >$ are isomorphic if and only if a strict monotone and bijective function $\varphi : I \to I'$ exists, such that $\forall\, i \in I$, $a_i = a'_{\varphi(i)}$.*

Proof. First, let us prove the implication "$\Longrightarrow$".

We denote by f the isomorphism between $(H, \circ_A)$ and $(H, \circ_B)$.

Similarly, with the finite case, we shall consider $p : H \to I$, $p(x) = j$, ($p' : H \to I'$, $p'(x) = j'$), where j (j', respectively) is the unique element of I (I', respectively) such that $x \in H_j$ ($x \in H'_{j'}$, respectively).

For $\{x,y\} \subset H_i$, we have $x \circ_A x = x \circ_A y = y \circ_A y = H_i$, so $f(x) \circ_B f(x) = f(x \circ_A y) = f(y) \circ_B f(y)$, that is $H'_{p'(f(x))} = f(H_i) = H'_{p'(f(y'))}$, whence $p'(f(x)) = p'(f(y))$, because for $\{r_1, r_2\} \subset I'$, $r_1 \neq r_2$, $H'_{r_1} \cap H'_{r_2} = \emptyset$.

Now, we can define the function $\varphi : I \longrightarrow I'$ in this manner: for $i = p(x)$, $x \in H$, $\varphi(i) = p'(f(x))$.

Let us remark that φ is an one-to-one function. For this, let us suppose $\exists\{j_1, j_2\} \subset I$, $j_1 \neq j_2$, such that $\varphi(j_1) = \varphi(j_2)$. It follows $f(H_{j_1}) = f(H_{j_2})$, which is absurd since $H_{j_1} \cap H_{j_2} = \emptyset$ and f is one-to-one.

φ is also onto. Indeed, since $f(H) = H$, we have:

$$H = \bigcup_{i' \in I'} H'_{i'} = f\left(\bigcup_{i \in I} H_i\right) = \bigcup_{i \in I} f(H_i) = \bigcup_{i \in I} H'_{\varphi(i)} = \bigcup_{i' \in \operatorname{Im}\varphi} H'_{i'},$$

so $I' = \operatorname{Im}\varphi$.

Therefore, φ is a bijection, hence $|I| = |I'|$.

f is an isomorphism, so $\forall\, i \in I$, $|f(H_i)| = |H_i|$, that is $\forall\, i \in I$, $a'_{\varphi(i)} = \left|H'_{\varphi(i)}\right| = |H_i| = a_i$.

Let us prove now the strict monotony of φ. We shall use the following notations: $u \wedge w = \min\{u, v\}$; $u \vee v = \max\{u, v\}$; $\forall\,(i,j) \in I^2$, $[i \wedge j, i \vee j] = \{t \in I \mid i \wedge j \leq t \leq i \vee j\}$ and $\forall\,(i', j') \in I^2$, $[i' \wedge j', i' \vee j'] = \{t' \in I' \mid i' \wedge j' \leq t' \leq i' \vee j'\}$.

Let $(i,j) \in I^2$, $i < j$ and let us consider $x \in H_i$ and $y \in H$. From $f(x \circ_A y) = f(x) \circ_B f(y)$ it follows

$$f\left(\bigcup_{i \leq r \leq j} H_r\right) = \bigcup_{\varphi(i) \wedge \varphi(j) \leq k \leq \varphi(i) \vee \varphi(j)} H'_k,$$

whence

$$\bigcup_{i \leq r \leq j} H'_{\varphi(r)} = \bigcup_{\varphi(i) \wedge \varphi(j) \leq k \leq \varphi(i) \vee \varphi(j)} H'_k;$$

therefore, $\{\varphi(r) \mid i \leq r \leq j\} = \{k \in I \mid \varphi(i) \wedge \varphi(j) \leq k \leq \varphi(i) \vee \varphi(j)\}$, that is:

$$(*) \quad \forall\,(i,j) \in I^2,\ i < j,\ \varphi[i,j]) = [\varphi(i) \wedge \varphi(j), \varphi(i) \vee \varphi(j)].$$

We have $\varphi(i) \neq \varphi(j)$, since $i < j$ and φ is a bijection; so, there are two possibilities: $\varphi(i) < \varphi(j)$ or $\varphi(j) < \varphi(i)$.

Case A. For $\varphi(i) < \varphi(j)$, φ is strict increasing on $[i,j]$. Indeed, let us observe that:

A_1. If $[i, j] = \{i, j\}$, that is obviously;

A_2. If there exists $r \in I$, such that $i < r < j$, then
$\varphi(i) = \varphi(i) \wedge \varphi(j) < \varphi(r) < \varphi(i) \vee \varphi(j) = \varphi(j)$, since $(*)$.

A_3. If there exists $(r, s) \in I^2$, such that $i < s < r < j$, we obtain, using $(*)$, that: $\varphi([i, r]) = [\varphi(i) \wedge \varphi(r), \varphi(i) \vee \varphi(r)]$; so, using A_2, $\varphi(i) = \varphi(i) \wedge \varphi(r) < \varphi(s) < \varphi(i) \vee \varphi(r) = \varphi(r) < \varphi(j)$.

Case B. For $\varphi(j) < \varphi(i)$, we can prove that φ is strict decreasing in a similar manner.

Moreover, we shall prove that for $\forall (i, j) \in I^2$, such that $i < j$, we have two situations:

1°. If $\varphi(i) < \varphi(j)$, then φ is strict increasing on I;

2°. If $\varphi(j) < \varphi(i)$, then φ is strict decreasing on I.

Indeed, in the first situation, we have already seen that φ is strict increasing on $[i, j]$. Let ℓ be an arbitrary element of I, such that $j < \ell$. So, $\varphi(j) \neq \varphi(\ell)$; if $\varphi(\ell) < \varphi(j)$, then $\varphi(\ell) < \varphi(i)$.

Indeed, $\varphi(\ell) \notin [\varphi(i), \varphi(j)] = \varphi([i, j])$, since φ is one-to-one.

But, from $\varphi(\ell) < \varphi(i)$, we obtain that φ is strict decreasing on $[i, \ell]$, as it follows from the case B. So that, φ is strict decreasing on $[i, j]$, too, which is not true. Therefore, $\forall \ell \in I$, $j < \ell$, $\varphi(j) < \varphi(\ell)$, that is φ is strict increasing on $[j, \ell]$, $\forall \ell > j$.

Similarly, we can prove that φ is strict increasing on $[\ell, i]$, for every $\ell \in I$, $\ell < i$. Therefore, φ is strict increasing on I.

Analogously, it follows 2°.

"$\Longleftarrow$" For $\forall i \in I$, we have $|K_i| = |H_i| = a_i = a'_{\varphi(i)} = \left|H'_{\varphi(i)}\right| =$
$= \left|K'_{\varphi(i)}\right|$, so we can suppose $K_i = K'_{\varphi(i)}$.

Let us define $f : (H, \circ_A) \to (H, \circ_B)$ in this manner: for $\forall i \in I$, $\forall k \in K_i$, $f(x_{x,i}) = x'_{k,\varphi(i)}$.

Hence, f is a bijection.

Let us verify now that f is a morphism.

For $\forall\,(i,j)\in I^2$ and $\forall\,(x,y)\in H_i{\times}H_j$, $\exists k\in K_i$, $\exists h\in K_i$, such that $x=x_{x,i}$ and $y=y_{h,j}$. We have:

$$f(x)\circ_B f(y)=f(x_{k,i})\circ_B f(y_{h,j})=x'_{k,\varphi(i)}\circ_B y'_{h,\varphi(j)}=\bigcup_{\varphi(i)\wedge\varphi(j)\leq t\leq\varphi(i)\vee\varphi(j)} H'_t;$$

$$f(x\circ_A y)=f\left(\bigcup_{i\wedge j\leq k\leq i\vee j} H_k\right)=\bigcup_{i\wedge j\leq k\leq i\vee j} H'_{\varphi(k)}.$$

If φ is strict increasing, then

$$\varphi(i\wedge j)=\varphi(i)\wedge\varphi(j) \text{ and } \varphi(i\vee j)=\varphi(i)\vee\varphi(j);$$

φ is also a bijection, so

$$f(x\circ_A y)=\bigcup_{\varphi(i\wedge j)\leq\varphi(k)\leq\varphi(i\vee j)} H'_{\varphi(k)}=\bigcup_{\varphi(i)\wedge\varphi(j)\leq s\leq\varphi(i)\vee\varphi(j)} H'_s,$$

whence $f(x\circ_A y)=f(x)\circ_B f(y)$.

If φ is strict decreasing, then $\varphi(i{\wedge}j)=\varphi(i){\vee}\varphi(j)$ and $\varphi(i{\vee}j)=$ $=\varphi(i)\wedge\varphi(j)$; φ is also a bijection, so

$$f(x\circ_A y)=\bigcup_{\varphi(i\vee j)\leq\varphi(k)\leq\varphi(i\wedge j)} H'_{\varphi(k)}=\bigcup_{\varphi(i)\wedge\varphi(j)\leq s\leq\varphi(i)\vee\varphi(j)} H'_s,$$

whence $f(x\circ_A y)=f(x)\circ_B f(y)$.

Therefore, we have obtained that f is a morphism and now the theorem is proved. ∎

Finally, we give other examples of hyperstructures associated with fuzzy subsets.

Let μ_A be a fuzzy subset on a universe H.

6. Example. Let us define the hyperoperation in the following manner:

$$y\otimes_A x=x\otimes_A y=\{z\in H\mid \mu_A(x)\leq\mu_A(z)\leq\mu_A(y)\}\cup$$
$$\cup\{z\in H\mid \mu_A(x)\leq 1-\mu_A(z)\leq\mu_A(y)\},$$

where we have supposed $\mu_A(x) \leq \mu_A(y)$.

$< H, \otimes >$ is not a join space, because from $a/b \cap c/d \neq \emptyset$ it results $\exists x$, such that $a \in b \otimes x$ and $c \in d \otimes x$.

If, for instance,

$$\bar{\mu}_A(a) = 1 - \mu_A(a) \in [\mu_A(b) \wedge \mu_A(x), \mu_A(b) \vee \mu_A(x)] \text{ and}$$
$$\mu_A(x) \in [\mu_A(d) \wedge \mu_A(x), \mu_A(d) \vee \mu_A(x)]$$

and if $\mu_A(x) \leq \mu_A(b)$ and $\mu_A(c) \leq \mu_A(d)$, a possible situation is the next:

$$\mu_A(x) \leq \bar{\mu}_A(a) \leq \mu_A(c) \leq \mu_A(b) \leq \mu_A(a) \leq \mu_A(d), \text{ whence:}$$

1. $[\mu_A(c), \mu_A(b)] \cap [\mu_A(a), \mu_A(d)]$ can be void (for $\mu_A(b) \neq \mu_A(a)$);

2. $\mu_A(d) \leq \mu_A(a) \geq \bar{\mu}_A(c) \geq \bar{\mu}_A(b)$,
   so $[\bar{\mu}_A(b), \bar{\mu}_A(c)] \cap [\mu_A(a), \mu_A(d)]$ can be void;

3. $\bar{\mu}_A(c) \geq \bar{\mu}_A(b) \geq \bar{\mu}_A(a) \geq \bar{\mu}_A(d)$,
   so $[\bar{\mu}_A(b), \bar{\mu}_A(c)] \cap [\bar{\mu}_A(a), \bar{\mu}_A(d)]$ can be void;

4. $\mu_A(b) \geq \bar{\mu}_A(c) \geq \bar{\mu}_A(a) \geq \bar{\mu}_A(d)$,
   so $[\bar{\mu}_A(d), \bar{\mu}_A(a)] \cap [\mu_A(b), \mu_A(c)]$ can be void.

So that,

$$\begin{aligned} a \otimes d \cap b \otimes c = (& [\mu_A(a) \wedge \mu_A(d), \mu_A(a) \vee \mu_A(d)] \cup \\ & \cup [\bar{\mu}_A(a) \wedge \bar{\mu}_A(d), \bar{\mu}_A(a) \vee \bar{\mu}_A(d)]) \cap \\ & \cap ([\mu_A(b) \wedge \mu_A(c), \mu_A(b) \vee \mu_A(c)] \cup \\ & \cup (\bar{\mu}_A(b) \wedge \bar{\mu}_A(c), \bar{\mu}_A(b) \vee \bar{\mu}_A(c)]) \end{aligned}$$

can be void.

7. Example. Let us consider

$$x \,\square_A\, y = \{z \in H \mid \mu_A(z) \in \{\mu_A(x), \mu_A(y), \bar{\mu}_A(x), \bar{\mu}_A(y)\}\}.$$

We have

$$\forall (x,y,z) \in H^3,\ (x \,\square_A\, y) \,\square_A\, z = x \,\square_A\, (y \,\square_A\, z) =$$
$$= \{\alpha \in H \mid \mu_A(\alpha) \in \{\mu_A(x), \mu_A(y), \mu_A(z), \bar{\mu}_A(x), \bar{\mu}_A(y), \bar{\mu}_A(z)\}\}$$

and $\forall (x,y) \in H^2,\ x \in x \,\square_A\, y$ so $x \,\square_A\, H = H \,\square_A\, x = H$.

Therefore, $(H, \square_A)$ is a hypergroup, with $\omega_H = H$. Moreover, $(H, \square_A)$ is regular and reversible. $(H, \square_A)$ is a join space, too.

Let $x \in a/b \cap c/d$,

$$\mu_A(a) \in \{\mu_A(b), \mu_A(x), \bar{\mu}_A(b), \bar{\mu}_A(x)\}$$
$$\mu_A(c) \in \{\mu_A(d), \mu_A(x), \bar{\mu}_A(d), \bar{\mu}_A(x)\}\}.$$

We can find $\alpha \in H$, such that

$$\mu_A(\alpha) \in \{\mu_A(a), \mu_A(d), \bar{\mu}_A(a), \bar{\mu}_A(d)\} \cap \{\mu_A(b), \mu_A(c), \bar{\mu}_A(b), \bar{\mu}_A(c)\}.$$

If $\mu_A(a)$ is $\mu_A(b)$ or $\bar{\mu}_A(b)$, we choose $\alpha = b$;
If $\mu_A(c)$ is $\mu_A(d)$ or $\bar{\mu}_A(d)$, we choose $\alpha = d$;
If $\{\mu_A(a), \mu_A(c)\} \subset \{\mu_A(x), \bar{\mu}_A(x)\}$, then we choose $\alpha = x$.
So, $(H, \square_A)$ is a join space.

Let us consider on H the equivalence relation:

$$x \approx y \iff \mu_A(y) \in \{\mu_A(x), 1 - \mu_A(x)\}$$

For $\forall (x,y) \in H^2,\ x \,\square_A\, y = \bar{x} \cup \bar{y}$. If $x \approx_A y,\ x \,\square_A\, y = \bar{x}$.

Let us consider now q_1 and q_2 be two fuzzy subsets on H.

Let $H/{\approx_{q_1}} = \{H_{\lambda_i} \mid i \in I\}$ and $H/{\approx_{q_2}} = \{H'_{\lambda'_{i'}} \mid i' \in I'\}$.

So, $\forall \lambda_i \in [0,1],\ H_{\lambda_i} = \{x \in H \mid q_1(x) = \lambda_i \text{ or } q_1(x) = 1 - \lambda_i\}$ and $\forall \lambda'_{i'} \in [0,1],\ H'_{\lambda'_{i'}} = \{x \in H \mid q_2(x) = \lambda'_{i'} \text{ or } q_2(x) = 1 - \lambda'_{i'}\}$.

Let us denote $a_i^{q_1} = |H_{\lambda_i}|$ and $a_{i'}^{q_2} = \left|H'_{\lambda'_{i'}}\right|$.

8. Proposition. *For q_1 and q_2 fuzzy subsets on a universe H, we have $(H, \square_{q_1}) \stackrel{\sim}{\longrightarrow} (H, \square_{q_2})$ if and only if $|I| = |I'|$ and $\{a_i^{q_1}\}_{i \in I} = \{a_{i'}^{q_2}\}_{i' \in I'}$.*

Proof. "$\Longrightarrow$" Let us denote the isomorphism by f. For $x \in H$, we have $f(x \,\square_{q_1}\, x) = f(x) \,\square_{q_2}\, f(x)$, that is $f(\bar{x}) = \overline{f(x)}$.

If $|\bar{x}| = a_{i_0}^{q_1}$ $(x \in H_{\lambda_{i_0}})$, then $\left|\overline{f(x)}\right| = a_{i_0}^{q_1}$.

But $\overline{f(x)} = \{y \in H \mid q_2(y) = q_2(f(x))) = \lambda'_{i'_0}$ or $q_2(y) = 1-\lambda'_{i'_0}\}$ and $\left|\overline{f(x)}\right| = a_{i'_0}^{q_2}$. So, $a_{i_0}^{q_1} = a_{i'_0}^{q_2}$.

Let us also prove that if $i_0 \neq j_0$, then $i'_0 = j'_0$.

Indeed, if we suppose $\exists (i_0, j_0) \in I^2$, $i_0 \neq j_0$ such that $i'_0 = j'_0$, that is $\exists x \in H_{\lambda_{i_0}}$ and $\exists y \in H_{\lambda_{j_0}}$, for which $\{f(x), f(y)\} \subset H'_{\lambda'_{i'_0}}$, then we have

$$
\begin{aligned}
H'_{\lambda'_{i'_0}} &= f(x) \,\square_{q_2}\, f(y) = f(x \,\square_{q_1}\, y) = \\
&= f(H_{\lambda_{i_0}} \cup H_{\lambda_{j_0}}) = f(H_{\lambda_{i_0}}) \cup f(H_{\lambda_{j_0}}),
\end{aligned}
$$

whence, $f(H_{\lambda_{i_0}}) = f(\bar{x}) = \overline{f(x)} = H'_{\lambda'_{i'_0}} = f(H_{\lambda_{i_0}}) \cup f(H_{\lambda_{j_0}})$, contradiction, since f is an isomorphism.

So, $|I| \leq |I'|$. Now, if we do the same reasoning for f^{-1}, we obtain $|I'| \leq |I|$, hence $|I| = |I'|$ and so we can consider $I = I'$.

"$\Longleftrightarrow$" First, let us denote $a_i = a_i^{q_1} = a_i^{q_2}$, for every $i \in I$ and let us define the bijection $f : (H, \square_{q_1}) \to (H, \square_{q_2})$, such that $\forall j \in I$, $f(H_{\lambda_j}) = H'_{\lambda'_j}$.

Let $\{x_{i,j}, x_{i',j}\} \subset H_{\lambda_j}$. we have

$$
f(x_{i,j} \,\square_{q_1}\, x_{i',j}) = f(\bar{x}_{i,j}) = f(H_{\lambda_j}) = H'_{\lambda'_j} = f(x_{i,j}) \,\square_{q_2}\, f(x_{i',j}).
$$

Let us consider now $x_{i,j} \in H_{\lambda_j}$ and $x_{\ell,k} \in H_{\lambda_k}$, where $j \neq k$.

$$
\begin{aligned}
f(x_{i,j} \,\square_{q_1}\, x_{\ell,k}) &= f(\bar{x}_{i,j} \cup \bar{x}_{\ell,k}) = f(H_{\lambda_j}) \cup (H_{\lambda_k}) = \\
&= f(H_{\lambda_j}) \cup f(H_{\lambda_k}) = H'_{\lambda'_j} \cup H'_{\lambda'_k} = f(x_{i,j}) \,\square_{q_2}\, f(x_{\ell,k}),
\end{aligned}
$$

whence $\forall (x, y) \in H^2$, $f(x \circ y) = f(x) \circ f(y)$. ■

Now, let $(I, \leq)$ be a totally ordered set.

9. Theorem. *Let $\pi = \{A_i\}_{i \in I}$ be a partition of a set H. Let us define*

$$\delta) \quad \forall (x,y) \in A_i^2,\ x \underset{\pi}{\circ} y = A_i, \ \text{if } i<j,\ x \in A_i,\ y \in A_j,\ x \underset{\pi}{\circ} y = \bigcup_{i \le s \le j} A_s.$$

Then $< H; \underset{\pi}{\circ} >$ is a hypergroup.

From Theorem 9 we obtain

10. Theorem. *For every function $\mu : H \to I$ such that $\forall x \in H$, $\mu^{-1}\mu(x) = A_{\mu(x)}$, the hypergroupoid defined*

$$\forall (x,y) \in H^2,\ x \underset{\mu}{\circ} y = \{z \mid \mu(x) \wedge \mu(y) \le \mu(z) \le \mu(x) \vee \mu(y)\}$$

is a join space which coincides with $< H; \circ >$.

So if $I = \mu(H) \subset [0,1]$, $< H; \circ >$ is the join space associated with the fuzzy set $< H; \mu >$.

11. Definition. We call $< H; \circ >$ a *I–pr–hypergroup*, if a partition $\pi = \{A_i\}_{i \in I}$ of H exists, which satisfies (δ).

12. Corollary. *A hypergroup $< H; \circ >$ is the join space associated with a fuzzy set if and only if it is an I–pr-hypergroup with $I \subset [0,1]$.*

Now, we consider the following generalization:

Let H be a nonempty set, $(L, \vee, \wedge)$ a lattice and $\mu : H \to L$.

We define on H the following hyperoperation:

$$\forall (x,y) \in H^2,\ x * y = \{a \mid \mu(x) \wedge \mu(y) \le \mu(a) \le \mu(x) \vee \mu(y)\}.$$

This hyperoperation has been studied by I. Tofan and A.C. Volf.

13. Theorem. *If $\mu(L)$ is a distributive sublattice of $(L, \vee, \wedge)$, then $(H, *)$ is a commutative hypergroup.*

Proof. First of all, we shall verify the associativity law. We shall check that

$\forall (x, y, z) \in H^3$,
$x * (y * z) = \{a \in H \mid \mu(x) \wedge \mu(y) \wedge \mu(z) \leq \mu(a) \leq \mu(a) \vee \mu(y) \vee \mu(z)\}$.

Let $u \in x * (y * z)$. Then there is $v \in y * z$, such that $u \in x * v$. We have $\mu(y) \wedge \mu(z) \leq \mu(v) \leq \mu(y) \vee \mu(z)$ and $\mu(x) \wedge \mu(v) \leq \mu(u) \leq \leq \mu(x) \vee \mu(v)$. Hence $\mu(x) \wedge \mu(y) \wedge \mu(z) \leq \mu(u) \leq \mu(x) \vee \mu(y) \vee \mu(z)$.

Now, let us consider $a \in H$, such that

$$\mu(x) \wedge \mu(y) \wedge \mu(z) \leq \mu(a) \leq \mu(x) \vee \mu(y) \vee \mu(z).$$

There exists $b \in \mu^{-1}[(\mu(y) \wedge \mu(z)) \vee (\mu(y) \wedge \mu(a)) \vee (\mu(z) \wedge \mu(a))]$, since $\mu(L)$ is a sublattice of L. We have $\mu(y) \wedge \mu(z) \leq \mu(b) \leq \leq \mu(y) \vee \mu(z)$, whence $b \in y * z$. On the other hand, we have

$$\begin{aligned}&\mu(x) \wedge \mu(b) = \mu(x) \wedge [(\mu(y) \wedge \mu(z)) \vee (\mu(y) \wedge \mu(a)) \vee (\mu(z) \wedge \mu(a))] = \\ &= (\mu(x) \wedge \mu(y) \wedge \mu(z)) \vee (\mu(x) \wedge \mu(y) \wedge \mu(a)) \vee (\mu(x) \wedge \mu(z) \wedge \mu(a)) = \\ &= (\mu(x) \wedge \mu(y) \wedge \mu(a)) \vee (\mu(x) \wedge \mu(z) \wedge \mu(a)) \leq \mu(a) \text{ and}\end{aligned}$$

$$\begin{aligned}&\mu(a) = \mu(a) \wedge (\mu(x) \vee \mu(y) \vee \mu(z)) = \\ &= (\mu(a) \wedge \mu(x)) \vee (\mu(a) \wedge \mu(y)) \vee (\mu(a) \wedge \mu(z)) \leq \\ &\leq \mu(x) \vee (\mu(a) \wedge \mu(y)) \vee (\mu(a) \wedge \mu(z)) \leq \\ &\leq \mu(x) \vee [(\mu(y) \wedge \mu(z)) \vee (\mu(a) \wedge \mu(y)) \vee (\mu(a) \wedge \mu(z))] = \mu(x) \vee \mu(b).\end{aligned}$$

Therefore, $\mu(x) \wedge \mu(b) \leq \mu(a) \leq \mu(x) \vee \mu(b)$, that is $a \in x * b \subset x * (y * z)$.

Thus,

$$x * (y * z) = \{a \in H \mid \mu(x) \wedge \mu(y) \wedge \mu(z) \leq \mu(a) \leq \mu(x) \vee \mu(y) \vee \mu(z)\}.$$

Similarly, it follows that

$$(x * y) * z = \{a \in H \mid \mu(x) \wedge \mu(y) \wedge \mu(z) \leq \mu(a) \leq \mu(x) \vee \mu(y) \vee \mu(z)\}.$$

Hence $\forall (x, y, z) \in H^3$, $x * (y * z) = (x * y) * z$. Moreover, $\forall (x, y) \in H^2$, we have $x \in x * y$ and $x * y = y * x$, whence $H = H * x = x * H$. Therefore, $(H, *)$ is a commutative hypergroup. ■

14. Proposition. *Let $(L, \vee, \wedge)$ be a lattice with final element, denoted by 1. If $\mu(L)$ is a sublattice of L, then there is $u \in H$, such that:*

(i) $x * u = y * u \Longrightarrow x * x = y * y$;

(ii) $\forall\, (x, y) \in H^2$, $\exists (m, M) \in H^2$ *such that* $\bigcap_{t \in x*y} t * u = M * u$ *and* $\bigcap_{x \in t*u \ni y} t * u = m * u$.

Proof. (i) Let $u \in \mu^{-1}(\{1\})$. We have $x * u = \{t \mid \mu(x) \leq \mu(t)\}$. Since $y \in y * u = x * u$, it follows $\mu(x) \leq \mu(y)$. Similarly, we obtain $\mu(y) \leq \mu(x)$, whence $x * x = y * y = \{t \mid \mu(x) = \mu(t) = \mu(y)\}$.

(ii) Let $m \in \mu^{-1}(\mu(x) \wedge \mu(y))$ and $M \in \mu^{-1}(\mu(x) \vee \mu(y))$. For any $t \in H$, we have the equivalence relations

$$t \in x * u \cap y * u \Longleftrightarrow \mu(x) \leq \mu(t) \text{ and }$$
$$\mu(y) \leq \mu(t) \Longleftrightarrow \mu(x) \vee \mu(y) \leq \mu(t) \Longleftrightarrow \mu(M) \leq \mu(t).$$

Hence, $x * u \cap y * u = M * u$.

Notice that if $t \in x * y$ then $\mu(t) \leq \mu(x) \vee \mu(y) = \mu(M)$ so $M * u \subseteq t * u$. Then $M * u \subseteq \bigcap_{t \in x*y} t * u \subseteq x * u \cap y * u = M * u$ (since $x \in x * y \ni y$), whence $M * u = \bigcap_{t \in x*y} t * u$.

On the other hand, notice that

$$x \in m * u \ni y, \quad \text{so} \quad \bigcap_{x \in t*u \ni y} t * u \subseteq m * u.$$

We also have $x \in t * u \ni y \Longrightarrow \mu(t) \leq \mu(x) \wedge \mu(y) = \mu(m) \Longrightarrow m \in t * u \Longrightarrow m * u \subseteq t * u$ and so $m * u \subseteq \bigcap_{x \in t*u \ni y} t * u$, whence it follows the equality. ■

15. Remark. Notice that the hypergroup $(H, *)$ satisfies the following properties for all $(x, y) \in H^2$:

1. $x \in x * y$;
2. $x * y = y * x$;
3. $x * (x * y) = x * y = (x * x) * y = (x * x) * (y * y) = (x * y) * y$.

Now, we consider a hyperstructure $(H, *)$ which satisfies the conditions 1, 2, 3 of the above remark and (i), (ii) of the above proposition.

We shall construct a lattice L and a map $\mu : H \to L$, such that "$*$" is exactly the hyperstructure induced by μ, that is

$$\forall\,(x, y) \in H^2,\ x * y = \{t \mid \mu(x) \wedge \mu(y) \leq \mu(t) \leq \mu(x) \vee \mu(y)\}.$$

Let us define on H the following equivalence relation:

$$x \sim y \Longleftrightarrow x * x = y * y$$

and let L be the quotient set $H/\!\sim$.

Let us define the following relation:

$$\hat{x} \leq \hat{y} \Longleftrightarrow y * y \subseteq x * u.$$

We shall verify that "$\leq$" is an order on L. Indeed, first of all, notice that "$\leq$" is well–defined:

$$\begin{aligned}&\text{if } \hat{x} = \hat{x}_1,\ \hat{y} = \hat{y}_1, \text{ and } \hat{x} \leq \hat{y}, \text{ then}\\ &y_1 * y_1 = y * y \subseteq x * u = x * x * u = x_1 * x_1 * u = x_1 * u.\end{aligned}$$

Moreover, we have $y * y \subseteq x * u \Longleftrightarrow y * u \subseteq x * u$.

Indeed, if $y*y \subseteq x*u$, then $y*u = (y*y)*u \subseteq (x*u)*u = x*u$. On the other hand, if $y * u \subseteq x * u$, then $y * y \subseteq y * u \subseteq x * u$.

Now, let us verify the antisymmetry. If $\hat{x} \leq \hat{y}$ and $\hat{y} \leq \hat{x}$, then $x \in y * u$ and $y \in x * u$, whence $x * u \subseteq (y * u) * u = y * u$ and similarly, we obtain $y * u \subseteq x * u$ so $\hat{x} = \hat{y}$.

In a similar way, the reflexivity and transitivity can be verified.

Therefore $(L, \leq)$ is an order set. For any $(\hat{x}, \hat{y}) \in L^2$, we have $\hat{m} = \inf(\hat{x}, \hat{y})$ and $\widehat{M} = \sup(\hat{x}, \hat{y})$. Indeed, we have $\hat{m} \leq \hat{x}$ and

$\hat{m} \leq \hat{y}$; moreover, if $\hat{t} \leq \hat{x}$ and $\hat{t} \leq \hat{y}$, then $\{x, y\} \subset t * u$ and by (ii) it follows $m * u \subseteq t * u$, so $\hat{m} \leq \hat{t}$. Therefore, $\hat{m} = \inf(\hat{x}, \hat{y})$.

On the other hand, $\hat{x} \leq \widehat{M}$ and $\hat{y} \leq \widehat{M}$; moreover, if $\hat{x} \leq \hat{z}$ and $\hat{y} \leq \hat{z}$ then $z * u \subseteq x * u \cap y * u = M * u$, so $\widehat{M} \leq \hat{z}$. Thus $\widehat{M} = \sup(\hat{x}, \hat{y})$.

Notice that the greatest element of L is $\hat{u}$.

Let us consider the canonical projection

$$\mu : H \to H/\!\sim\; = L, \quad \mu(x) = \hat{x}.$$

In the above conditions, it follows the following:

16. Theorem. *For any* $(x, y) \in H^2$, *we have*

$$x * y = \{t \in H \mid \mu(x) \wedge \mu(y) \leq \mu(t) \leq \mu(x) \vee \mu(y)\}.$$

Proof. Let m and M be the elements which appear in (ii). We have

$$\mu(x) \wedge \mu(y) = \mu(m) \text{ and } \mu(x) \vee \mu(y) = \mu(M).$$

We shall verify the equivalence relation:

$$t \in x * y \Longleftrightarrow M * u \subseteq t * u \subseteq m * u.$$

"$\Longrightarrow$: We have $x \in m * u$, $y \in m * u$ and $x * y \subseteq m * u$, so we obtain $x * y * u \subseteq (m * u) * u = m * u$, whence $t * u \subseteq m * u$. By (ii) it follows that $M * u \subseteq t * u$.

"$\Longleftarrow$: We have $t * u \subseteq m * u = \bigcap_{x \in s * u \ni y} s * u = \bigcap_{x * y \subset s * u} s * u$, so $t \in t * u \subseteq x * y$. Therefore

$$t \in x * y \Longleftrightarrow M * u \subseteq t * u \subseteq m * u \Longleftrightarrow \mu(m) \leq \mu(t) \leq \mu(M).$$

■

§2. Direct limit and inverse limit of join spaces associated with fuzzy subsets

In the first part of this paragraph, the direct limit of a direct family of join spaces is studied; in particular, join spaces associated with fuzzy subsets are considered.

The second part of the paragraph is dedicated to the study of the inverse limit of an inverse family of hypergroups. It is again analysed the case of join spaces associated with fuzzy subsets. These results have been by obtained by V. Leoreanu.

I). In [322], G. Romeo introduced the notion of the direct limit of a direct family of semihypergroups. First, let us recall some definitions:

17. Definition. We say that (see [447]) a family $\{(H_i, \otimes_i)\}_{i\in I}$ of join spaces is a *direct family* if:

1) $(I, \leq)$ is a directed partially ordered set;

2) $\forall (i,j) \in I^2$, we have $i \neq j \Longleftrightarrow H_i \cap H_j = \emptyset$;

3) $\forall (i,j) \in I^2$, $i \leq j$, there is a homomorphism $\varphi_{ij} : H_i \longrightarrow H_j$, such that $\varphi_{jk} \circ \varphi_{ij} = \varphi_{ik}$, if $i \leq j \leq k$ and φ_{ii} is the identity mapping for all $i \in I$.

Let $H = \cup_{i\in I} H_i$. Let us define, as in [322], on H the following equivalence relation:

$x \sim y$ if and only if the following implication is satisfied:

$$(x,y) \in H_i \times H_j \Longrightarrow \text{there is } k \in I;\ k \geq i,\ k \geq j, \text{ such that } \varphi_{ik}(x) = \varphi_{jk}(y).$$

If $x_i \in H_i$ and $i \leq j$, we denote $\varphi_{ij}(x_i)$ by x_j. We also denote by $\bar{x}$ the equivalence class of x and by $\overline{H}$ the set of equivalence classes.

$\overline{H}$ is a hypergroup, respect to the following hyperoperation:

$$\bar{x} * \bar{y} = \{\bar{z} \mid \exists i \in I,\ \exists x_i \in \bar{x} \cap H_i,\ \exists y_i \in \bar{y} \cap H_i,\\ \exists z_i \in \bar{z} \cap H_i,\ \text{such that } z_i \in x_i \otimes_i y_i\}$$

(see [322]).

18. Proposition. *If $\{(H_i, \otimes_i)\}_{i \in I}$ is a direct family of semihypergroups, such that $\forall i \in I$, $\exists k \in I$, $i \leq k$, for which $(H_k, \otimes_k)$ is a join space, then $(\overline{H}, *)$ is a join space.*

Proof. We only need to check the implication (see Theorem 4, [322]):

$$\forall (\bar{x}, \bar{y}, \bar{z}, \bar{w}) \in \overline{H}^4,\ \bar{x}/\bar{y} \cap \bar{z}/\bar{w} \neq \emptyset \Longrightarrow \bar{x} * \bar{w} \cap \bar{y} * \bar{z} \neq \emptyset$$

From $\bar{x}/\bar{y} \cap \bar{z}/\bar{w} \neq \emptyset$ it follows that there is $\bar{u} \in \overline{H}$, such that $\bar{x} \in \bar{y} * \bar{u}$ and $\bar{z} \in \bar{w} * \bar{u}$; so, there is $(i, j) \in I^2$, for which $x_i \in y_i \otimes_i u_i$ and $z_j \in w_j \otimes_j u_j$. Since I is directed partially ordered, it follows that $\exists k \in I$, such that $i \leq k$ and $j \leq k$. Moreover, we can suppose that $(H_k, \otimes_k)$ is a join space, by the hypothesis. So, we have:

$$\varphi_{ik}(x_i) = x_k \in \varphi_{ik}(y_i) \otimes_k \varphi_{ik}(u_i) = y_k \otimes_k u_k$$

and similarly, $z_k \in w_k \otimes_k u_k$, whence $u_k \in x_k/y_k \cap z_k/w_k$ and it follows that $x_k \otimes_k w_k \cap y_k \otimes_k z_k \neq \emptyset$, because $(H_k, \otimes_k)$ is a join space.

Hence, $\bar{x} * \bar{w} \cap \bar{y} * \bar{z} \neq \emptyset$. ■

We shall consider now $\mathcal{F} = \{(H_i, \mu_i)\}_{i \in I}$ a family of fuzzy subsets.

In §2, it is introduced a join space associated with a fuzzy subset, in the following manner: $\forall (x_i, y_i) \in H_i^2$, we have:

$$x_i \circ_i y_i = \{z_i \in H_i \mid \min\{\mu_i(x_i), \mu_i(y_i)\} \leq \mu_i(z_i) \leq\\ \leq \max\{\mu_i(x_i), \mu_i(y_i)\}\}.$$

19. Definition. Let (H, μ) and (H', μ') be fuzzy sets. The function $f{:}H \to H'$ is called a *f.s. homomorphism* if

$$\forall (x,y) \in H^2, \text{ such that } \mu(x) < \mu(y), \text{ we have } \mu'(f(x)) < \mu'(f(y))$$
$$\text{and if } \mu(x) = \mu(y), \text{ then } \mu'(f(x)) = \mu'(f(y)).$$

20. Definition. Let $\mathcal{F} = \{(H_i, \mu_i)\}_{i \in I}$ be a family of fuzzy subsets. We say that $\mathcal{F}$ is a *direct family of fuzzy subsets* if:

1) $(I, \leq)$ is a directed partially ordered set;

2) $\forall (i,j) \in I^2$, we have $i \neq j \Longleftrightarrow H_i \cap H_j \neq \emptyset$;

3) $\forall (i,j) \in I^2$, $i \leq j$, there is a f.s. homomorphism $\varphi_{ij} : H_i \to H_j$, such that: if $i \leq j \leq k$, we have $\varphi_{jk} \circ \varphi_{ij} = \varphi_{ik}$ and φ_{ii} is the identity mapping for all $i \in I$.

Let $\{(H_i, \mu_i)\}_{i \in I}$ be a direct family of fuzzy subsets and let us consider now $\bar{\mu}{:}\overline{H} \to [0,1]$, such that the following condition holds: $\forall (\bar{x}, \bar{y}) \in \overline{H}^2$, $\bar{\mu}(\bar{x}) < \bar{\mu}(\bar{y})$ if and only if $\exists i \in I$, $\exists x_i \in \bar{x} \cap H_i$, $\exists y_i \in \bar{y} \cap H_i$, such that $\mu_i(x_i) < \mu_i(y_i)$.

21. Proposition. *The following equivalence relation holds:*

$$[\exists i \in I,\ \exists x_i \in \bar{x} \cap H_i,\ \exists y_i \in \bar{y} \cap H_i,\ \textit{such that } \mu_i(x_i) < \mu_i(y_i)]$$
$$\Longleftrightarrow [\forall j \in I,\ \forall x_j \in \bar{x} \cap H_j,\ \forall y_j \in \bar{y} \cap H_j : \mu_j(x_j) < \mu_j(y_j)].$$

Proof. "$\Longrightarrow$" First, we show that $\forall \{x_i, x_i'\} \subset \bar{x} \cap H_i$, we have $\mu_i(x_i) = \mu_i(x_i')$.

Indeed, since $x_i \sim x_i'$ it follows that there is $k \in I$, $i \leq k$, such that $\varphi_{ik}(x_i) = \varphi_{ik}(x_i')$, that is $x_k = x_k'$.

Suppose that $\mu_i(x_i) < \mu_i(x_i')$. Then $\mu_k(\varphi_{ik}(x_i)) < \mu_k(\varphi_{ik}(x_i'))$, that is $\mu_k(x_k) < \mu_k(x_k')$ contradiction with $x_k = x_k'$.

We shall check now that $\forall j \in I$, we have $\mu_j(x_j) < \mu_j(y_j)$.

For $j \in I$, $i \leq j$, we have $\mu_j(\varphi_{ij}(x_i)) < \mu_j(\varphi_{ij}(y_i))$, that is $\mu_j(x_j) < \mu_j(y_j)$.

Let us suppose that there is $k \in I$, such that $\mu_k(x_k) > \mu_k(y_k)$. Since $(I, \leq)$ is directed partially ordered, it follows that there is $t \in I$, $k \leq t$, $i \leq t$. Since $\mu_k(x_k) > \mu_k(y_k)$ it follows $\mu_t(x_t) > \mu_t(y_t)$ and since $\mu_i(x_i) < \mu_i(y_i)$ it follows $\mu_t(x_t) < \mu_t(y_t)$, contradiction!

Therefore, for any $j \in I$, we have $\mu_j(x_j) < \mu_j(y_j)$. ∎

22. Corollary. *We have*

$$\begin{aligned}
&\bar{\mu}(\bar{x}) = \bar{\mu}(\bar{y}) \\
&\Longleftrightarrow [\exists i \in I,\ \exists x_i \in \bar{x} \cap H_i,\ \exists y_i \in \bar{y} \cap H_i,\ \textit{such that } \mu_i(x_i) = \mu_i(y_i)] \\
&\Longleftrightarrow [\forall j \in I,\ \forall x_i \in \bar{x} \cap H_i,\ \forall y_i \in \bar{y} \cap H_i,\ \mu_j(x_j) = \mu_j(y_j)].
\end{aligned}$$

23. Remark. Let (H, μ) and (H', μ') be fuzzy subsets. If $f: H \to H'$ is a f.s. homomorphism, then the following implication holds:

$$\forall (x, y) \in H^2,\ f(x) = f(y) \Longrightarrow \mu(x) = \mu(y).$$

Proof. Indeed, since $f(x) = f(y)$ it follows $\mu'(f(x)) = \mu'(f(y))$. If we suppose now that $\mu(x) < \mu(y)$, then $\mu'(f(x)) < \mu'(f(y))$, contradiction! Therefore, $\mu(x) = \mu(y)$. ∎

So, we can define the function:

$$g : \operatorname{Im} f \longrightarrow [0, 1],\quad g(f(x)) = \mu(x).$$

We can choose $\bar{\mu}$ in many manners.

24. Examples.

1. Let $i_0 \in I$ and let us define $\mu'(\bar{x}) = \mu_{i_0}(x_{i_0})$, $\forall \bar{x} \in \overline{H}$. Then we can consider $\bar{\mu}(\bar{x}) = \mu'(\bar{x})$, $\forall \bar{x} \in \overline{H}$.

2. Let F be a finite subset of I and $|F|$ the cardinal of F

$$\mu''(\bar{x}) = \sum_{i \in F} \mu_i(x_i) / |F|.$$

Remember that $\forall\{x_i, x_i'\} \subset \bar{x} \cap H_i$, we have $\mu_i(x_i) = \mu_i(x_i')$.

If $\mu''(\bar{x})<\mu''(\bar{y})$, that is $\sum\limits_{i\in F} \mu_i(x_i)/|F|<\sum\limits_{i\in F} \mu_i(y_i)/|F|$, then $\exists i_0 \in F$, such that $\mu_{i_0}(x_{i_0}) < \mu_{i_0}(y_{i_0})$. So, we can consider $\bar{\mu}(\bar{x}) = \mu''(\bar{x})$, $\forall \bar{x} \in \overline{H}$.

25. Proposition. *Let $\{(H_i, \mu_i)\}_{i\in I}$ be a direct family of fuzzy subsets and let $\{(H_i, \circ_i)\}_{i\in I}$ be the family of join spaces associated with the previous fuzzy subsets. Then $\{(H_i, \circ_i)\}_{i\in I}$ is a direct family of join spaces.*

Proof. We only need to prove that for $\forall(i,j) \in I^2$, $i \leq j$, $\varphi_{ij} : H_i \longrightarrow H_j$ is a homomorphism of join spaces, that is $\forall(x_i, y_i) \in H_i^2$, $\forall z_i \in x_i \circ_i y_i$, we have $\varphi_{ij}(z_i) \in \varphi_{ij}(x_i) \circ_j \varphi_{ij}(y_i)$, that is $z_j \in x_j \circ_j y_j$.

Indeed, by $z_i \in x_i \circ_i y_i$ it follows

$$\min\{\mu_i(x_i), \mu_i(y_i)\} \leq \mu_i(z_i) \leq \max\{\mu_i(x_i), \mu_i(y_i)\}.$$

Suppose $\mu_i(x_i) \leq \mu_i(y_i)$; we have $\mu_i(x_i) \leq \mu_i(z_i) \leq \mu_i(y_i)$. Since for $i \leq j$, φ_{ij} is a f.s. homomorphism, we obtain

$$\mu_j(\varphi_{ij}(x_i)) \leq \mu_j(\varphi_{ij}(z_i)) \leq \mu_j(\varphi_{ij}(y_i)),$$

that is $\mu_j(x_j) \leq \mu_j(z_j) \leq \mu_j(y_j)$, whence $z_j \in x_j \circ_j y_j$. ■

26. Theorem. *Let $\{(H_i, \mu_i)\}_{i\in I}$ be a direct family of fuzzy subsets and $\{(H_i, \circ_i)\}_{i\in I}$ the direct family of join spaces associated with the previous fuzzy subsets. Let $(\overline{H}, *)$ be the direct limit of the direct family of join spaces.*

Then $(\overline{H}, *)$ is also a join space, associated with a fuzzy subset.

Proof. Let $(\overline{H}, \circ)$ be the join space associated with a fuzzy subset $\bar{\mu}$, which satisfies the following condition:

$$\bar{\mu}(\bar{x}) < \bar{\mu}(\bar{y}) \Longleftrightarrow [\exists i \in I,\ \exists x_i \in \bar{x}\cap H_i,\ \exists y_i \in \bar{y}\cap H_i : \mu_i(x_i) < \mu_i(y_i)].$$

Then $\bar{x}\circ\bar{y} = \{\bar{z} \in \overline{H} \mid \min\{\bar{\mu}(\bar{x}), \bar{\mu}(\bar{y})\}\} \leq \bar{\mu}(\bar{z}) \leq \max\{\bar{\mu}(\bar{x}), \bar{\mu}(\bar{y})\}$. Suppose that $\bar{\mu}(\bar{x}) \leq \bar{\mu}(\bar{y})$. Then $\bar{x} \circ \bar{y} = \{\bar{z} \mid \bar{\mu}(\bar{x}) \leq \bar{\mu}(\bar{x}) \leq$

$\leq \bar{\mu}(\bar{y})\} = \{\bar{z} \mid \exists i \in I,\ \exists x_i \in \bar{x} \cap H_i,\ \exists z_i \in \bar{z} \cap H_i : \mu_i(x_i) \leq \mu_i(z_i)$ and $\exists j \in I,\ \exists z_j \in \bar{z} \in H_j,\ \exists y_j \in \bar{y} \cap H_j : \mu_j(z_j) \leq \mu_j(y_j)\}$.

Since I is a directed partially ordered set, it follows that there is $k \in I,\ i \leq k,\ j \leq k$. We have $\mu_k(\varphi_{ik}(x_i)) \leq \mu_k(\varphi_{ik}(z_i))$ that is $\mu_k(x_k) \leq \mu_k(z_k)$ and similarly, $\mu_k(z_k) \leq \mu_k(y_k)$. Therefore, $\bar{x} \circ \bar{y} =$ $= \{\bar{z} \mid \exists k \in I,\ \exists x_k \in \bar{x} \cap H_k, \exists z_k \in \bar{z} \cap H_k, \exists y_k \in \bar{y} \cap H_k : \mu_k(x_k) \leq$ $\leq \mu_k(z_k) \leq \mu_k(y_k)\} = \{\bar{z} \mid \exists k \in I : z_k \in x_k \circ_k y_k\} = \bar{x} * \bar{y}$. Then the join spaces $(\overline{H}, \circ)$ and $(\overline{H}, *)$ coincide. ■

II) First, we shall introduce the notion of *inverse limit* of hypergroups and then we shall study it for an inverse family of join spaces associated with fuzzy subsets.

27. Definition. We say that a family of hypergroups $\{(H_i, \otimes_i)\}_{i\in I}$ is an *inverse family* if:

1. $(I, \leq)$ is a directed partially ordered set;
2. $\forall (i,j) \in I^2$, we have $H_i \cap H_j = \emptyset \Longleftrightarrow i \neq j$;
3. $\forall (i,j) \in I^2,\ i \geq j$, there is a homomorphism of hypergroups $\psi_{ij} : H_i \longrightarrow H_j$, such that: if $i \geq j \geq k,\ \psi_{jk} \circ \psi_{ij} = \psi_{ik}$ and $\forall i \in I,\ \psi_{ii}$ is the identity mapping.

Let us consider now $\left(\prod\limits_{i\in I} H_i, \otimes\right)$ the direct product and let

$$\widetilde{H} = \left\{p \in \prod_{i\in I} H_i \mid \psi_{ij}(p_i) = p_j,\ \forall i \geq j\right\},$$

where $p = (p_i)_{i\in I}$. If $\widetilde{H} \neq \emptyset$, we define on $\widetilde{H}$ the hyperoperation:

$$\widetilde{x} \circ \widetilde{y} = \{\widetilde{z} \in \widetilde{H} \mid \widetilde{z} \in \widetilde{x} \otimes \widetilde{y}\} = \widetilde{x} \otimes \widetilde{y} \cap \widetilde{H}.$$

The assumption $\widetilde{H} \neq \emptyset$ is really necessary. In [447], G. Grätzer presents an example of an inverse family of nonvoid sets, whose inverse limit is void. The following theorem shows that this cannot happen if all the sets are finite and nonvoid.

28. Theorem. [[447], Th.1, p.132] *The inverse limit of a family of nonvoid finite sets is always nonvoid.*

Another situation for which the inverse limit of a family of nonvoid sets $\{H_i\}_{i\in I}$ is nonvoid is the following one:

If I has a maximum element, then $\widetilde{H} \neq \emptyset$.

Indeed, if $s = \max I$, then $\forall p = (p_i)_{i\in I}$, $\exists \widetilde{p} \in \widetilde{H}$. $\widetilde{p}_i = \psi_{si}(p_s)$, because $\forall (i,j) \in I^2$, $i \geq j$, we have: $\psi_{ij}\ (\psi_{si}(p_s)) = \psi_{sj}(p_s)$, that is $\psi_{ij}(\widetilde{p}_i) = \widetilde{p}_j$.

In the following, we shall consider $(I, \leq)$ a partially ordered set, with a maximum element.

29. Theorem. *Let I be a partially ordered set, with a maximum element s. If $\{(H_i, \otimes_i)\}_{i\in I}$ is an inverse family of hypergroups, then $(\widetilde{H}, \circ)$ is a hypergroup. Moreover, if $\forall i \in I$, $(H_i, \otimes_i)$ is a join space, then $(\widetilde{H}, \circ)$ is also a join space.*

Proof. Let us verify first that $(\widetilde{H}, \circ)$ is a hypergroup.

The associativity. We shall check that $\forall(\widetilde{x}, \widetilde{y}, \widetilde{z}) \in \widetilde{H}^3$, $(\widetilde{x}\circ\widetilde{y})\circ\widetilde{z}=(\widetilde{x}\otimes\widetilde{y})\otimes\widetilde{z}\cap\widetilde{H}$. We have to verify only the inclusion "$\supset$". Let $\widetilde{t} \in (\widetilde{x} \otimes \widetilde{y}) \otimes \widetilde{z} \cap \widetilde{H}$. There is $u \in \widetilde{x} \otimes \widetilde{y}$, such that $\widetilde{t} \in u \otimes \widetilde{z}$, so $\forall i \in I$, $\widetilde{t}_i \in u_i \otimes_i \widetilde{z}_i$, particularly $\widetilde{t}_s \in u_s \otimes_s \widetilde{z}_s$. For all $j \in I$, we have $\psi_{sj}(\widetilde{t}_s) \in \psi_{sj}(u_s) \otimes_j \psi_{sj}(\widetilde{z}_s)$, that means $\widetilde{t}_j \in \psi_{sj}(u_s) \otimes_j \widetilde{z}_j$. Let $\widetilde{u} \in \widetilde{H}$, defined in this manner: $\widetilde{u}_j = \psi_{sj}(u_s)$, $\forall j \in I$. We have: $\widetilde{t}_j \in \widetilde{u}_j \otimes_j \widetilde{z}_j$, $\forall j \in I$, whence $\widetilde{t} \in \widetilde{u} \circ \widetilde{z}$.

Since $u \in \widetilde{x} \otimes \widetilde{y}$, it follows $u_j \in \widetilde{x}_j \otimes_j \widetilde{y}_j$, $\forall j \in I$; so, $u_s \in \widetilde{x}_s \otimes_s \widetilde{y}_s$, whence $\psi_{sj}(u_s) \in \psi_{sj}(\widetilde{x}_s) \otimes_j \psi_{sj}(\widetilde{y}_s)$, $\forall j \in I$, that is $\widetilde{u}_j \in \widetilde{x}_j \otimes_j \widetilde{y}_j$, $\forall j \in I$, hence $\widetilde{u} \in \widetilde{x} \otimes \widetilde{y} \cap \widetilde{H} = \widetilde{x} \circ \widetilde{y}$. Then, $\widetilde{t} \in \widetilde{u} \circ \widetilde{z} \subset (\widetilde{x} \circ \widetilde{y}) \circ \widetilde{z}$.

Similarly, we prove that $\widetilde{x} \circ (\widetilde{y} \circ \widetilde{z}) = \widetilde{x} \otimes (\widetilde{y} \otimes \widetilde{z}) \cap \widetilde{H}$. Therefore $(\widetilde{x} \circ \widetilde{y}) \circ \widetilde{z} = (\widetilde{x} \otimes \widetilde{y}) \otimes \widetilde{z} \cap \widetilde{H} = \widetilde{x} \otimes (\widetilde{y} \otimes \widetilde{z}) \cap \widetilde{H} = \widetilde{x} \circ (\widetilde{y} \circ \widetilde{z})$, $\forall(\widetilde{x}, \widetilde{y}, \widetilde{z}) \in \widetilde{H}^3$.

The reproducibility. For any $(\widetilde{x}, \widetilde{y}) \in \widetilde{H}^2$, there is $z \in \prod_{i \in I} H_i$ such that $\widetilde{x} \in \widetilde{y} \otimes z$, whence $\forall i \in I$, we have $\widetilde{x}_i \in \widetilde{y}_i \otimes_i z_i$. From $\widetilde{x}_s \in \widetilde{y}_s \otimes_s z_s$, it follows $\psi_{sj}(\widetilde{x}_s) \in \psi_{sj}(\widetilde{y}_s) \otimes_j \psi_{sj}(z_s)$, $\forall j \in I$, that is $\widetilde{x}_j \in \widetilde{y}_j \otimes_j \psi_{sj}(z_s)$. Let us consider $\widetilde{z} \in \widetilde{H}$, such that $\widetilde{z}_j = \psi_{sj}(z_s)$, $\forall j \in I$. So, $\forall i \in I$, we have $\widetilde{x}_j \in \widetilde{y}_j \otimes_j \widetilde{z}_j$, whence $\widetilde{x} \in \widetilde{y} \circ \widetilde{z}$. Therefore $\widetilde{y} \circ \widetilde{H} = \widetilde{H}$ and similarly, we have $\widetilde{H} \circ \widetilde{y} = \widetilde{H}$.

Therefore, $(\widetilde{H}, \circ)$ is a hypergroup.

Let us suppose now that $\forall i \in I$, $(H_i, \otimes_i)$ is a join space. We shall prove the following implication:

$$\forall (\widetilde{x}, \widetilde{y}, \widetilde{z}, \widetilde{t}) \in \widetilde{H}^4,\ \widetilde{x} \underset{\circ}{/} \widetilde{y} \cap \widetilde{t} \underset{\circ}{/} \widetilde{z} \neq \emptyset \Longrightarrow \widetilde{x} \circ \widetilde{z} \cap \widetilde{y} \circ \widetilde{t} \neq \emptyset.$$

From $\widetilde{x} \underset{\circ}{/} \widetilde{y} \cap \widetilde{t} \underset{\circ}{/} \widetilde{z} \neq \emptyset$, it follows that $\exists \widetilde{u} \in \widetilde{H} : \widetilde{x} \in \widetilde{y} \circ \widetilde{u} \in \widetilde{y} \otimes \widetilde{u}$ and $\widetilde{t} \in \widetilde{z} \circ \widetilde{u} \subset \widetilde{z} \otimes \widetilde{u}$. Then, $\widetilde{x} \underset{\otimes}{/} \widetilde{y} \cap \widetilde{t} \underset{\otimes}{/} \widetilde{z} \neq \emptyset$ in $\left(\prod_{i \in I} H_i, \otimes\right)$, which is a join space, so $\widetilde{x} \otimes \widetilde{z} \cap \widetilde{y} \otimes \widetilde{t} \neq \emptyset$. Hence $\exists v \in \widetilde{x} \otimes \widetilde{z}$ and $v \in \widetilde{y} \otimes \widetilde{t}$, that means $\forall i \in I$, $v_i \in \widetilde{x}_i \otimes_i \widetilde{z}_i$ and $v_i \in \widetilde{y}_i \otimes_i \widetilde{t}_i$. From $v_s \in \widetilde{x}_s \otimes_s \widetilde{z}_s$, it follows that $\forall j \in I$, $\psi_{sj}(v_s) \in \psi_{sj}(\widetilde{x}_s) \otimes_j \psi_{sj}(\widetilde{z}_s) = \widetilde{x}_j \otimes_j \widetilde{z}_j$.

Let us consider $\widetilde{v} \in \widetilde{H}$, such that $\widetilde{v}_j = \psi_{sj}(v_s)$, $\forall j \in I$. We have $\widetilde{v}_j \in \widetilde{x}_j \otimes_j \widetilde{z}_j$, $\forall j \in I$, that means $\widetilde{v} \in \widetilde{x} \circ \widetilde{z}$. Similarly, from $v_s \in \widetilde{y}_s \otimes_s \widetilde{t}_s$, it follows $\widetilde{v} \in \widetilde{y} \circ \widetilde{t}$.

Therefore, $\widetilde{x} \circ \widetilde{z} \cap \widetilde{y} \circ \widetilde{t} \neq \emptyset$, so $(\widetilde{H}, \circ)$ is a join space. ∎

30. Definition. $(\widetilde{H}, \circ)$ is called the *inverse limit of the inverse family* $\{(H_i, \otimes_i)\}_{i \in I}$.

Finally, we shall analyse the inverse limit of an inverse family of join spaces associated with fuzzy subsets.

31. Definition. Let $\mathcal{F} = \{(H_i, \mu_i)\}_{i \in I}$ be a family of fuzzy subsets. We say that $\mathcal{F}$ is an *inverse family* of fuzzy subsets if:

1. $(I, \leq)$ is a directed partially ordered set;
2. $\forall (i, j) \in I^2$, we have $i \neq j \Longleftrightarrow H_i \cap H_j = \emptyset$;

3. $\forall(i,j)\in I^2$, $i \geq j$, there is a f.s. homomorphism $\psi_{ij} : H_i \rightarrow H_j$, such that: if $i \geq j \geq k$, we have $\psi_{jk} \circ \psi_{ij} = \psi_{ik}$ and φ_{ii} is the identity mapping for all $i \in I$.

32. Proposition. *Let $\mathcal{F} = \{(H_i, \mu_i)\}_{i\in I}$ be an inverse family of fuzzy subsets. Then the family $\{(H_i, \circ_i)\}_{i\in I}$ of join spaces, associated with the previous fuzzy subsets, is an inverse family.*

Proof. We shall check that $\forall(i,j) \in I^2$, $i \geq j$, ψ_{ij} is a homomorphism of join spaces, that means: if $z_i \in x_i \circ_i y_i$, then $\psi_{ij}(z_i) \in \in \psi_{ij}(x_j) \circ_j \psi_{ij}(y_i)$.

Suppose $\mu_i(x_i) \leq \mu_i(y_i)$. From $z_i \in x_i \circ_i y_i$, it follows $\mu_i(x_i) \leq \leq \mu_i(z_i) \leq \mu_i(y_i)$ and since ψ_{ij} is a f.s. homomorphism, we obtain

$$\mu_j(\psi_{ij}(x_i)) \leq \mu_j(\psi_{ij}(z_i)) \leq \mu_j(\psi_{ij}(y_i)),$$

that is $\psi_{ij}(z_i) \in \psi_{ij}(x_i) \circ_j \psi_{ij}(y_i)$. ∎

33. Proposition. *Let $\{(H_i, \mu_i)\}_{i\in I}$ be an inverse family of fuzzy subsets and $\{(H_i, \circ_i)\}_{i\in I}$ the associated inverse family of join spaces. If $\widetilde{H} \neq \emptyset$ and $\exists i \in I : \mu_i(\widetilde{x}_i) < \mu_i(\widetilde{y}_i)$, where $\widetilde{x} = (\widetilde{x}_i)_{i\in I} \in \widetilde{H}$ and $\widetilde{y} = (\widetilde{y}_i)_{i\in I} \in \widetilde{H}$, then $\forall j \in I$, we have $\mu_j(\widetilde{x}_j) < \mu_j(\widetilde{y}_j)$.*

Proof. Let us consider $j \in I$, $j \leq i$. Since ψ_{ij} is a f.s. homomorphism, from $\mu_i(\widetilde{x}_i) < \mu_i(\widetilde{y}_i)$ it results $\mu_j(\psi_{ij}(\widetilde{x}_i)) < \mu_j(\psi_{ij}(\widetilde{y}_i))$, that is $\mu_j(\widetilde{x}_j) < \mu_j(\widetilde{y}_j)$.

Let us suppose that $\exists p \in I$, such that $\mu_p(\widetilde{x}_p) \geq \mu_p(\widetilde{y}_p)$. Since I is a directed partially ordered set, it follows that $\exists t \in I$, $t \geq i$, $t \geq p$.

If $\mu_t(\widetilde{x}_t) < \mu_t(\widetilde{y}_t)$ it follows $\mu_p(\psi_{tp}(\widetilde{x}_t)) < \mu_p(\psi_{tp}(\widetilde{y}_t))$ that is $\mu_p(\widetilde{x}_p) < \mu_p(\widetilde{y}_p)$, contradiction with the made assumption.

If $\mu_t(\widetilde{x}_t) \geq \mu_t(\widetilde{y}_t)$ it follows $\mu_i(\psi_{ti}(\widetilde{x}_t)) \geq \mu_i(\psi_{ti}(\widetilde{y}_t))$ that is $\mu_i(\widetilde{x}_i) \geq \mu_i(\widetilde{y}_i)$, contradiction with the hypothesis.

Therefore, $\forall j \in I$, we have $\mu_j(\widetilde{x}_j) < \mu_j(\widetilde{y}_j)$. ∎

34. Corollary. *In the hypothesis of the previous proposition, we have that if $\exists i \in I$, such that $\mu_i(\widetilde{x}_i) = \mu_i(\widetilde{y}_i)$, then $\forall j \in I$, $\mu_j(\widetilde{x}_j) = = \mu_j(\widetilde{y}_j)$.*

35. Theorem. *Let $\{(H_i, \mu_i)\}_{i\in I}$ be an inverse family of fuzzy subsets, $\{(H_i, \circ_i)\}_{i\in I}$ the associated inverse family of join spaces and let suppose $\widetilde{H} \neq \emptyset$. Then the inverse limit $(\widetilde{H}, \circ)$ is also a join space associated with a fuzzy subset.*

Proof. Let $(\widetilde{H}, \bullet)$ be the join space associated with a fuzzy subset $\widetilde{\mu}$, which satisfies the following condition: if $(\widetilde{x}, \widetilde{y}) \in \widetilde{H}^2$, then

$$\widetilde{\mu}(\widetilde{x}) < \widetilde{\mu}(\widetilde{y}) \iff \exists i \in I : \mu_i(\widetilde{x}_i) < \mu_i(\widetilde{y}_i).$$

We have

$$\widetilde{x} \bullet \widetilde{y} = \{\widetilde{z} \in \widetilde{H} \mid \min\{\widetilde{\mu}(\widetilde{x}), \widetilde{\mu}(\widetilde{y})\}\} \leq \widetilde{\mu}(\widetilde{z}) \leq \max\{\widetilde{\mu}(\widetilde{x}), \widetilde{\mu}(\widetilde{y})\}.$$

Suppose $\widetilde{\mu}(\widetilde{x}) \leq \widetilde{y}(\widetilde{y})$. Then $\widetilde{x} \bullet \widetilde{y} = \{\widetilde{z} \mid \widetilde{\mu}(\widetilde{x}) \leq \widetilde{\mu}(\widetilde{z}) \leq \widetilde{\mu}(\widetilde{y})\}$.

From $\widetilde{\mu}(\widetilde{x}) \leq \widetilde{\mu}(\widetilde{z})$ and the previous proposition, it follows that $\mu_i(\widetilde{x}_i) \leq \mu_i(\widetilde{z}_i)$, $\forall i \in I$. Therefore, $\widetilde{x} \bullet \widetilde{y} = \{\widetilde{z} \mid \mu_i(\widetilde{x}_i) \leq \mu_i(\widetilde{z}_i) \leq \leq \mu_i(\widetilde{y}_i),\ \forall i \in I\} = \{\widetilde{z} \mid \widetilde{z}_i \in \widetilde{x}_i \circ_i \widetilde{y}_i,\ \forall i \in I\} = \widetilde{x} \circ \widetilde{y}$.

Then the join spaces $(\widetilde{H}, \bullet)$ and $(\widetilde{H}, \circ)$ coincide. ■

36. Remark. We can choose $\widetilde{\mu}$ is many manners. For instance,

1. $\forall \widetilde{x} \in \widetilde{H},\ \widetilde{\mu}(\widetilde{x}) = \mu_{i_0}(\widetilde{x}_{i_0})$ for some $i_0 \in I$.
2. $\forall \widetilde{x} \in \widetilde{H},\ \widetilde{\mu}(\widetilde{x}) = \sum\limits_{i\in F} \mu_i(\widetilde{x}_i)/|F|$, where F is a finite subset of I, and $|F|$ is the cardinal of F.

Indeed, we have $\widetilde{\mu}(\widetilde{x}) < \widetilde{\mu}(\widetilde{y}) \iff \exists i \in I$, such that $\mu_i(\widetilde{x}) < \mu_i(\widetilde{y}_i)$.

§3. Rough sets, fuzzy subsets and join spaces

Let H be a set and R be an equivalence relation on H. Let A be a subset of H.

The *main question* addressed by rough sets (Pawlak, 1982) is:

How to represent A by means of H/R?

Denote by $R(x)$ the equivalence class of $x \in H$.

37. Definition. A *rough set* is a pair of subsets $(\overline{R}(A), \underline{R}(A))$ of H, which approximate as close as possible A from outside and inside, respectively:

$$\overline{R}(A) = \bigcup_{R(x)\cap A \neq \emptyset} R(x);$$

$$\underline{R}(A) = \bigcup_{R(x)\subseteq A} R(x).$$

Rough sets have been utilized as an instrument to study in deep the *theory of knowledge* (Artificial Intelligence) by Pawlak (a Polish mathematician) and many others.

One can remark (Biswas, 1999) that Rough Sets can be considered a special case of Fuzzy subsets, letting correspond to $(\overline{R}(A), \underline{R}(A))$ the membership function μ_A, defined

$$\mu_A(x) = \frac{|R(x) \cap A|}{|R(x)|}.$$

Now, let us see how join spaces can be associated with rough sets.

The results presented in this paragraph belong to P. Corsini.

38. Theorem. *The partial hyperoperation*

$$\forall (x, y) \in H^2, \; x \circ y = \overline{R}(\{x, y\}) - \underline{R}(\{x, y\})$$

is defined everywhere if and only if

$$(\varepsilon) \qquad \forall x \in H, \; |R(x)| \geq 3$$

Proof. Let us prove now the implication $\Longleftarrow$.

Set $\forall x \; |R(x)| \geq 3$. Then we have

$$\underline{R}(\{x, y\}) = \bigcup_{R(z)\subset\{x,y\}} R(z) = \emptyset$$

whence $x \circ y = R(x) \cup R(y) \neq \emptyset$.

Let us prove the implication $\Longrightarrow$.

Let us suppose x exists such that $R(x) = \{x, x'\}$ and $x \neq x'$. Then

$$x \circ x' = \bigcup_{R(z) \cap \{x,x'\} \neq \emptyset} R(z) - \bigcup_{R(z) \subset \{x,x'\}} R(z) = R(x) - R(x) = \emptyset.$$

Let us suppose x exists such that $R(x) = \{x\}$.

By the same way one finds $x \circ x = R(x) - R(x) = \emptyset$. Therefore, $\forall x,\ |R(x)| \geq 3$.

Then $\langle H; \circ \rangle$ is a hypergroupoid if and only if $\forall x,\ |R(x)| \geq 3$.■

39. Theorem. *$< H; \circ >$ is a join space if and only if*

$$\forall x \in H,\ |R(x)| \geq 3.$$

Proof. Set $\forall (x, y),\ x \otimes y = \overline{R}(\{x, y\}) = R(x) \cup R(y)$. By Theorem 38 it is sufficient to prove that if $< H; \circ >$ is a hypergroupoid, then it is a join space.

Let us remark that the hypothesis $|R(x)| \geq 3$ implies: $< \circ > = < \otimes >$, so $x \circ y = R(x) \cup R(y)$. It follows that $< H; \circ >$ is a commutative semi–hypergroup.

Moreover, since every x is an identity, it follows that $< H; \circ >$ is a hypergroup.

It remains to prove that the implication $a/b \cap c/d \neq \emptyset \Longrightarrow a \circ d \cap b \circ c \neq \emptyset$ is satisfied.

Set (I): $a/b \ni x \in c/d$, that is $a \in b \circ x$, $c \in d \circ x$, whence

$$a \in R(b) \cup R(x),\ c \in R(d) \cup R(x)$$

moreover $a \circ d = R(a) \cup R(d),\ b \circ c = R(b) \cup R(c)$.

We have $a \in a \circ d$, so, if $a \in R(b) \subset b \circ c$, it follows $a \in a \circ d \cap b \circ c$.

By the same way, $c \in R(d)$ implies $c \in a \circ d \cap b \circ c$.

Let us suppose now $a \notin R(b)$ and $c \notin R(d)$. Then it follows $a \in R(x)$ whence $x \in R(a) \subset a \circ d$, $c \in R(x)$, whence $x \in R(c) \subset b \circ c$. Therefore (I) implies $a \circ d \cap b \circ c \neq \emptyset$, so $< H; \otimes >$ is a join space. ■

Let us suppose now $|H| < \aleph_0$.

40. Theorem. *Let* $< \mathcal{P}^*(H); \mu >$ *be a fuzzy subset. There is a knowledge* $< H; R >$ *such that*

$$\forall X \in \mathcal{P}^*(H),\ \mu(X) = \mu_R(X) = \frac{|\underline{R}(X)|}{|\overline{R}(X)|}$$

if and only if the following condition is satisfied

(D) *An integer* $m > 0$, *and a partition of* H, $\{A_i\}_{i \in I(m)}$ *exist so that, for all non empty subsets* S *and* J *of* $I(m)$ *such that* $S \cap J = \emptyset$, *for every family* $\{A'_s\}_{s\in S}$ *of subsets,* $A'_s \subsetneq A_s$, *setting* $\forall i \in I(m)$, $a_i = |A_i|$, *we have:*

1) $\mu\left(\bigcup_{s\in S} A'_s\right) = 0.$

2) $\mu\left(\bigcup_{j\in J} A_j \cup \bigcup_{s\in S} A'_s\right) = \dfrac{\sum_{j\in J} a_j}{\sum_{j\in J} a_j + \sum_{s\in S} a_s}.$

Proof. (D) is sufficient.

Let R be the equivalence relation on H such that $H/R = \{A_i \mid i \in I(m)\}$.

$\forall X \in \mathcal{P}^*(H)$, we can represent X as the union

$$X = \bigcup_{j\in J} A_j \cup \bigcup_{s\in S} A'_s$$

where

$$J \cup S \subset I(m),\ J \cap S = \emptyset$$
$$J = \{j \in I(m) \mid A_j \subset X\},$$
$$S = \{s \in I(m) \mid A_s \not\subset X,\ A'_s = X \cap A_s \neq \emptyset\}$$

So, we have:

$$\underline{R}(X) = \bigcup_{j\in J} A_j,$$
$$\overline{R}(X) = \bigcup_{A_t \cap X \neq \emptyset} A_t = \bigcup_{j\in J} A_j \cup \bigcup_{s\in S} A_s$$

Therefore

$$\mu(X) = \frac{\sum\limits_{j\in J} a_j}{\sum\limits_{j\in J} a_j + \sum\limits_{s\in S} a_s} = \frac{|\underline{R}(X)|}{|\overline{R}(X)|} = \mu_R(X)$$

The condition (D) is necessary.
Let $\{A_i\}_{i\in I(m)}$ be the set of equivalence classes of R.
Then $\forall X \in \mathcal{P}^*(H)$, if we set

$$J = \{j \in I(m) \mid A_j \subset X\}$$
$$S = \{s \in I(m) \mid A_s \not\subset X\},\ \forall s \in S,\ A'_s = A_s \cap X,$$

we have

$$\underline{R}(X) = \bigcup_{j\in J} A_j$$
$$\overline{R}(X) = \bigcup_{A_i\cap X\neq\emptyset} A_i = \bigcup_{j\in J} A_j \cup \bigcup_{\substack{A_s\not\subset X\\ A_s\cap X\neq\emptyset}} A_s$$

So we obtain:

$$\mu\left(\bigcup_{s\in S} A'_s\right) = 0,\quad \mu(X) = \frac{\sum\limits_{j\in J} a_j}{\sum\limits_{j\in J} a_j + \sum\limits_{s\in S} a_s}$$

■

§4. Direct limits and inverse limits of join spaces associated with rough sets

The results of this paragraph have been obtained by V. Leoreanu.

I) First of all we establish necessary and sufficient or only sufficient conditions for direct limits and products of models associated with rough sets to be join spaces.

Let us recall some definitions. A *model* is a pair $< H, \rho >$, where H is a nonempty set and ρ is a binary relation on H.

Let us recall what a *rough set* is.

Let H be a nonempty set and R an equivalence relation on H. For every $X \subset H$, $X \neq \emptyset$, set

$$\underline{R}(X) = \bigcup_{R(y) \subset X} R(y) \text{ and } \overline{R}(X) = \bigcup_{R(z) \cap X \neq \emptyset} R(z) = \bigcup_{w \in X} R(w).$$

The pair $(\overline{R}(X), \underline{R}(X))$ is a *rough set*. We have seen that a join space is associated with a rough set in the following manner:

41. Theorem. *Let R be an equivalence relation defined on a nonempty set H and $< \circ >$ the partial hyperoperation defined*

$$(*) \qquad x \circ y = \overline{R}(\{x, y\}) - \underline{R}(\{x, y\}).$$

Then $< \circ >$ is defined everywhere $\Longleftrightarrow \forall x \in H,\ |R(x)| \geq 3 \Longleftrightarrow$ $< H, \circ >$ is a join space.

If $< H'; \rho' >$ is another model, we say that a function $f{:}H \to H'$ is a *homomorphism of the models* if for every $(x, y) \in \rho$, we have $(f(x), f(y)) \in \rho'$. A family of models $\{< H_i, \rho_i >\}_{i \in I}$ is *direct* if the following conditions holds:

(i) $(I, \leq)$ is a directed partially ordered set;

(ii) $\forall (i, j) \in I^2,\ i \neq j \Longrightarrow H_i \cap H_j = \emptyset$;

(iii) $\forall (i, j) \in I^2$, if $i \leq j$, a homomorphism of models $\varphi_j^i : H_i \to H_j$ is defined, such that if $i \leq j \leq k$, we have $\varphi_k^j \varphi_j^i = \varphi_k^i$ and $\forall i \in I$, $\varphi_i^i = \mathrm{Id}(H_i)$.

On $H = \bigcup_{i \in I} H_i$ the following binary relation is defined as follows:

$$\forall (x_i, y_i) \in H_i \times H_j, \quad x_i \sim y_j \Longleftrightarrow \exists k \in I,$$

$k \geq i,\ k \geq j$, such that $\varphi_k^i(x_i) = \varphi_k^j(y_j)$.

The relation "$\sim$" is an equivalence relation.

$\varphi_j^i(x_j)$ is denoted by x_j.

Set $\overline{H} = H/\sim$.

On $\overline{H}$ is defined the binary relation $\bar{\rho}$ as follows

$$(\bar{x}, \bar{y}) \in \bar{\rho} \Longleftrightarrow \exists q \in I,\ \exists x_q \in \bar{x} \cap H_q,$$
$$\exists z_q \in \bar{z} \cap H_q, \text{ such that } (x_q, z_q) \in \rho_q.$$

Let $< \overline{H}, \odot >$ $(H_i, \odot_i))$ be the partial hypergroupoid corresponding to $\bar{\rho}$ (ρ_i, respectively) and defined by $(*)$.

42. Theorem. *If $< \overline{H}, \odot >$ is a join space, then there is $\ell \in I$, such that $(H_\ell, \odot_\ell)$ is a join space.*

Moreover, for every $i \in I$, $i \geq \ell$, we have that $(H_i, \odot_i)$ is a join space.

Proof. Since $< \overline{H}, \odot >$ is a join space, then $\forall\, \bar{x} \in \overline{H}$, we have $|\bar{\rho}(\bar{x})| \geq 3$. Let $\{\bar{x}, \bar{y}_1, \bar{y}_2\}$ be three different elements of $\bar{\rho}(\bar{x})$. We have

$$(1) \qquad \begin{array}{ll} \forall\, i \in I,\ \forall\, x_i \in \bar{x} \cap H_i, & \forall\, y_{1i} \in \bar{y}_1 \cap H_i, \\ \forall\, y_{2i} \in \bar{y}_2 \cap H_i, & x_i \neq y_{1i} \neq y_{2i} \neq x_i, \end{array}$$

otherwise $\bar{x}, \bar{y}_1, \bar{y}_2$ would not be different. Since $\bar{x}\, \bar{\rho}\, \bar{y}_1$, it follows that

$$\exists j \in I,\ \exists x_j \in \bar{x} \cap H_j,\ \exists y_{1j} \in \bar{y}_1 \cap H_j : (x_j, y_{1j}) \in \rho_j.$$

Similarly, since $\bar{x}\, \bar{\rho}\, \bar{y}_2$, it follows that

$$\exists k \in I,\ \exists x_k \in \bar{x} \cap H_k,\ \exists y_{2k} \in \bar{y}_2 \cap H_k : (x_k, y_{2k} \in \rho_k.$$

But $\bar{x}_k = \bar{x} = \bar{x}_j$, so $x_k \sim x_j$, that means $\exists \ell \in I$, $\ell \geq k$, $\ell \geq j$, such that $\varphi_\ell^k(x_k) = \varphi_\ell^j(x_j) = x_\ell$. Using now the fact that $\forall\, (i,j) \in I^2$, $i \leq j$, $\varphi_j^i : H_i \to H_j$ is a homomorphism of models, we have the implications:

$$(x_j, y_{1j}) \in \rho_j \Longrightarrow (\varphi_\ell^j(x_j), \varphi_\ell^j(y_{1j})) \in \rho_\ell$$

and

$$(x_k, y_{2k}) \in \rho_k \Longrightarrow (\varphi_\ell^k(x_k), \varphi_\ell^k(y_{2k})) \in \rho_\ell.$$

Therefore, $(x_\ell, y_{1\ell}) \in \rho_\ell \ni (x_\ell, y_{2\ell})$.

By (1), we have $x_\ell \neq y_{1\ell} \neq y_{2\ell} \neq x_\ell$, so $|\rho_\ell(x_\ell)| \geq 3$.

Since $\bar{x}$ is whichever in $\overline{H}$, it follows that x_ℓ is whichever in H_ℓ. So, by Theorem 41, it follows that $(H_\ell, \odot_\ell)$ is a join space.

Now, since $(x_\ell, y_{1\ell}) \in \rho_\ell \ni (x_\ell, y_{2\ell})$ it follows that $\forall\, i \in I$, $i \geq \ell$, we have $(\varphi_i^\ell(x_\ell), \varphi_i^\ell(y_{1\ell})) \in (\varphi_i^\ell(x_\ell), \varphi_i^\ell(y_{2\ell}))$ that is

$$(x_i, y_{1i}) \in \rho_i \ni (x_i, y_{2i}).$$

Moreover, by (1), it follows $x_i \neq y_{1i} \neq y_{2i} \neq x_i$ and since x_i is whichever in H_i, we have that $(H_i, \odot_i)$ is a join space, by Theorem 41. ∎

43. Theorem. $< \overline{H}, \odot >$ *is a join space if and only if* $\exists \ell \in I$, $\forall\, x_\ell \in H_\ell$, $\exists \{y_{1\ell}, y_{2\ell}\} \subset \rho_\ell(x_\ell)$ *such that* $\bar{x}_\ell \neq \bar{y}_{1\ell} \neq \bar{y}_{2\ell} \neq \bar{x}_\ell$.

Proof. "$\Longrightarrow$" By the previous theorem we have that

$$\exists \ell \in I,\ \forall\, x_\ell \in H_\ell,\ \exists (y_{1\ell}, y_{2\ell}) \in H_\ell^2,$$

such that $x\ell \neq y_{1\ell} \neq y_{2\ell} \neq x_\ell$ and $\forall\, i \in I$, $i \geq \ell$, $\varphi_i^\ell(x_\ell) = x_i \neq$ $\neq \varphi_i^\ell(y_{1\ell}) = y_{1i} \neq \varphi_i^\ell(y_{2\ell}) = y_{2i} \neq x_i$ and $(x_i, y_{1i}) \in \rho_i \ni (x_i, y_{2i})$, so it results the thesis.

"$\Longleftarrow$" Let us suppose that $\exists \ell \in I$, $\forall\, x_\ell \in H_\ell$, $\exists \{y_{1\ell}, y_{2\ell}\} \subset \rho_\ell(x_\ell) : \bar{x}_\ell \neq \bar{y}_{1\ell} \neq \bar{y}_{2\ell} \neq \bar{x}_\ell$. So, $\forall\, \bar{x}_\ell \in \overline{H}$, $|\bar{\rho}(\bar{x}_\ell)| \geq 3$, whence $< \overline{H}, \odot >$ is a join space. ∎

44. Proposition. *Let* $\{< H_i, \rho_i >\}_{i \in I}$ *be a direct family of models. If there is* $k \in I$, *such that* $\forall\, t \in I$, $t \geq k$, φ_t^k *is injective, and such that* $< H_k, \odot_k >$ *is a join space, then* $< \overline{H}, \odot >$ *is a join space.*

Proof. By Theorem 41, $< H_k, \odot_k >$ is a join space if and only if $\forall\, (x, y) \in H_k^2$,

$$x \odot_k y = \overline{\rho_k}(\{x, y\}) - \underline{\rho_k}(\{x, y\}) \neq \emptyset$$

if and only if $\forall x \in H_k$, $|\rho_k(x)| \geq 3$. We have the implication $(x,y) \in \rho_k \Longrightarrow (\bar{x},\bar{y}) \in \bar{\rho}$.

Let us remark that if $(y_1, y_2) \in H_k^2$, $y_1 \neq y_2$, then $\forall t \in I$, $t \geq k$, we have $\varphi_t^k(y_1) \neq \varphi_t^k(y_2)$, that means $\bar{y}_1 \neq \bar{y}_2$. Therefore, if $\forall x \in H_k$ we have $|\rho_k(x)| \geq 3$, then $\forall \bar{x} \in \overline{H}$, $|\bar{\rho}(\bar{x})| \geq 3$, so, by Theorem 41, $< \overline{H}, \odot >$ is a join space. ■

45. Remark. If I has a maximum M and φ_M^k is injective, then $\forall t \in I$, $t \geq k$, we have φ_t^k is injective.

Proof. We have $\varphi_M^t \circ \varphi_t^k = \varphi_M^k$ and since φ_M^k is injective, it follows φ_t^k is injective. ■

Direct products

Let $< H_1, \otimes_1 >$ and $< H_2, \otimes_2 >$ be two hyperstructures, where for all $i \in \{1,2\}$, $\forall (x,y) \in H_i^2$,

$$x \otimes_i y = \overline{\rho_i}(\{x,y\}) - \underline{\rho_i}(\{x,y\}).$$

Let $\rho_1 \times \rho_2$ be the binary relation defined on $H = H_1 \times H_2$ as follows:

$$((a_1, x_1), (a_2, x_2)) \in \rho_1 \times \rho_2 \Longleftrightarrow (a_1, a_2) \in \rho_1 \text{ and } (x_1, x_2) \in \rho_2.$$

Let $\otimes$ be the hyperoperation defined on H as follows:

$$\forall (\alpha, \beta) \in H^2,\ \alpha \otimes \beta = \overline{\rho_1 \times \rho_2}(\{\alpha,\beta\}) - \underline{\rho_1 \times \rho_2}(\{\alpha,\beta\}).$$

46. Proposition. *If $< H_1, \otimes_1 >$ or $< H_2, \otimes_2 >$ is a join space, then $< H, \otimes >$ is a join space.*

Proof. Let us suppoose $< H_1, \otimes_1 >$ is a join space. Then $\forall a_1 \in H_1$, $|\rho_1(a_1)| \geq 3$. Let $\{a_1, a_2, a_3\} \subset \rho_1(a_1)$, $a_1 \neq a_2 \neq a_3 \neq a_1$. Then $\forall x_1 \in H_2$, it follows that $((a_1, x_1), (a_1, x_1))$, $((a_1, x_1), (a_2, x_1))$ and $((a_1, x_1), (a_3, x_1))$ are different elements of $\rho_1 \times \rho_2$, whence $\forall (a_1, x_1) \in H$, $|(\rho_1 \times \rho_2)((a_1, x_1))| \geq 3$, so $< H, \otimes >$ is a join space.

■

47. Proposition. *If* $< H_1, \otimes_1 >$ *and* $< H_2, \otimes_2 >$ *are partial hypergroupoids defined as in Theorem* 37, *such that*

$$\forall\, a_1 \in H_1, \quad |\rho_1(a_1)| = 2 \quad \textit{and}$$
$$\forall\, x_1 \in H_2, \quad |\rho_2(x_1)| = 2,$$

then $< H, \otimes >$ *is a join space.*

Proof. For every $a_1 \in H_1$ and $x_1 \in H_2$, set $\rho_1(a_1) = \{a_1, a_2\}$ and $\rho_2(x_1) = \{x_1, x_2\}$. So, $((a_1, x_1), (a_1, x_1))$, $((a_1, x_1), (a_2, x_1))$, $((a_1, x_1), (a_2, x_2))$ are different elements of $\rho_1 \times \rho_2$, that is $\forall (a_1, x_1) \in H$, $|(\rho_1 \times \rho_2)(a_1, x_1)| \geq 3$, that means $< H, \otimes >$ is a join space.

48. Remark. By the proof of the previous proposition, it follows that if $< H, \otimes >$ is a join space and (a_1, x_1) is whichever in H, we have:

(i) if $|\rho_1(a_1)| = 1$ then $|\rho_2(x_1)| \geq 3$;

(i) if $|\rho_2(a_1)| = 2$ then $|\rho_2(x_1)| \geq 2$;

(i) if $|\rho_2(a_1)| = 3$ then $|\rho_2(x_1)|$ can be whichever nonzero natural number. ■

II) In the following, it is shown that the direct (inverse) limit of a direct (inverse) family of join spaces associated with rough sets is a join space associated with a rough set.

II.1) Direct limit of a direct family of join spaces associated with rough sets

Let $\{< H_i, \rho_i >\}_{i \in I}$ be a direct family of models, $H = \bigcup_{i \in I} H_i$ and let consider on H the following equivalence relation (see [322]): $x \sim y$ if and only if the following implication is satisfied:

$$(x, y) \in H_i \times H_j \Longrightarrow \exists k \in I;\ k \geq i; k \geq j, \text{ such that } \varphi_{ik}(x) = \varphi_{jk}(y).$$

If $x_i \in H_i$ and $i \leq j$, we shall denote $\varphi_{ij}(x_i)$ by x_j and by $\overline{H}$ the quotient set $H/\sim = \{\bar{x} \mid x \in H\}$.

We define on $\overline{H}$ the following binary relation (see [232]):

$$\begin{aligned} &\forall\,(\bar{x},\bar{y})\in\overline{H}^2,\ \bar{x}\rho^*\bar{y} \text{ if and only if } \exists i\in I, \\ &\exists x_i\in\bar{x}\cap H_i,\ \exists y_i\in\bar{y}\cap H_i,\ \text{such that } x_i\rho_i y_i. \end{aligned} \tag{2}$$

49. Definition. $(\overline{H},\rho^*)$ is called the *direct limit* of the direct family of models $\{H_i,\rho_i\}_{i\in I}$.

50. Proposition. *If* $\forall\, i \in I$, ρ_i *is an equivalence relation on* H_i, *then* ρ^* *is an equivalence relation on* $\overline{H}$.

Proof. The reflexivity and symmetry result directly by the definition of ρ^*. Let's suppose now $\bar{x}\rho^*\bar{y}$ and $\bar{y}\rho^*\bar{z}$. It follows there are $(i,j) \in I^2$, $x_i \in \bar{x}\cap H_i$, $y_i \in \bar{y}\cap H_i$, $y_j \in \bar{y}\cap H_j$ and $z_j \in \bar{z}\cap H_j$, such that $x_i\rho_i y_i$ and $y_j\rho_j z_j$. We have $y_i \sim y_j$, so there is $k \in I$, $k \geq i$, $k \geq j$, such that $\varphi_{ik}(y_i) = \varphi_{jk}(y_j) = y_k$. Since φ_{ik} and φ_{jk} are homomorphisms of models, it follows: $\varphi_{ik}(x_i)\rho_k\varphi_{ik}(y_i)$ and $\varphi_{jk}(y_j)\rho_k\varphi_{jk}(z_j)$, that is $x_k\rho_k y_k$ and $y_k\rho_k z_k$, whence $x_k\rho_k z_k$, so $\bar{x}\rho^*\bar{z}$. Therefore, ρ^* is transitive, hence it is an equivalence relation. ∎

51. Proposition. *Let* $\{< H_i,\rho_i >\}_{i\in I}$ *be a direct family of models, where* $\forall\, i \in I$, ρ_i *is an equivalence relation, such that* $\forall\, x_i \in H_i$, $|\rho_i(x_i)| \geq 3$. *For any* $i \in I$, *let us consider the hyperoperation* "$\circ_{\rho_i}$" *defined on* H_i *as in* $(*)$ (*Theorem* 41), *that is:*

$$x_i\circ_{\rho_i}y_i = \overline{\rho_i}(\{x_i,y_i\}) - \underline{\rho_i}(\{x_i,y_i\}).$$

Then $\{< H_i,\circ_{\rho_i} >\}_{i\in I}$ *is a direct family of join spaces.*

Proof. It is sufficient to notice that if $(i,j) \in I^2$, $i \leq j$ and φ_{ij} is a homomorphism of models, then φ_{ij} is a homomorphism of join spaces. Indeed, $\forall\, i \in I$, $\forall\,(x_i,y_i) \in H_i^2$, we have

$$x_i\circ_{\rho_i}y_i = \overline{\rho_i}(\{x_i,y_i\}) - \underline{\rho_i}(\{x_i,y_i\}) = \overline{\rho_i}(\{x_i,y_i\}) = \rho_i(x_i)\cup\rho_i(y_i)$$

because ρ_i is an equivalence relation and so, any equivalence class has at least three elements.

On the other hand, if $x'_i \in \rho_i(x_i)$, then $\forall j \in I,\ i \leq j$,

$$\varphi_{ij}(x'_i) = x'_j \in \rho_j(\varphi_{ij}(x_i)) = \rho_j(x_j),$$

since φ_{ij} is a homomorphism of models. Therefore,

$$\varphi_{ij}(\rho_i(x_i)) = \varphi_{ij}(\{x'_i \in H_i \mid x'_i \rho_i x_i\}) = \{x'_j \in H_j \mid x'_j \rho_j x_j\} = \rho_j(x_j),$$

whence

$$\begin{aligned}\varphi_{ij}(x_i \circ_{\rho_i} y_i) = \varphi_{ij}(\rho_i(x_i) \cup \rho_i(y_i)) = \varphi_{ij}(\rho_i(x_i)) \cup \varphi_{ij}(\rho_i(y_i)) = \\ = \rho_j(x_j) \cup \rho_j(y_j) = x_j \circ_{\rho_j} y_j = \varphi_{ij}(x_i) \circ_{\rho_j} \varphi_{ij}(y_i),\end{aligned}$$

hence φ_{ij} is a homomorphism of join spaces. ∎

Let us consider on $\overline{H}$ the following hyperoperation (see [322]):

$$\begin{aligned}\bar{x} * \bar{y} = \{\bar{z} \mid \exists i \in I,\ \exists x_i \in \bar{x} \cap H_i,\ \exists y_i \in \bar{y} \cap H_i, \\ \exists z_i \in \bar{z} \cap H_i,\ \text{such that } z_i \in x_i \circ_{\rho_i} y_i\}.\end{aligned}$$

52. Definition. $(\overline{H}, *)$ is called the *direct limit* of the direct family of join spaces $\{< H_i, \circ_{\rho_i} >\}_{i \in I}$.

53. Theorem. *Let $\{< H_i, \rho_i >\}_{i \in I}$ be a direct family of models, where $\forall i \in I$, ρ_i is an equivalence relation, such that $\forall x_i \in H_i$, $|\rho_i(x_i)| \geq 3$ and let $\{< H_i, \circ_{\rho_i} >\}_{i \in I}$ be the corresponding direct family of join spaces. Then the direct limit $(\overline{H}, *)$ of the previous direct family of join spaces is a join space, associated with the model $(\overline{H}, \rho^*)$, where ρ^* is the equivalence relation defined by* (2).

Proof. First, notice that $(\overline{H}, *)$ is a join space, being a direct limit of a direct family of join spaces (Prop. 1, [235]).

Let " $\circ_{\rho^*}$ " be the hyperoperation, associated with the equivalence relation ρ^*, defined as in $(*)$, Theorem 41, on the set $\overline{H}$:

$$\bar{x} \circ_{\rho^*} \bar{y} = \overline{\rho^*}(\{\bar{x}, \bar{y}\}) - \underline{\rho}^*(\{\bar{x}, \bar{y}\}) = \overline{\rho^*}(\{\bar{x}, \bar{y}\}) = \rho^*(\bar{x}) \cup \rho^*(\bar{y}),$$

since ρ^* is an equivalence relation and so, $\forall \bar{z} \in \overline{H}$, $|\rho^*(\bar{z})| \geq 3$. On the other hand,

$$\bar{x} * \bar{y} = \{\bar{z} \in \overline{H} \mid \exists i \in I,\ \exists x_i \in \bar{x} \cap H_i,\ \exists y_i \in \bar{y} \cap H_i,\\ \exists z_i \in \bar{z} \cap H_i,\ \text{such that } z_i \in x_i \circ_{\rho_i} y_i\}.$$

We have $x_i \circ_{\rho_i} y_i = \overline{\rho_i}(\{x_i, y_i\}) - \underline{\rho_i}(\{x_i, y_i\}) = \rho_i(x_i) \cup \rho_i(y_i)$, so since $z_i \in x_i \circ_{\rho_i} y_i$, one obtains $z_i \rho_i x_i$ or $z_i \rho_i y_i$, whence $\bar{z}\rho^*\bar{x}$ or $\bar{z}\rho^*\bar{y}$, that is $\bar{z} \in \rho^*(\bar{x}) \cup \rho^*(\bar{y})$.

Therefore, $\bar{x} * \bar{y} = \{\bar{z} \mid \bar{z} \in \rho^*(\bar{x}) \cup \rho^*(\bar{y})\} = \bar{x} \circ_{\rho^*} \bar{y}$, that means the join spaces $(\overline{H}, *)$ and $(\overline{H}, \circ_{\rho^*})$ coincide. ■

II.2) Direct products of join spaces associated with rough sets and inverse limit of an inverse family of join spaces associated with rough sets

Let $\{< H_i, \rho_i >\}_{i \in I}$ be a family of models, where $\forall i \in I$, ρ_i is an equivalence relation.

54. Remark. *The direct product* $\rho = \prod_{i \in I} \rho_i$ *of the family* $\{\rho_i\}_{i \in I}$ *is an equivalence relation on* $H = \prod_{i \in I} H_i$. *We recall that if* $x = (x_i)_{i \in I}$ *and* $y = (y_i)_{i \in I}$ *are in* H, *then* $x\rho y$ *if and only if* $\forall i \in I$, $x_i \rho_i y_i$.

Let us denote by" $\circ_\rho$ " and by " $\circ_{\rho_i}$ ", where $i \in I$, the hyperoperations induced by ρ, respectively, by ρ_i, defined as in $(*)$, Theorem 41, on the set H, respectively on the set H_i.

55. Proposition. *If* $\{< H_i, \circ_{\rho_i} >\}_{i \in I}$ *is a family of partial hypergroupoids* ($\forall i \in I$, ρ_i *is an equivalence relation*), *such that at least one is a join space, then* $< H, \circ_\rho >$ *is a join space.*

Proof. We shall verify that $\forall x = (x_i)_{i \in I} \in H$, $|\rho(x)| \geq 3$. We have $\rho(x) = \{y \in (y_i)_{i \in I} \in H \mid \forall i \in I,\ x_i \rho_i y_i\}$. Since $\exists i_0 \in I$, such that $< H_{i_0}, \circ_{\rho_{i_0}} >$ is a join space, it follows that $\forall x_{i_0} \in H_{i_0}$, $|\rho_{i_0}(x_{i_0})| \geq 3$, whence $\forall x \in H$, $|\rho(x)| \geq 3$, therefore $< H, \circ_\rho >$ is a join space. ■

56. Proposition. *If* $\{< H_i, \circ_{\rho_i} >\}_{i \in I}$ *is a family of partial hypergroupoids* ($\forall\, i \in I$, ρ_i *is an equivalence relation*), *such that there are* i_0 *and* j_0 *in* I, $i_0 \neq j_0$, *for which* $\forall\, x_{i_0} \in H_{i_0}$, $|\rho_{i_0}(x_{i_0})| \geq 2$ *and* $\forall\, x_{j_0} \in H_{j_0}$, $|\rho_{j_0}(x_{j_0})| \geq 2$, *then* $< H, \circ_\rho >$ *is a join space.*

Proof. By hypothesis, for any $x = (x_i)_{i \in I}$, we have

$$|\rho(x)| \geq |\rho_{i_0}(x_{i_0})| \cdot |\rho_{j_0}(x_{j_0})| \geq 4,$$

so $< H, \circ_\rho >$ is a join space. ■

57. Proposition. *Let* $\{< H_i, \circ_{\rho_i} >\}_{i \in I}$ *be a family of join spaces* ($\forall\, i \in I$, ρ_i *is an equivalence relation, such that* $\forall\, x_i \in H$, $|\rho_i(x_i)| \geq 3$) *and let* $< H, \otimes >$ *be the direct product of this family, that is* $\forall\, x = (x_i)_{i \in I} \in H$, $\forall\, y = (y_i)_{i \in I} \in H$ *we have* $x \otimes y = (x_i \circ_{\rho_i} y_i)_{i \in I}$. *Then the join space* $< H, \otimes >$ *is an enlargement of the join space* $< H, \circ_\rho >$.

Proof. For any $x = (x_i)_{i \in I}$ and $y = (y_i)_{i \in I}$ elements of H, we have:

$$x \otimes y = (x_i \circ_{\rho_i} y_i)_{i \in I} = (\rho_i(x_i) \cup \rho_i(y_i))_{i \in I}.$$

On the other hand,

$$\begin{aligned} x \circ_\rho y = \rho(x) \cup \rho(y) = \{z \in H \mid \forall\, i \in I,\ z_i \in \rho_i(x_i)\} \cup \\ \cup \{z \in H \mid \forall\, i \in I,\ z_i \in \rho_i(y_i)\} \subseteq (\rho_i(x_i) \cup \rho_i(y_i))_{i \in I} = x \otimes y, \end{aligned}$$

that means that the join space $< H, \otimes >$ is an enlargement of the join space $< H, \circ_\rho >$. ■

Let us study now the inverse limit of an inverse family of join spaces associated with rough sets.

58. Proposition. *Let* $\{< H_i, \rho_i >\}_{i \in I}$ *be an inverse family of models, where* $\forall\, i \in I$, ρ_i *is an equivalence relation, such that* $\forall\, x_i \in H_i$, $|\rho_i(x_i)| \geq 3$. *Then the family* $\{< H, \circ_{\rho_i} >\}_{i \in I}$ *of join spaces, where* "$\circ_{\rho_i}$" *is defined as in* (*), *Theorem* 41, *on* H_i, *is an inverse family of join spaces.*

Proof. The proof is similar to that one for direct families. ∎

Let $\{< H_i, \circ_{\rho_i} >\}_{i\in I}$ be an inverse family of join spaces, where $\forall\, i\in I$, the hyperoperation " $\circ_{\rho_i}$ " is defined on H_i, as in $(*)$, Theorem 41.

We consider the following subset of the direct product $H = \prod_{i\in I} H_i$:

$$\widetilde{H} = \{p \in H \mid f_{ij}(p_i) = p_j,\ \forall\, i \geq j\}, \text{ where } p = (p_i)_{i\in I}.$$

If $\widetilde{H} \neq \emptyset$, we define on $\widetilde{H}$ the hyperoperation:

$$\widetilde{x} \circ \widetilde{y} = \widetilde{x} \otimes \widetilde{y} \cap \widetilde{H}. \tag{3}$$

59. Remark. *If I has a maximum, then* $\widetilde{H} \neq \emptyset$ (see [235]).

60. Definition. Let $\{< H_i, \circ_i >\}_{i\in I}$ be an inverse family of join spaces and let $< H = \prod_{i\in I} H_i, \otimes >$ be the direct product of this family. Suppose $\widetilde{H} \neq \emptyset$. Then $< \widetilde{H}, \circ >$ is called the *inverse limit* of this inverse family of join spaces, where "$\circ$" is defined on $\widetilde{H}$, as in (3).

61. Theorem. *Let $\{< H_i, \circ_{\rho_i} >\}_{i\in I}$ be a family of join spaces ($\forall\, i\in I$, ρ_i is an equivalence relation, and let's suppose that $\forall\, x_i \in H_i$, $|\rho_i(x_i)| \geq 3$), such that $(I, \leq)$ has a maximum s. Let $< H, \otimes >$ be the direct product of this family. Then the hyperoperations "$\otimes$" and "$\circ_\rho$" coincide, where $\rho = \prod_{i\in I} \rho_i$.*

Proof. Let x and y be two arbitrary elements of H and let $z \in x \otimes y$. It follows that $\forall\, i \in I$, we have $z_i \in x_i \circ_{\rho_i} y_i = \rho_i(x_i) \cup \rho_i(y_i)$; particularly, $z_s \in \rho_s(x_s) \cup \rho_s(y_s)$. Suppose $z_s \in \rho_s(x_s)$, that is $z_s \rho_s x_s$; hence, $\forall\, i \in I$, $f_{si}(z_s) = z_i \rho_i f_{si}(x_s) = x_i$, whence $z \in \rho(x)$. It results $z \in \rho(x) \cup \rho(y) = x \circ_\rho y$, so, $x \otimes y \subset x \circ_\rho y$ and since $x \circ_\rho y \subset x \otimes y$, one obtains that the hyperoperations " $\circ_\rho$" and "$\otimes$" coincide. ∎

62. Theorem. *Let* $\{<H_i, \rho_i>\}_{i\in I}$ *be an inverse family of models, where* $\forall i \in I$, ρ_i *is an equivalence relation, such that* $\forall x_i \in H_i$, $|\rho_i(x_i)| \geq 3$ *and let* $\{< H_i, \circ_{\rho_i} >\}_{i\in I}$ *be the associated inverse family of join spaces.*

If I *has a maximum* s, *then the inverse limit of the inverse family of join spaces is a join space associated with a rough set.*

Proof. Let $(\widetilde{H}, \circ)$ be the inverse limit of the inverse family of join spaces.

Let $\rho = \prod_{i\in I} \rho_i$ and $\widetilde{\rho} = \rho \cap \widetilde{H} \times \widetilde{H}$. The relation $\widetilde{\rho}$ is an equivalence relation on $\widetilde{H}$. We shall verify that the hyperoperations "$\circ$" and "$\circ_{\widetilde{\rho}}$" coincide.

Let $\widetilde{x}$ and $\widetilde{y}$ be two arbitrary elements of $\widetilde{H}$. We have $\widetilde{x} \circ_\rho \widetilde{y} = \rho(\widetilde{x}) \cup \rho(\widetilde{y})$, because $< H, \circ_\rho >$ is a join space, so $\forall z \in H$, $|\rho(z)| \geq 3$. So,

$$\widetilde{x} \circ \widetilde{y} = \widetilde{x} \otimes \widetilde{y} \cap \widetilde{H} = \widetilde{x} \circ_\rho \widetilde{y} \cap \widetilde{H} =$$
$$= (\rho(\widetilde{x}) \cap \widetilde{H}) \cup (\rho(\widetilde{y}) \cap \widetilde{H}) = \widetilde{\rho}(\widetilde{x}) \cup \widetilde{\rho}(\widetilde{y}).$$

On the other hand, $\forall \widetilde{u} \in \widetilde{H}, \widetilde{\rho}(\widetilde{u}) = \{\widetilde{v} \in \widetilde{H} \mid \widetilde{v}\widetilde{\rho}\widetilde{u}\} = \{\widetilde{v} \in \widetilde{H} \mid \forall i \in I,\ v_i \rho_i u_i\}$, where $\widetilde{v} = (v_i)_{i\in I}$ and $\widetilde{u} = (u_i)_{i\in I}$. So, $\widetilde{v} \in \widetilde{\rho}(\widetilde{u})$ implies $v_s \rho_s u_s$. Conversely, if $v_s \rho_s u_s$, then $\forall i \in I$, $f_{si}(v_s) \rho_i f_{si}(u_s)$ (because $\widetilde{u}$ and $\widetilde{v}$ are in $\widetilde{H}$ and $\forall i \in I$, $s \geq i$), that means $\forall i \in I$, $v_i \rho_i u_i$, whence $\widetilde{v} \in \widetilde{\rho}(\widetilde{u})$.

Therefore, $\widetilde{v} \in \widetilde{\rho}(\widetilde{u})$ if and only if $v_s \in \rho_s(u_s)$. Since $\forall u_s \in H_s$, $|\rho_s(u_s)| \geq 3$, it follows that $\forall \widetilde{u} \in \widetilde{H}$, $|\widetilde{\rho}(\widetilde{u})| \geq 3$, because for every $v_s \in \rho_s(u_s)$, there exists $\widetilde{v} = (f_{si}(v_s))_{i\in I} \in \widetilde{\rho}(\widetilde{u})$ and this correspondence is injective. Hence, $\widetilde{x} \circ_{\widetilde{\rho}} \widetilde{y} = \widetilde{\rho}(\widetilde{x}) \cup \widetilde{\rho}(\widetilde{y}) = \widetilde{x} \circ \widetilde{y}$, therefore the join spaces $< \widetilde{H}, \circ >$ and $< \widetilde{H}, \circ_{\widetilde{\rho}} >$ coincide. ∎

§5. Hyperstructures and Factor Spaces

Another application of hyperstructures, again in the setting of Fuzzy Set Theory and in particular of Decision Making is that one to *Factor Spaces.*

Factor Space Theory was introduced in 1981 by Pei–Zuang Wang. Hong Xing Li and Vincent C. Yen have applied Factor Spaces to Fuzzy Decision Making. Every Factor Space can be considered a generalization of the physical coordinate space.

We have a *universe* U of objects, where $|U| < \aleph_0$, for instance the universe of living beings, a *set of concepts* (as the concepts of being either a man, or a mammalian or an insect or a plant etc.) and a *set of factors*, that is a set of functions $f : U \to X(f)$ from the universe U to $X(f)$, the set of states of f, for instance the height which sets in correspondence with every object u the size of the height of u (when the height is definable for u, otherwise $f(u) = \theta$, where θ is the empty state). So, every object u of the universe can be represented by the $\{f(u)\}_{f\in F}$-ple, where its coordinates are the elements $f(u) \in X(f)$, for every factor $f \in F$.

A *description frame* is just a triple (U, C, F), where C is the set of concepts. Now, let us suppose that a concept $\alpha \in C$ has as extension, not simply a crisp set (that is a subset of U), but a fuzzy subset $\widetilde{A}$.

If (U, C, F) is given and $f \in F$ is a factor, a *hypergroupoid* $< U, \hat{\circ} >$ can be associated as follows:

$$x \,\hat{\circ}\, y = \left\{ a \mid \widetilde{A}(a) \in \left[\bigvee_{f(z)=f(x)} \widetilde{A}(z), \bigvee_{f(v)=f(y)} \widetilde{A}(v) \right] \right\}.$$

Let "$\circ$" be the following hyperoperation defined on U:

$$x \circ y = \{\lambda \mid \widetilde{A}(\lambda) \in [\widetilde{A}(x), \widetilde{A}(y)\}.$$

The following results have been obtained by P. Corsini.

63. Theorem. *$< U; \hat{\circ} >$ is a semi–hypergroup.*

Proof. Since U is finite, there is $x_0 \in U$ such that $\bigvee_{f(z)=f(x)} \widetilde{A}(z) = \widetilde{A}(x_0)$, and there is $y_0 \in U$ such that $\bigvee_{f(v)=f(y)} \widetilde{A}(v) = \widetilde{A}(y_0)$. So we have $x \,\hat{\circ}\, y = x_0 \circ y_0$.

If $t \in U$ and $\bigvee_{f(u)=f(t)} \widetilde{A}(u) = \widetilde{A}(t_0)$, we have clearly

$$(x \,\hat{\circ}\, y) \,\hat{\circ}\, t = (x_0 \circ y_0) \circ t_0 = x_0 \circ (y_0 \circ t_0) = x \,\hat{\circ}\, (y \,\hat{\circ}\, t),$$

whence $< U; \hat{\circ} >$ is a semi–hypergroup. ∎

Since U is finite, there is $p \in U$ such that

$$\widetilde{A}(p) = \min\{\widetilde{A}(z) \mid z \in U\}.$$

64. Theorem. *Let us suppose* $f^{-1}f(p) \subset \widetilde{A}^{-1}(\widetilde{A}(p))$. *Then* $< U; \hat{\circ} >$ *is a hypergroup.*

Proof. It is enough, by Theorem 63, to prove that $< U; \hat{\circ} >$ is a quasi–hypergroup.

Let us prove, first, that for every $a, b \in U$, if $\widetilde{A}(a) \geq \bigvee_{f(z)=f(b)} \widetilde{A}(z)$, then x exists such that $a \in b \,\hat{\circ}\, x$.

It is enough to set $x = a$. Indeed, we have

$$\bigvee_{f(z)=f(b)} \widetilde{A}(z) \leq \widetilde{A}(a) \leq \bigvee_{f(v)=f(a)} \widetilde{A}(v)$$

whence

$$\widetilde{A}(a) \in \left[\bigvee_{f(z)=f(b)} \widetilde{A}(z), \bigvee_{f(v)=f(x)} \widetilde{A}(v)\right]$$

therefore $a \in b \,\hat{\circ}\, a = b \,\hat{\circ}\, x$.

Let us suppose now $\widetilde{A}(a) \leq \bigvee_{f(z)=f(b)} \widetilde{A}(z)$.

Set $y = p$. Hence

$$\bigvee_{f(v)=f(p)} \widetilde{A}(v) = \bigvee_{v \in f^{-1}f(p)} \widetilde{A}(v) \leq \bigvee_{v \in \widetilde{A}^{-1}\widetilde{A}(p)} \widetilde{A}(v) = \widetilde{A}(p),$$

therefore

$$\bigvee_{f(v)=f(y)} \widetilde{A}(v) = \widetilde{A}(p) \leq \widetilde{A}(a) \leq \bigvee_{f(z)=f(b)} \widetilde{A}(z),$$

so it follows $a \in y \,\hat{\circ}\, b = p \,\hat{\circ}\, b$. ∎

65. Corollary. *With every factor* $f \in F$ *endowed with an extension satisfying the condition*

$$f^{-1}f(p) \subset \widetilde{A}^{-1}\widetilde{A}(p),$$

a join space $< U; \hat{\circ} >$ *is associated.*

Proof. It follows straight off, from Theorem 64 and from Theorem 4 [70]. ∎

66. Theorem. *Let* $< U; \hat{\circ} >$ *be a hypergroup. If* $x \in p/U$ (see Definition 156 [437]), *we have* $f^{-1}(f(x)) \subset \widetilde{A}^{-1}\widetilde{A}(p)$ *whence* $f^{-1}(f(p)) \subset \widetilde{A}^{-1}(\widetilde{(p)})$.

Proof. $\forall (a, b) \in U^2$, x exists such that $a \in x \,\hat{\circ}\, b$. If $\widetilde{A}(a) \leq \leq \bigvee_{f(v)=f(b)} \widetilde{A}(v)$, we have

$$\bigvee_{f(z)=f(x)} \widetilde{A}(z) \leq \widetilde{A}(a) \leq \bigvee_{f(v)=f(b)} \widetilde{A}(v).$$

So, if we set $a = p$, it follows

$$\widetilde{A}(p) \leq \bigvee_{f(z)=f(x)} \widetilde{A}(z) \leq \widetilde{A}(p),$$

whence $\forall z \in f^{-1}(f(x))$ we have $\widetilde{A}(z) = \widetilde{A}(p)$, therefore

$$f^{-1}(f(x)) \subset \widetilde{A}^{-1}(\widetilde{A}(p)).$$

Since $q \in U$ exists such that $p \in p \,\hat{\circ}\, q$, one obtains

$$f^{-1}f(p) \subset \widetilde{A}^{-1}(\widetilde{A}(p)).$$

∎

65. Corollary. $< U; \hat{\circ} >$ *is a join space if and only if*

$$f^{-1}(f(p)) \subset \widetilde{A}^{-1}(\widetilde{A}(p)).$$

Proof. It follows straight off, from Theorem 66 and Corollary 65. ∎

§6. Hypergroups induced by a fuzzy subset. Fuzzy hypergroups

R. Ameri and M.M. Zahedi have considered an interesting hyperstructure $(G, \circ_\mu)$, associated with a fuzzy subset μ. Notice that μ is a fuzzy subset on a group $(G, \cdot)$. They have proved that if μ is subnormal, then the hyperstructure $(G, \circ_\mu)$ is a hypergroup and under suitable conditions, it is a join space.

We mention here some of their results.

68. Definition. Let $(G, \cdot)$ be a group and μ a fuzzy subset on G. We say that μ is a *fuzzy subgroup* on G if and only if :

1) $\forall\,(x, y) \in G^2,\ \mu(xy) \geq \min(\mu(x), \mu(y))$;

2) $\forall\,x \in G,\ \mu(x^{-1}) = \mu(x)$.

Let $X \neq \emptyset$. Denote by $FS(X)$ the set of all nonzero fuzzy subsets on X. From now on, we shall denote by e the identity of the group G.

The concept of fuzzy subgroup was introduced by Rosenfeld [328]. Notice that if $\mu \in FS(G)$, then μ is a fuzzy subgroup of G if and only if any nonempty set $\mu_t = \{x \mid \mu(x) \geq t\}$ is a subgroup of G, where $t \in [0, 1]$.

69. Definition. Let $\mu \in FS(G)$. We say that μ is

1) *symmetric* if $\forall\,x \in G$, we have $\mu(x) = \mu(x^{-1})$;

2) *invariant* if $\forall\,(x, y) \in G^2$, we have $\mu(xy) = \mu(yx)$;

3) *subnormal* if it is both symmetric and invariant.

70. Definition. Let $\mu \in FS(G)$ and $x \in G$. The *left fuzzy coset* $x\mu \in FS(G)$ of μ is defined by:

$$\forall\,g \in G,\ (x\mu)(g) = \mu(x^{-1}g).$$

Similarly, the *right fuzzy coset* $\mu x \in FS(G)$ of μ is defined by:

$$\forall\, g \in G, \;\; (\mu x)(g) = \mu(gx^{-1}).$$

71. Definition. Let $(H, \circ)$ be a hypergroup and μ a fuzzy subset on H.

We say that μ is a *fuzzy subhypergroup* on G if the following conditions hold:

1) $\forall\,(x, y) \in H^2, \;\; \inf\limits_{z \in x \circ y} \mu(z) \geq \inf\{\mu(x), \mu(y)\};$

2) $\forall\,(x, a) \in H^2, \;\; \exists (y, z) \in H^2$, such that $x \in a \circ y \cap z \circ a$ and $\inf\{\mu(y), \mu(z)\} \geq \inf\{\mu(a), \mu(x)\}$.

The following result can be easily proved:

72. Proposition. *Let $\mu \in FS(G)$ and $(x, y, z) \in G^3$. Then we have:*

1) $x\mu = y\mu \Longleftrightarrow zx\mu = zy\mu;$

2) $x\mu = y\mu \Longleftrightarrow xz\mu = yz\mu$, *if μ is subnormal.*

Let us make the following notations: if $\mu \in FS(G)$ and $(a, b) \in G^2$, then we denote ${}^a\mu = \{x \in G \mid x\mu = a\mu\}$, $\mu^a = \{x \in G \mid \mu x = \mu a\}$, $a\mu^e = \{ax \mid x \in \mu^e\}$ and $\mu^a\mu^b = \{xy \mid x \in \mu^a,\, y \in \mu^b\}$. If μ is invariant, then $\forall\, a \in G$, we have ${}^a\mu = \mu^a$.

Another result which can be easily proved is the following one:

73. Proposition. *Let $\mu \in FS(G)$ be subnormal. Then:*

1) $\forall\,(x, y) \in G^2$, *we have* $x\mu = y\mu \Longleftrightarrow xy^{-1} \in \mu^e;$

2) μ^e *is a normal subgroup of* $G;$

3) $\forall\, a \in G, \;\; \mu^a = a\mu^e;$

4) $\forall\,(a, b) \in G^2, \;\; \mu^a\mu^b = \mu^{ab}.$

Now, let us consider on G the following hyperoperation:

$$\circ_\mu : G \times G \to \mathcal{P}^*(G), \ \circ_\mu((a,b)) = \mu^a \mu^b$$

So, "$\circ_\mu$" is the hyperoperation induced by μ.

74. Theorem. *Let $\mu \in FS(G)$.*

1) *Then $(G, \circ_\mu)$ is a quasi–hypergroup.*

2) *If μ is subnormal, then $(G, \circ_\mu)$ is a hypergroup.*

Proof. 1) We shall verify that $\forall a \in G$, $a \circ_\mu G = G = G \circ_\mu a$. Let $b \in G$. We have $b \in a \circ_\mu (a^{-1}b) \cap (ba^{-1}) \circ_\mu a$. Hence, $(G, \circ_\mu)$ is a quasi–hypergroup.

2) Let us check the associativity. Let $(a, b, c) \in G^3$. Since μ is subnormal, we have

$$(a \circ_\mu b) \circ_\mu c = \bigcup_{x \in \mu^a \mu^b} \mu^x \mu^c = \bigcup_{x \in \mu^{ab}} \mu^{xc} = \mu^{(ab)c}.$$

We have used that $x \in \mu^{ab}$ implies $\mu x = \mu ab$.

Similarly, we obtain $a \circ_\mu (b \circ_\mu c) = \mu^{a(bc)}$.

Therefore, $(a \circ_\mu b) \circ_\mu c = a \circ_\mu (b \circ_\mu c)$. ■

75. Proposition. *Let $\mu \in FS(G)$ be subnormal. Then $(G, \circ_\mu)$ is a commutative hypergroup if and only if $[G, G]$, the commutator subgroup of G, is included in μ^e.*

Proof. Let $(a, b) \in G^2$. We have $a \circ_\mu b = b \circ_\mu a \iff \mu^{ab} = \mu^{ba} \iff ab\mu = ba\mu \iff aba^{-1}b^{-1} \in \mu^e$, therefore $(G, \circ_\mu)$ is commutative if and only if $[G, G] \subseteq \mu^e$. ■

76. Theorem. *Let $\mu \in FS(G)$ be subnormal. Then $(G, \circ_\mu)$ is a quasi–canonical hypergroup.*

Moreover, there exists a good homomorphism from $(G, \cdot)$ to $(G, \circ_\mu)$.

Proof. Let $x \in G$. We have $x \in \mu^x = \mu^{ex} = \mu^e\mu^x = e \circ_\mu x$ and, similarly, we have $x \in \mu^x = \mu^{xe} = x \circ_\mu e$.

Moreover, $e \in x \circ_\mu x^{-1} \cap x^{-1} \circ_\mu x$. Now, let $z \in x \circ_\mu y = \mu^{xy}$. It follows $\mu z = \mu xy$ whence $\mu x = \mu zy^{-1}$, that is $x \in z \circ_\mu y^{-1}$. On the other hand, $\mu y = \mu x^{-1} z$ implies that $y \in x^{-1} \circ_\mu z$. Therefore, $(G, \circ_\mu)$ is a quasi–canonical hypergroup.

Let $\varphi : G \longrightarrow \mathcal{P}^*(G)$, $f(a) = \mu^a$. We have

$$f(ab) = \mu^{ab} = \bigcup_{xy \in \mu^{ab}} \mu^{xy} = \bigcup_{x \in \mu^a, y \in \mu^b} x \circ_\mu y = \mu^a \circ_\mu \mu^b = f(a) \circ_\mu f(b).$$

Hence, f is a good homomorphism. ■

77. Theorem. *Let $\mu \in FS(G)$ be subnormal. Then $(G, \circ_\mu)$ is a join space if and only if $[G, G] \subseteq \mu^e$.*

Proof. $\Longleftarrow$" Let $(a, b, c, d) \in G^4$. We have to verify only the implication:

$$a/b \cap c/d \neq \emptyset \Longrightarrow a \circ_\mu d \cap b \circ_\mu c \neq \emptyset.$$

If $x \in a/b \cap c/d$, then $a \in x \circ_\mu b$ and $c \in x \circ_\mu d$, whence $\mu a = \mu xb$ and $\mu c = \mu xd$. It results $\mu ad = \mu xbd$ and $\mu bc = \mu bxd$. Therefore, we have the equivalence relations: $a \circ_\mu d \cap b \circ_\mu c \neq \emptyset \Longleftrightarrow \mu ad = \mu bc \Longleftrightarrow \mu xbd = \mu bxd \Longleftrightarrow (xbd)(bxd)^{-1} \in \mu^e \Longleftrightarrow xbx^{-1}b^{-1} \in \mu^e$. So, if $[G, G] \subset \mu^e$, then $a \circ_\mu d \cap b \circ_\mu c \neq \emptyset$. Therefore, $(G, \circ_\mu)$ is a join space.

"$\Longrightarrow$" Conversely, if $(G, \circ_\mu)$ is a join space, then according to the previous calculations, $\forall (x, b) \in G^2$, we have $xbx^{-1}b^{-1} \in \mu^e$, whence $[G, G] \subseteq \mu^e$. ■

In the following, we mention here some results on fuzzy hypergroups, obtained by P. Corsini and I. Tofan.

Let M be a nonempty set. An application

$$\square : M \times M \longrightarrow \mathcal{P}(M)^* = \mathcal{P}(M) - \{\emptyset\} =$$
$$= \{0,1\}^M - \{0 : M \longrightarrow \{0,1\} \mid \forall x \in M,\ 0(x) = 0\},$$

is called *hyperoperation* on M.

78. Definition. An application

$$\bullet : M\times M \rightarrow \mathcal{F}^*(M)=[0,1]^M\backslash\{0 : M \rightarrow [0,1] | \forall x\in M,\ 0(x)=0\}$$

is called *f-hyperoperation* (fuzzy hyperoperation) on M.

For any $(a,b) \in M^2$, $H \subseteq M$, $H \neq \emptyset$ and $\varepsilon \in (0,1]$, we denote:

$$a \oplus b = \{x \in M \mid (a \bullet b)(x) \neq 0\},$$
$$a \oplus H = \bigcup_{h\in H} a \oplus h,\ H \oplus a = \bigcup_{h\in H} h \oplus a$$
$$a \otimes b = \{x \in M \mid (a \bullet b)(x) = 1\},$$
$$a \otimes H = \bigcup_{h\in H} a \otimes h,\ H \otimes a = \bigcup_{h\in H} h \otimes a$$
$$a \bullet_\varepsilon b = \{x \in M \mid (a \bullet b)(x) \geq \varepsilon\},$$
$$a \bullet_\varepsilon H = \bigcup_{h\in H} a \bullet_\varepsilon h,\ H \bullet_\varepsilon a = \bigcup_{h\in H} h \bullet_\varepsilon a.$$

The following situations are possible:

R1) $\forall a \in M,\ a \bullet M = \chi_M = M \bullet a$,
where $\chi_M : M \longrightarrow [0,1]$ and $\forall x \in M,\ \chi_M(x) = 1$;

R2) $\forall a \in M,\ a \oplus M = M = M \oplus a$;

R3) $\forall a \in M,\ a \otimes M = M = M \otimes a$;

R4) $\forall \varepsilon \in (0,1],\ \forall a \in M,\ a \bullet_\varepsilon M = M = M \bullet_\varepsilon a$.

79. Definition. A nonempty set M, on which is defined a f-hyperoperation $\bullet : M\times M \longrightarrow \mathcal{F}^*(M)$ which satisfy the associativity law and the reproductibility R_i is called a *f_i-hypergroup* (for $i \in \{1,2,3,4\}$).

80. Proposition. *Let* $(M,\Box)$ *be a hypergroup. If one defines*

$$\bullet : M\times M \longrightarrow \mathcal{F}^*(M)$$

by $a \bullet b : M \longrightarrow [0,1]$ *: if* $x \in a \,\square\, b$, $(a \bullet b)(x) = 1$, *if* $x \notin a \,\square\, b$, $(a \bullet b)(x) = 0$, *then one obtains a* f_i*–hypergroup* ($i \in \{1,2,3,4\}$).

Proof. For any $(a,b,c) \in M^3$, we have

$$[(a \bullet b) \bullet c](x) = \sup_{(a \bullet b)(h) \neq 0} \{(h \bullet c)(x)\} = \sup_{h \in a \square b} \{(h \bullet c)(x)\},$$

so if $x \in (a \square b) \square c$, $[(a \bullet b) \bullet c](x)=1$, if $x \notin (a \square b) \square c$, $[(a \bullet b) \bullet c](x)=0$. Similarly, one obtains: if $x \in a \,\square\, (b \,\square\, c)$, $[a \bullet (b \bullet c)](x) = 1$, if $x \notin a \,\square\, (b \,\square\, c)$, $[a \bullet (b \bullet c)](x) = 0$.

Therefore, the associativity of "$\bullet$" holds.

Moreover, we have:

$$a \otimes M = \bigcup_{m \in M} a \otimes m = \bigcup_{m \in M} \{x \mid (a \bullet m)(x) = 1\} = \bigcup_{m \in M} a \,\square\, m = M.$$

In a similar way, one proves: $M = M \otimes a$, then $< M, \bullet >$ is a f_3–hypergroup. Similarly, we can verify the statement for the other hyperoperations. ■

81. Proposition. *If* $< M, \bullet >$ *is a* f_i*–hypergroup (for an* $i \in$ $\in \{1,2,3,4\}$*), then* $< M, \oplus >$ *is a hypergroup.*

Proof. It is enough to prove the associativity of "$\oplus$".

For any $(a,b,c) \in M^3$, we have:

$$(a \oplus b) \oplus c = \bigcup_{t \in a \oplus b} t \oplus c = \bigcup_{(a \bullet b)(t) \neq 0} \{x \mid (t \bullet c)(x) \neq 0\} =$$

$$= \{x \mid \sup_{(a \bullet b)(t) \neq 0} \{(t \bullet c)(x) \neq 0\}\} = \{x \mid [(a \bullet b) \bullet c](x) \neq 0\} =$$

$$= \{x \mid [a \bullet (b \bullet c)](x) \neq 0\} = a \oplus (b \oplus c).$$

■

§7. Fuzzy subhypermodules over fuzzy hyperrings

Let R be a commutative hyperring with identity, M an R-hypermodule and L a completely distributive lattice.

If X is a nonempty set, then we denote by $F(X)$ the set of all fuzzy subsets of X, that is $F(X) = \{\mu \mid \mu : X \longrightarrow L \text{ is a function}\}$.

We present here some results, about fuzzy subhypermodules over fuzzy hyperrings, obtained by M.M. Zahedi and R. Ameri.

82. Definition. Let $\mu \in F(R)$. We say that μ is a *fuzzy subhyperring* of R if the following conditions hold:

(i) $\forall (x, y) \in R^2,\ \forall z \in x + y,\ \mu(z) \geq \mu(x) \wedge \mu(y)$;

(ii) $\forall x \in R,\ \mu(-x) \geq \mu(x)$;

(iii) $\forall (x, y) \in R^2,\ \mu(xy) \geq \mu(x) \wedge \mu(y)$.

We denote by $FR(R)$ the set of all fuzzy hyperrings of R.

83. Theorem. *Let $\mu \in F(R)$. then μ is a fuzzy hyperring if and only if any nonempty level subset $\mu_\alpha = \{x \in R \mid \mu(x) \geq \alpha\}$ is a subhyperring of R, where $\alpha \in L$.*

Proof. Let $\mu \in FR(R)$ and μ_α be nonempty. Let x, y be in μ_α and $z \in x - y$. Since $\mu(z) \geq \mu(x) \wedge \mu(y) \geq \alpha$, it results $z \in \mu_\alpha$, whence $x - y \subseteq \mu_\alpha$.

Similarly, from $\mu(xy) \geq \mu(x) \wedge \mu(y) \geq \alpha$, it follows $xy \in \mu_\alpha$. Therefore, μ_α is a subhyperring of R.

Conversely, let us suppose that any nonempty μ_α is a subhyperring. For $(x, y) \in R^2$ and $z \in x + y$, set $\alpha = \mu(x) \wedge \mu(y)$. Then $x + z \subseteq \mu_\alpha$. Hence, $\forall z \in x + y,\ \mu(z) \geq \mu(x) \wedge \mu(y)$.

Similarly, we obtain $\mu(xy) \geq \mu(x) \wedge \mu(y)$. For $x \in R$, set $\alpha = \mu(x)$. Then $x \in \mu_\alpha$, so $-x \in \mu_\alpha$, that is $\mu(-x) \geq \alpha = \mu(x)$.

Therefore, μ is a fuzzy subhyperring of R. ■

84. Definition. Let $\mathcal{V} \in F(R)$. We say that $\mathcal{V}$ is a *fuzzy hyperideal* of R if it satisfies the following conditions:

(i) $\forall (x,y) \in R^2,\ \forall z \in x+y,\ \mathcal{V}(z) \geq \mathcal{V}(x) \wedge \mathcal{V}(y)$;

(ii) $\forall x \in R,\ \mathcal{V}(-x) \geq \mathcal{V}(x)$;

(iii) $\mathcal{V}(xy) \geq \mathcal{V}(x) \vee \mathcal{V}(y)$.

We denote by $FI(R)$ the set of all fuzzy hyperideals of R.

The following theorem can be proved in a similar way as the above theorem.

85. Theorem. *Let $\mathcal{V} \in F(R)$. Then $\mathcal{V}$ is a fuzzy hyperideal of R if and only if any nonempty $\mathcal{V}_\alpha = \{x \in R \mid \mathcal{V}(x) \geq \alpha\}$ is a hyperideal of R, where $\alpha \in L$.*

86. Definition. Let $\theta \in F(M)$ and $\mathcal{V} \in FI(R)$. We say that θ is a $\mathcal{V}$*–fuzzy subhypermodule* of M if and only if the following conditions hold:

1°) $\forall (x,y) \in M^2,\ \forall z \in x+y,\ \theta(z) \geq \theta(x) \wedge \theta(y)$;

2°) $\forall x \in M,\ \mu(-x) \geq \theta(x)$;

3°) $\forall x \in M,\ \forall r \in R,\ \theta(rx) \geq \mathcal{V}(r) \wedge \theta(x)$.

We denote by $Fm_R^{\mathcal{V}}(M)$ the set of all $\mathcal{V}$-fuzzy subhypermodules of M.

87. Theorem. *Let $\theta \in F(M)$ and $\mathcal{V} \in FI(R)$. Then $\theta \in Fm_R^{\mathcal{V}}(R)$ if and only if any $\theta_\alpha = \{x \in M \mid \theta(x) \geq \alpha\}$ is a canonical hypergroup of M, where $\alpha \in L$. Particularly, if $\mathcal{V}_\alpha$ is nonempty, then θ_α is a $\mathcal{V}_\alpha$–subhypermodule of M.*

Proof. Let any nonempty θ_α be a subhypergroup of M. By Theorem 83, the conditions 1°) and 2°) are satisfied. Thus θ is a subhypergroup of M. If $\mathcal{V}_\alpha$ is nonempty and $x \in \theta_\alpha$, $r \in \mathcal{V}_\alpha$, then $\theta(rx) \geq \mathcal{V}(r) \wedge \theta(x) \geq \alpha$. Hence θ is a fuzzy subhypergroup of M.

Conversely, let $\theta \in Fm_R^{\mathcal{V}}(M)$ and θ_α be nonempty, where $\alpha \in L$. By Theorem 83, it follows that θ_α is a subhypergroup of M. Moreover, if $\mathcal{V}_\alpha$ is nonempty, then for $x \in \theta_\alpha$, $r \in \mathcal{V}_\alpha$, we have: $\theta(rx) \geq \mathcal{V}(r) \wedge \theta(x) \geq \alpha$. Hence $rx \in \theta_\alpha$. Therefore θ_α is a $\mathcal{V}_\alpha$-subhypermodule of M.

88. Definition. Let a θ be a nonconstant $\mathcal{V}$-fuzzy subhypermodule of M. We say that θ is *weakly fuzzy primary (prime) subhypermodule* if $\theta(ra) > \theta(a)$ implies that $\exists n \geq 1$, such that $\forall x \in M$, $\theta(r^n x) \geq \theta(ra)$ (respectively, $\theta(rx) \geq \theta(ra)$).

89. Proposition. *Let N be a proper subhypermodule of M. Then χ_N is weakly fuzzy primary (prime) $\mathcal{V}$-fuzzy subhypermodule of M if and only if N is a primary (prime) subhypermodule of M.*

Proof. Let N be a prime subhypermodule of M. Thus $N \neq M$ and χ_N is nonconstant. Let $a \in M$ and $r \in R$, be such that $\chi_N(ra) > \chi_N(a)$. Then $\chi_N(ra) = 1$ and $\chi_N(a) = 0$. Hence, $\forall x \in M$, $\chi_N(r^n x) \geq \chi_N(ra)$. So χ_N is weakly fuzzy prime.

The proof is similar if N is primary.

The converse is immediate. ∎

90. Theorem. *Let θ be a $\mathcal{V}$-fuzzy subhypermodule of M, $\theta(0) = = \mathcal{V}(0) = 1$ and $L = [0,1]$. Then θ is a weakly fuzzy primary (prime) subhypermodule if and only if every θ_t is a primary (prime) $\mathcal{V}_t$-subhypermodule of M, $\forall t \in [0,1]$.*

Proof. Let θ be a weakly fuzzy primary (prime) subhypermodule of M. Let $r \in \mathcal{V}_t$ and $a \in M$ such that $ra \in \theta_t$ and $a \notin \theta_t$. Then $\theta(ra) \geq t$ and $\theta(a) < t$. Since θ is weakly fuzzy primary (prime), it follows that $\exists n \geq 1$, such that $\forall x \in M$, $\theta(r^n x) \geq \theta(ra) \geq t$ (respectively, $\forall x \in M$, $\theta(rx) \geq \theta(ra) \geq t$).

Conversely, suppose that θ_t is primary (prime) $\mathcal{V}_t$–subhypermodule, $\forall t \in [0,1]$. Let $r \in R$ and $a \in M$, such that $\theta(ra) > \theta(a)$. Set $t = \theta(ra)$. Thus $ra \in \theta_t$, but $a \notin \theta_t$. Since θ_t is primry (prime), it follows that $\exists n \geq 1$ such that $\forall x \in M$, $\theta(r^n x) \geq t = \theta(ra)$ (respectively, $\forall x \in M$, $\theta(rx) \geq t = \theta(ra)$).

Therefore, θ is weakly $\mathcal{V}$-fuzzy primary (prime).

§8. On Chinese hyperstructures

Some Chinese mathematicians (see for instance [244], [435]) have developed an interesting theory of the groups which have as supports, subsets of the set of non empty subsets of a group G. These Chinese structures have been derived from the fuzzy subset theory.

In this paragraph another connection between these groups (called HX–groups) and hyperstructures is established and analyzed. From every HX-group a hypergroupoid is obtained (by P. Corsini) which is always a H_v-group and in some case, a join space. Another connection had already been emphasized in [70].

Let us remind some definitions.

I. An H_v*-group* is a hypergroupoid $< H; \circ >$ such that

a) $\forall\,(a,x,y) \in H^3, \quad (x \circ y) \circ z \cap x \circ (y \circ z) \neq \emptyset$

b) it is a quasi-hypergroup, that is
$\forall\, a \in H^2,\ a \circ H = H \circ H = H.$

II. An H_b*-group* is an H_v-group $< H; \circ >$ such that there is a group operation $< \cdot >$ so that $\forall\,(x,y) \in H^2$, we have

$$x \cdot y \in x \circ y.$$

III. Let $< G; \cdot >$ be a group and $\mathcal{P}^*(G)$ the set of non empty subsets of G. An HX*-group* is a non empty subset H of $\mathcal{P}^*(G)$ which is a group with respect to the operation:

$$\forall\,(A,B) \in \mathcal{P}^*(G) \times \mathcal{P}^*(G),\ A \cdot B = \{xy \mid x \in A,\ y \in B\}.$$

91. Definition. Let $\mathcal{G}$ be an HX-group with G as support and E as identity. We call *Chinese hypergroupoid* the hyperstructure $< G^*; \hat{\circ} >$, where $G^* = \bigcup_{A\in\mathcal{G}} A$, and $\forall (x,y) \in G^*\times G^*$, $x \,\hat{\circ}\, y = \bigcup_{\substack{A\ni x, B\ni y\\ \{A,B\}\subset\mathcal{G}}} A \cdot B$.

Set $\forall x \in G^*$, $\alpha(x) = \{A \mid A \in \mathcal{G},\ A \ni x\}$ and $A(x) = \bigcup_{A\in\alpha(x)} A$.

92. Lemma. $\forall (x,y) \in G^* \times G^*$, *we have*

$$x \,\hat{\circ}\, y = A(x) \cdot A(y).$$

Proof. Indeed, if $z \in x \,\hat{\circ}\, y$, then A, B exist in $\mathcal{G}$ such that $z \in A\cdot B$, $A \ni x$, $B \ni y$. We clearly have $A \subset A(x)$, $B \subset A(y)$ whence $x \,\hat{\circ}\, y \subset A(x) \cdot A(y)$.

On the converse, set $w \in A(x) \cdot A(y)$. Then A, B exist in $\mathcal{G}$ such that $w \in A \cdot B$, $A \ni x$, $B \ni y$ whence $A(x) \cdot A(y) \subset x \,\hat{\circ}\, y$. ∎

93. Theorem. *The hypergroupoid* $< G^*; \hat{\circ} >$ *is an* H_v*-group. Moreover, it is clearly an* H_b*-group* [440].

Proof. Let us see first that it is a quasi-hypergroup. Indeed, $\forall (a,b) \in G^{*2}$, $\forall (A,B) \in \mathcal{G}\times\mathcal{G}$ such that $A \ni a$, $B \ni b$, there exists $X \in \mathcal{G}$ such that $A = B \cdot X$; follows $a \in B \cdot X$. Therefore $\forall x \in X$, we have $a \in b\hat{\circ}x$.

By the same way we find $y \in \mathcal{G}$ such that $y \in Y$. We have $a \in y \,\hat{\circ}\, b$.

Let us prove now that $<G^*; \hat{\circ}>$ satisfies the condition

$$(x \,\hat{\circ}\, y) \,\hat{\circ}\, z \cap x \,\hat{\circ}\, (y \,\hat{\circ}\, z) \neq \emptyset.$$

$\forall\,(x,y,z)\in G^{*3}$, we have

$$(x\,\hat{\circ}\,y)\,\hat{\circ}\,z=(A(x)\cdot A(y))\,\hat{\circ}\,z=$$

$$=\left(\left(\bigcup_{A_1\ni x}A_1\right)\cdot\left(\bigcup_{A_2\ni y}A_2\right)\right)\,\hat{\circ}\,z=\left(\bigcup_{A_1\ni x,A_2\ni y}A_1\cdot A_2\right)\,\hat{\circ}\,z=$$

$$=\bigcup_{A_u\ni u\in A_1'A_2',\,A_1'\ni x,\,A_2'\ni y,\,A_3\ni z}A_uA_3\supset(A_1A_2)A_3$$

$\forall\,(A_1,A_2,A_3)\in\mathcal{G}^3$ such that $A_1\ni x,\ A_2\ni y,\ A_3\ni z$

$$x\,\hat{\circ}\,(y\,\hat{\circ}\,z)=\bigcup_{A_1\ni x,\ A_v\ni v\in A_2'A_3'}A_1A_v\supset A_1(A_2A_3)$$

for $\forall\,(A_1,A_2,A_3)\in\mathcal{G}^3$ such that $A_1\ni x,\ A_2\ni y,\ A_3\ni z$.

Therefore $< G^*;\,\hat{\circ} >$ is an H_v-group. ■

94. Proposition. *If $\mathcal{G}$ is an HX-group such that*

(1) $$\forall\,(A,B)\in\mathcal{G}\times\mathcal{G},\ A\cap B\neq\emptyset\Longrightarrow A=B,$$

then $< G^*;\,\hat{\circ} >$ is a hypergroup.

Proof. It is enough to remark the condition (1) implies $\forall\,x\in G^*$, $|\alpha(x)|=1$.

So $<\,\hat{\circ}\,>$ is associative.

Therefore, since $< G^*;\,\hat{\circ} >$ is a quasi–hypergroup, it is a hypergroup. ■

95. Proposition. *Let $\mathcal{G}$ be an HX-group, then $< G^*;\,\hat{\circ} >$ is a regular reversible H_v-group, moreover it is feebly quasi–canonical.*

Proof. Indeed, $\forall\,p\in E$, $\forall\,x\in G^*$, we have

$$x\,\hat{\circ}\,p=A(x)\cdot A(p)\supset A(x)\cdot E\supset A(x)\ni x.$$

Moreover, $\forall x \in G^*$, $\forall y \in A^{-1}$ for $x \in A \in \mathcal{G}$ we have

$$x \hat{\circ} y \supset A(x) \cdot A^{-1} \supset A \cdot A^{-1} = E.$$

Finally, if $a \in b \,\hat{\circ}\, c = A(b) \cdot A(c)$ then $\exists b' \in \bigcup_{B \in \alpha(b)} B$, $\exists c' \in \bigcup_{C \in \alpha(c)} C$ such that $a = b'c'$. Follows $\exists A \in \alpha(a)$, $\exists B' \in \alpha(b)$, $\exists C' \in \alpha(a)$ such that $A = B'C'$; follows $B' = AC'^{-1}$ where $C'C'^{-1} = C^{-1}C = E$.

Hence $b' \in A(a) \cdot A(c'')$, $\forall c'' \in C'^{-1}$ from which $b \in a \,\hat{\circ}\, c''$ and $c'' \in i(c)$. ■

96. Proposition. *Let $\mathcal{G}$ be an abelian HX-group such that the condition (1) is satisfied. Then $< G^*; \hat{\circ} >$ is a join space and a feebly quasi–canonical hypergroup.*

Proof. We shall denote the operation of G by "+".

Set $x \in a/b \cap c/d$, then $x \in G^*$ exists such that $a \in x \hat{\circ} b = x + B$ where $\alpha(b) = \{B\}$, $\alpha(x) = \{X\}$ and $c \in x \,\hat{\circ}\, d = X + D$, where $\alpha(d) = \{D\}$. Then if $\alpha(a) = \{A\}$, $\alpha(c) = \{C\}$, we have $A = X + B$, $C = X + D$, whence $D = -X + C$. Follows

$$\begin{aligned} A + D = X + B + (-X) + C = B + C + X + (-X) = \\ = B + C + E = B + C. \end{aligned}$$

$< G^*; \hat{\circ} >$ is feebly canonical by Proposition 4. ■

97. Proposition. *Let $< G_1^*; \hat{\circ}_1 >$ and $< G_2^*; \hat{\circ}_2 >$ be Chinese hypergroupoids. Then the cartesian product $< G_1^* \times G_2^*; \circ >$ with a product defined*

$$(x_1, x_2) \circ (y_1, y_2) = (x_1 \,\hat{\circ}_1\, y_1, x_1 \,\hat{\circ}_2\, y_2)$$

is again a Chinese hypergroupoid.

Proof. Indeed,

$$(x_1 \,\hat{\circ}_1\, y_1, x_2 \,\hat{\circ}_2\, y_2) =$$

$$= \bigcup_{A_1 \in \alpha(x_1),\, B_1 \in \alpha(y_1)} (A_1 \cdot B_1) \times \bigcup_{A_2 \in \alpha(x_2),\, B_2 \in \alpha(y_2)} (A_2 \cdot B_2) =$$

$$= \bigcup_{\forall i,\, A_i \in \alpha(x_i),\, B_i \in \alpha(y_i)} (A_1, A_2) \circ (B_1, B_2) = (x_1, x_2) \,\hat{\circ}\, (y_1, y_2)$$

where $< \hat{\circ} >$ is the hyperoperation corresponding to the structure of HX-group in $G_1 \times G_2$ defined by $\mathcal{G} = \{(A_1, A_2) \mid A_i \in \mathcal{G}_i\}$. ■

§8. Hyperstructures associated with ordered sets

In this paragraph some new hyperoperations are introduced in a different context and analyzed. The setting is as follows. In the place of the membership function $\mu : U \rightarrow [0, 1]$, we have a function $\widetilde{A}$ from a finite universe H to a totally ordered set $< V; \leq >$.

We shall suppose in the following U to be a non empty finite set, $\widetilde{A}$ to be a function from U to a totally orderd set $< V, \leq >$. We shall denote, for $\forall\, (x, y) \in U^2$

$\widetilde{A}(x) \vee \widetilde{A}(y)$ the maximum between $\widetilde{A}(x)$ and $\widetilde{A}(y)$, and
$\widetilde{A}(x) \wedge \widetilde{A}(y)$ the minimum between $\widetilde{A}(x)$ and $\widetilde{A}(y)$.

These results have been obtained by P. Corsini.

We consider the following hyperoperations $\eta_1, ..., \eta_4$ and we shall analyze their properties.

Let be $\forall\, (a, b) \in U^2$

$\eta_1)$ $a \overset{<}{\underset{\vee}{\circ}} b = \{x \in U \mid \widetilde{A}(x) \leq \widetilde{A}(a) \vee \widetilde{A}(b)\}$

$\eta_2)$ $a \overset{>}{\underset{\vee}{\circ}} b = \{y \in U \mid \widetilde{A}(y) \geq \widetilde{A}(a) \vee \widetilde{A}(b)\}$

$\eta_3)$ $a \overset{<}{\underset{\wedge}{\circ}} b = \{u \in U \mid \widetilde{A}(u) \leq \widetilde{A}(a) \wedge \widetilde{A}(b)\}$

$\eta_4)$ $a \underset{\wedge}{\overset{>}{\circ}} b = \{v \in U \mid \widetilde{A}(v) \geq \widetilde{A}(a) \wedge \widetilde{A}(b)\}$

98. Theorem.

1) *The hypergroupoids* $\eta_1), \eta_2), \eta_3), \eta_4)$ *are associative.*

2) $\eta_1), \eta_2), \eta_3), \eta_4)$ *are endowed with identities.*

3) $\eta_1), \eta_4)$ *are hypergroups.*

Proof. 1) We can remark that $\forall\,(a,b,c) \in U^3$ we have

$$
\begin{aligned}
(a \underset{\vee}{\overset{<}{\circ}} b) \underset{\vee}{\overset{<}{\circ}} c &= \{x \in U \mid \widetilde{A}(x) \leq \widetilde{A}(a) \vee \widetilde{A}(b) \vee \widetilde{A}(c)\} = \\
&= a \underset{\vee}{\overset{<}{\circ}} (b \underset{\vee}{\overset{<}{\circ}} c) \\
(a \underset{\vee}{\overset{>}{\circ}} b) \underset{\vee}{\overset{>}{\circ}} c &= \{y \in U \mid \widetilde{A}(y) \geq \widetilde{A}(a) \vee \widetilde{A}(b) \vee \widetilde{A}(c)\} = \\
&= a \underset{\vee}{\overset{>}{\circ}} (b \underset{\vee}{\overset{>}{\circ}} c) \\
(a \underset{\wedge}{\overset{>}{\circ}} b) \underset{\wedge}{\overset{>}{\circ}} c &= \{z \in U \mid \widetilde{A}(z) \geq \widetilde{A}(a) \wedge \widetilde{A}(b) \wedge \widetilde{A}(c)\} = \\
&= a \underset{\wedge}{\overset{>}{\circ}} (b \underset{\wedge}{\overset{>}{\circ}} c) \\
(a \underset{\wedge}{\overset{<}{\circ}} b) \underset{\wedge}{\overset{<}{\circ}} c &= \{u \in U \mid \widetilde{A}(u) \leq \widetilde{A}(a) \wedge \widetilde{A}(b) \wedge \widetilde{A}(c)\} = \\
&= a \underset{\wedge}{\overset{<}{\circ}} (b \underset{\wedge}{\overset{<}{\circ}} c)
\end{aligned}
$$

2) Set

$$
\begin{aligned}
x_0 &: \widetilde{A}(x_0) = \min\{\widetilde{A}(x) \mid x \in U\} \\
x_1 &: \widetilde{A}(x_1) = \max\{\widetilde{A}(x) \mid x \in U\}
\end{aligned}
$$

We have clearly that x_0 is an identity for both $< U; \underset{\vee}{\overset{<}{\circ}} >$ and $< U; \underset{\vee}{\overset{>}{\circ}} >$.

x_1 is an identity for both $< U; \underset{\wedge}{\overset{<}{\circ}} >$ and $< U; \underset{\wedge}{\overset{>}{\circ}} >$.

3) $\forall\,(a,b) \in U$ we have $a \in a \underset{\vee}{\overset{<}{\circ}} b$ and $a \in a \underset{\wedge}{\overset{>}{\circ}} b$. ∎

99. Theorem.

1) *U coincides with the set of identities of* $< U; \overset{<}{\underset{\vee}{\circ}} >$ *and* $< U; \overset{>}{\underset{\wedge}{\circ}} >$.

2) *The set of identities of* $< U; \overset{>}{\underset{\vee}{\circ}} >$ *is* $\widetilde{A}^{-1}\widetilde{A}(x_0)$.
The set of identities of $< U; \overset{<}{\underset{\wedge}{\circ}} >$ *is* $\widetilde{A}^{-1}\widetilde{A}(x_1)$.

3) $< U; \overset{<}{\underset{\vee}{\circ}} >$ *and* $< U; \overset{>}{\underset{\wedge}{\circ}} >$ *are regular reversible hypergroups.*

4) $< U; \overset{>}{\underset{\vee}{\circ}} >$ *and* $< U; \overset{<}{\underset{\wedge}{\circ}} >$ *are not hypergroups.*

Proof. 1) Indeed $\forall x \in U,\ \forall u \in U$, we have

$$\widetilde{A}(x) \leq \widetilde{A}(x) \vee \widetilde{A}(u),\ \ \widetilde{A}(x) \geq \widetilde{A}(x) \wedge \widetilde{A}(u)$$

whence $x \in x \overset{<}{\underset{\vee}{\circ}} u,\ \ x \in x \overset{>}{\underset{\wedge}{\circ}} u$.

2) Set $z \in A(x_0),\ \ v \in A(x_1),\ \ x \in U$. Then

$$\widetilde{A}(x) = \widetilde{A}(x) \vee \widetilde{A}(z) \text{ whence } x \in x \overset{>}{\underset{\vee}{\circ}} z$$

$$\widetilde{A}(x) = \widetilde{A}(x) \wedge \widetilde{A}(v) \text{ whence } x \in x \overset{<}{\underset{\wedge}{\circ}} v.$$

On the other hand, if $y \notin A(x_0)$ we have

$$x_0 \notin x_0 \overset{>}{\underset{\vee}{\circ}} y$$

It follows that y is not an identity for η_2). Hence $\widetilde{A}^{-1}\widetilde{A}(x_0) =$ $= E^{>}_{\vee}$ = the set of identities of η_2). By a similar proof one sees that $\widetilde{A}^{-1}\widetilde{A}(x_1) = E^{<}_{\wedge}$.

3) It follows from 1).

4) $< U; \overset{>}{\underset{\vee}{\circ}} >,\ < U; \overset{<}{\underset{\wedge}{\circ}} >$ are not hypergroups.

Indeed, if $\widetilde{A}(b) < \widetilde{A}(a)$, x does not exist such that $b \in a \overset{>}{\underset{\vee}{\circ}} x$ that is $\widetilde{A}(b) \geq \widetilde{A}(a) \vee \widetilde{A}(x)$. Similarly, if $\widetilde{A}(b) > \widetilde{A}(a)$, y does not exist such that $\widetilde{A}(b) \leq \widetilde{A}(a) \wedge \widetilde{A}(y)$, that is $a \in b \overset{<}{\underset{\wedge}{\circ}} y$. ■

100. Definition.

I. We call *quasi–join space* a commutative semi–hypergroup which satisfies the condition

j) $a/b \cap c/d \neq \emptyset \Longrightarrow a \circ d \cap b \circ c \neq \emptyset.$

II. We call *semi–join space* a semi–hypergroup which satisfies the condition (j).

101. Theorem. $< U; \overset{<}{\underset{\vee}{\circ}} >$, $< U; \overset{>}{\underset{\wedge}{\circ}} >$ *are join spaces,* $< U; \overset{>}{\underset{\vee}{\circ}} >$, $< U; \overset{<}{\underset{\wedge}{\circ}} >$ *are semi–join spaces.*

Proof. Set $\forall\,(a,b) \in U^2$

$$\begin{array}{lll} a \wedge b = a & \text{if and only if} & A(a) \leq A(b) \\ a \wedge b = b & \text{if and only if} & A(b) \leq A(a) \\ a \vee b = a & \text{if and only if} & A(a) \geq A(b) \\ a \vee b = b & \text{if and only if} & A(b) \geq A(a) \end{array}$$

Follows $\widetilde{A}(a \wedge b) = \widetilde{A}(a) \wedge \widetilde{A}(b)$, $\widetilde{A}(a \vee b) = \widetilde{A}(a) \vee \widetilde{A}(b)$.

ε_1) Set $(a,b,c,d) \in U^4$, $y = a \wedge c$. We have

$$\begin{array}{lll} y \leq a \leq a \vee d & \text{whence} & y \in a \overset{<}{\underset{\vee}{\circ}} d \\ y \leq c \leq b \vee c & \text{whence} & y \in b \overset{<}{\underset{\vee}{\circ}} c \end{array}$$

So $a \overset{<}{\underset{\vee}{\circ}} d \cap b \overset{<}{\underset{\vee}{\circ}} c \neq \emptyset$.

Hence $< U; \overset{<}{\underset{\vee}{\circ}} >$ by 3) Theorem 98 is a join space.

ε_4) By a similar proof one sees that

$$a \vee c \in a \overset{>}{\underset{\wedge}{\circ}} d \cap b \overset{>}{\underset{\wedge}{\circ}} c.$$

So, by 3) Theorem 98 $< U; \overset{>}{\underset{\wedge}{\circ}} >$ is a join space.

ε_2) Set $x \in a/b \cap c/d$. Then

$$a \in b \overset{>}{\underset{\vee}{\circ}} x = \{z \mid \widetilde{A}(z) \geq \widetilde{A}(b) \vee \widetilde{A}(x)\} \text{ and}$$
$$c \in d \overset{>}{\underset{\vee}{\circ}} x = \{z \mid \widetilde{A}(z) \geq \widetilde{A}(d) \vee \widetilde{A}(x)\}$$

Follows

$$\widetilde{A}(a \vee c) = \widetilde{A}(a) \vee \widetilde{A}(c) \geq \widetilde{A}(b) \vee \widetilde{A}(d) \vee \widetilde{A}(x) \geq \widetilde{A}(b) \vee \widetilde{A}(d).$$

Hence

$$\widetilde{A}(a \vee c) \geq \widetilde{A}(a) \vee \widetilde{A}(d) \quad \text{whence} \quad a \vee c \in a \overset{>}{\underset{\vee}{\circ}} d$$
$$\widetilde{A}(a \vee c) \geq \widetilde{A}(b) \vee \widetilde{A}(c) \quad \text{whence} \quad a \vee c \in b \overset{>}{\underset{\vee}{\circ}} c.$$

Therefore $< U; \overset{>}{\underset{\vee}{\circ}} >$, by Theorem 98 is a quasi–join space.

ε_3) By a similar proof one sees that if in $< U; \overset{<}{\underset{\wedge}{\circ}} >$, $a/b \cap c/d \neq \emptyset$ then $a \wedge c \in a \overset{<}{\underset{\wedge}{\circ}} d \cap b \overset{<}{\underset{\wedge}{\circ}} c \neq \emptyset$. Therefore $< U; \overset{<}{\underset{\wedge}{\circ}} >$ is by Theorem 98 a semi–join space. ■

Chapter 6

Automata

The definition of an automaton, we shall present here, has its origins in a paper of Kleene (1956). The title "Representation of events in nerve sets and finite automata" of Kleene's paper gives an idea of its motivation.

The concept of automaton had led to important results, both in mathematics and in theoretical computer science.

Automata are in fact very familiar objects, in the shape of coin machines.

The last twenty years have developed a body of research known under the names of Automaton Theory and Formal Language Theory.

We mention Biology between the fields which have significant connections with Automaton Theory.

Here, we have presented the connections of Automaton Theory and Language Theory with another field, known as Hyperstructure Theory.

Using tools and methods of Hyperstructure Theory, G.G. Massouros gave a new proof of the famous Kleene's Theorem, which states that:

> *"A subset of the set of words M^* is acceptable from an automaton $\mathcal{M}$ if and only if it is defined by a regular expression."*

As follows, an association is attempted between Automaton and Language Theory and Hyperstructure Theory.

In the following sections, we shall present some important results on these topics, obtained by G.G. Massouros and by G.G. Massouros & J. Mittas (see §1, §2) and then by J. Chvalina & L. Chvalinová (see §3).

§1. Language theory and hyperstructures

Let M be an alphabet, M^* the set of words defined over M (M^* is the closure of M), λ the empty word. The set M^* endowed with the operation of *concatenation* of the words, that is $x \cdot y = xy$, is a monoid, with neutral element the empty word.

The *length* $\ell(x)$ of a word $x \in M^*$ is the number of its letters, so $\ell(\lambda) = 0$ and $\forall\,(x, y) \in M^* \times M^*$, $\ell(xy) = \ell(x) + \ell(y)$.

Let us define on M^* the following hyperoperation:

$$\forall\,(x, y) \in M^* \times M^*,\ x + y = \{x, y\}.$$

1. Proposition. *$< M^*, + >$ is a join space.*

Proof. Indeed, $< M^*, + >$ is a commutative hypergroup and moreover $\forall\,(x, y) \in M^* \times M^*$, we have

$$x/y = \{z \in M^* \mid x \in y + z\} = \begin{cases} x, & \text{if } x \neq y \\ M^*, & \text{if } x = y \end{cases},$$

whence it is clear that $\forall\,(x, y, z, w) \in M^{*4}$, the following implication holds:

$$x/y \cap z/w \neq \emptyset \Longrightarrow x + w \cap y + z \neq \emptyset.$$

■

2. Definition. A *hyperringoid* is a structure $< H, +, \cdot >$, where $< H, + >$ is a join space, $< H, \cdot >$ is a semigroup and the multiplication "$\cdot$" is bilaterally distributive with respect to the hyperoperation "$+$".

3. Remark. $< M^*, +, \cdot >$ is a unitary hyperringoid.

Indeed, $< M^*, + >$ is a join space, $< M^*, \cdot >$ is a monoid and the concatenation is bilaterally distributive with respect to the addition.

Let us consider now the following binary relation L on M^*:

$$xLy \Longleftrightarrow \ell(x) = \ell(y).$$

This is an equivalence, called *length equivalence.*

It is possible to verify the following:

4. Proposition.

(i) *$< M^*/L, \oplus >$ is a join space, where* $\forall\, (\bar{x}, \bar{y}) \in (M^*/L)^2,\ \bar{x} \oplus \bar{y} = \{\bar{x}, \bar{y}\}$.

(ii) *$< M^*/L, \oplus, \odot >$ is a unitary hyperringoid, where $\forall\, (\bar{x}, \bar{y}) \in (M^*/L)^2$, $\bar{x} \odot \bar{y} = \overline{xy}$ and the multiplicatively neutral element is $\overline{\lambda} = \{\lambda\}$.*

(iii) *If we set $\forall\, (x, y) \in M^{*2}$, $x \boxplus y = \bar{x} \cup \bar{y}$, then $< M^*, \boxplus >$ is a join space.*

(iv) *$< M^*, \boxplus, \cdot >$ is a unitary hyperringoid, where "$\cdot$" is the concatenation and the multiplicatively neutral element is λ.*

5. Definition. A join space $< H, + >$ is called *fortified join space* if the following conditions hold:

(i) there is a unique neutral element denoted by 0 (the *zero* of H), that is $\exists 0 \in H$, such that $\forall x \in H$, $x \in 0 + x$;

(ii) every element x of H has exactly one inverse $-x$, that is $\forall x \in H$, $\exists! -x \in H$, such that $0 \in x + (-x) = x - x$;

(iii) the hypergroup $< H, + >$ is *partially reversible*, that is: $\forall\, (x, y, z) \in H^3$, if $z \in x+y$, then either $y \in z-x$ or $x \in z-y$.

6. Definition. Let $< H, + >$ be a join space. If the following axioms are satisfied:

(i) there exists a unique neutral element 0, such that every nonzero element x of H has a nonempty set $i(x)$ of nonzero inverses of x in H (with respect to 0);

(ii) the hypergroup $< H, + >$ is partially reversible, that is:
$a \in x+y \Longrightarrow (\exists x' \in i(x),\ y \in a+x' \text{ or } \exists y' \in i(y),\ x \in a+y')$

then $< H, + >$ is called *polysymmetrical fortified join space.*

7. Definition. A hyperringoid $(H, +, \cdot)$ is called *fortified* if its additive structure is fortified and its zero element is a bilaterally absorbing element for the multiplication, that is

$$\forall\, x \in H,\ 0x = x0 = 0.$$

Let us adjoin to the set M^* an element 0, considering it as a *zero element*, with the properties:

$$\forall\, x,\ 0x = x0 = 0,\ 0 + x = \{0, x\},\ x + x = \{0, x\}.$$

Let $\overline{M} = M^* \cup \{0\}$. We obtain the following

8. Proposition. *$< \overline{M}, +, \cdot >$ is a fortified unitary hyperringoid.*

We notice that if the length $\ell(0)$ of the zero word were the natural number 0, then the length equivalence in $\overline{M}$ would not be compatible with respect to the multiplication. Indeed, $\forall\, x \in \overline{M}$, since $\hat{0} = \{0, \lambda\}$, we would have $\hat{0} \cdot \hat{x} = \widehat{0x} = \widehat{\lambda x}$, so $\hat{0} = \widehat{\lambda x}$, which is absurd (where $\forall\, x \in \overline{M}, \hat{x}$, is the equivalence class of x).

But, we can define the *order* of a word x (ord x) on $\overline{M}$ in the following manner:

$$\forall\, x \in M^*,\ \operatorname{ord} x = \ell(x) + 1 \text{ and } \operatorname{ord} 0 = 0.$$

Let $\sim$ be the following relation on $\overline{M}$:

$$x \sim y \Longleftrightarrow \operatorname{ord} x = \operatorname{ord} y.$$

"$\sim$" is an equivalence relation, which is called *order equivalence*. Its restriction on M^* coincides with the length equivalence on M^*. Similarly as in Proposition 4, we can define $\oplus$ and $\odot$ on $\overline{M}/\sim$. The relation "$\sim$" is compatible with respect to both the hyperoperation and the operation of $< \overline{M}, \oplus, \odot >$. Thus we have the

9. Proposition. $< \overline{M}/\sim, \oplus, \odot >$ *is a unitary fortified hyperringoid.*

§2. Automata and hyperstructures

10. Definition. An *automaton* is a 5-tuple (S, M, S_0, F, t), where S is a finite set of states, M is an alphabet of input letters, S_0 and F are the set of the start and final states, respectively and t is a state transition function.

If the automaton is *deterministic*, then t has the domain $S \times M$ and range S. If the automaton is *nondeterministic*, then t has the domain $S \times M$ and range $P(S)$.

We shall define on S several hyperoperations, such that we obtain hypergroup structures on S.

I. The attached order hypergroup

We suppose that there exists a *conventional* start state $s_{0'}$, so that every state $s \in S$ is connected to $s_{0'}$ (see Definition 14).

11. Definition. The *order* of a state $s \in S$ is the natural number $\ell + 1$, where ℓ is the minimum of the lengths of words which lead from the conventional start state $s_{0'}$ to s.

We denote the order of s by $\operatorname{ord} s$.

We define $\operatorname{ord} s_{0'} = 0$.

Let us define now on S the following *order* equivalence: if $(s_1, s_2) \in S^2$, $s_1 \sim s_2 \Longleftrightarrow \operatorname{ord} s_1 = \operatorname{ord} s_2$.

For any $s \in S$, let $\hat{s} = \{s' \in S \mid s \sim s'\}$.

Let us consider the following commutative hyperoperations on S:

$$1^\circ\ \forall\,(s_1, s_2) \in S^2,\ s_1 + s_2 = \begin{cases} s_2, & \text{if } \operatorname{ord} s_1 < \operatorname{ord} s_2 \\ \bigcup\limits_{\operatorname{ord} s < \operatorname{ord} s_1} \hat{s}, & \text{if } \operatorname{ord} s_1 = \operatorname{ord} s_2 \neq 0 \\ s'_0, & \text{if } s_1 = s_2 = s_{0'}. \end{cases}$$

$$2^\circ\ \forall\,(s_1, s_2) \in S^2,\ s_1 + s_2 = \begin{cases} s_2, & \text{if } \operatorname{ord} s_1 < \operatorname{ord} s_2 \\ \bigcup\limits_{\operatorname{ord} s < \operatorname{ord} s_1} \hat{s}, & \text{if } \operatorname{ord} s_1 = \operatorname{ord} s_2 \text{ and } \\ & s_{0'} \neq s_1 \neq s_2 \neq s_{0'} \\ \bigcup\limits_{\operatorname{ord} s \leq \operatorname{ord} s_1} \hat{s}, & \text{if } s_1 = s_2. \end{cases}$$

$$3^\circ\ \forall\,(s_1, s_2) \in S^2,\ s_1 + s_2 = \begin{cases} \hat{s}_2, & \text{if } 0 \neq \operatorname{ord} s_1 < \operatorname{ord} s_2 \\ \bigcup\limits_{0 \neq \operatorname{ord} s \leq \operatorname{ord} s_1} \hat{s}, & \text{if } \operatorname{ord} s_1 = \operatorname{ord} s_2 \text{ and } \\ & s_{0'} \neq s_1 \neq s_2 \neq s_{0'} \\ \bigcup\limits_{\operatorname{ord} s \leq \operatorname{ord} s_1} \hat{s}, & \text{if } s_{0'} \neq s_1 = s_2 \\ s_2 = s_2 + s_{0'}, & \text{if } s_1 = s_{0'}. \end{cases}$$

In each case, $< S, + >$ is a canonical hypergroup.

II. The attached grade hypergroup

Let (S, M, S_0, F, t) be a deterministic automaton.

12. Definition. We call *grade* of a state $s \in S$ and we denote it by grad s, the set $\{x \in M^* \mid t^*(s,x) \in F\}$, where $t^* : S \times M^* \longrightarrow S$ is the extended state transition function, which is defined recursively as follows:

$$\forall\, s \in S,\ \forall\, a \in M,\ t^*(s,a) = t(s,a);$$
$$\forall\, s \in S,\ t^*(s,\lambda) = s;$$
$$\forall\, s \in S,\ \forall\, x \in M^*,\ \forall\, a \in M,\ t^*(s,ax) = t^*(t(s,a),x).$$

We define the relation R on the set of states S, as follows:

$$s_1 R s_2 \Longleftrightarrow \operatorname{grad} s_1 = \operatorname{grad} s_2.$$

This relation is an equivalence relation on S, called *grade equivalence.*

Let us denote by $\widetilde{s}_1$ the equivalence class of s_1, with respect to R.

Let us define on S the following hyperoperation

$$s_1 + s_1 = \widetilde{s}_1 \cup \widetilde{s}_2.$$

Then $(S,+)$ is a join space.

Now, let us suppose that the automaton (S, M, S_0, F, t) has only one final state, the state s_T, otherwise we endow it with a conventional one.

We define on S the following hyperoperation "+":

$$s_1 + s_2 = \begin{cases} \widetilde{s}_1 \cup \widetilde{s}_2, & \text{if } \widetilde{s}_1 \neq \widetilde{s}_2 \text{ and } s_1 \neq s_T \neq s_2 \\ \widetilde{s}_1 \cup \{s_T\}, & \text{if } \widetilde{s}_1 = \widetilde{s}_2. \end{cases}$$

Then $< S, + >$ is a polysymmetrical fortified join space, called the *attached grade hypergroup* of the automaton.

13. Remark. If in a polysymmetrical fortified join space H, the family $\{S(x)\}_{x \in H}$ forms a partition of H, then the relation ρ defined by: $x\rho y \Longleftrightarrow S(x) = S(y)$ is an equivalence relation on H and the factor set H/ρ, endowed with the hyperoperation

$$\overline{C}_x + \overline{C}_y = \{\overline{C}_z \mid \overline{C}_x \cup \overline{C}_y\}$$

becomes a fortified join space.

The grade notion is very important for the creation of the minimum automaton which accepts the same language as the initial one. If in an automaton there exist two states of the same grade, then it makes no difference, for the process of reaching the final state, whether we are on one or on the other. By Remark 13, if the attached grade hypergroup is polysymmetrical, then we can construct a fortified join space and so, its corresponding automaton has less states than the original one, but it accepts exactly the same language as it.

III. The attached hypergroup of the paths

14. Definition. The state s_2 of S will be called *connected* to the state s_1 of S if there exists $x \in M^*$, such that $s_2 = t^*(s_1, x)$.

If x consists only in one letter, then the state s_2 is called *successive* to s_1.

Notice that if s_2 is connected (successive) to s_1, this does not imply that s_1 is connected (succesive) to s_2.

We define the following hyperoperation on the set of states S:

$\forall\,(x_1, x_2) \in S^2$,

$$s_1 \circ s_2 = \begin{cases} \{s \in S \mid \exists (x,y) \in M^{*2} \text{ such that } s = t^*(s_1, x) \text{ and} \\ \qquad\qquad s_2 = t^*(s, y)\}, \quad \text{if } s_2 \text{ is connected to } s_1 \\ \{s_1, s_2\}, \qquad\qquad \text{if } s_2 \text{ is not connected to } s_1. \end{cases}$$

Then $(S, \circ)$ is a non–commutative hypergroup.

Using this hypergroup, an important theorem of Languages and Automaton Theory can be proved by tools and methods of

Hypercompositional Structure Theory: the Theorem of Kleene (see [258]).

IV. The attached hypergroup of the operation

Now, we shall point out that an automaton can be in a certain state in a certain moment (clock pulse). In other words, we consider "time" as one of the factors that are involved.

Therefore, it is convenient to consider the cartesian product $S \times \mathbb{N}$ (S being the set of states).

If the automaton is in the state s during the clock pulse t, we write (s, t).

15. Definition. An element (s, t) of $S \times \mathbb{N}$ is called *activated* if after t clock pulses, the automaton can be found in the state s. We say that (s_2, r) is *succesive* to (s_1, t) if s_2 is succesive to s_1 and $r = t + 1$.

We say that (s_2, r) is *connected* to (s_1, t) if s_2 is connected to s_1 and $t < r$.

Let $A \subseteq S \times \mathbb{N}$ be the set of activated elements and tA^* the generalization of the extended state transition function t^*, that is $tA^* : (S \times \mathbb{N}) \times M^* \longrightarrow S \times \mathbb{N}$, $tA^*((s,t), x) = (t^*(s,x), t + |x|)$, where $|x|$ is the length of the word x. We define on A the following hyperoperation:

$$(s_1, m) \circ (s_2, n) = \begin{cases} \{tA^*((s_1, m), x) \mid & x \in \text{Prefix}\, r \text{ and} \\ & tA^*((s_1, m), r) = (s_2, n)\}, \\ & \text{if } (s_2, n) \text{ is connected} \\ & \text{to } (s_1, m) \\ \{(s_1, m), (s_2, n)\} & \text{otherwise.} \end{cases}$$

16. Proposition. *$(A, \circ)$ is a non commutative hypergroup.*

Proof. "$\circ$" is associative. Indeed, if (s_j, n) is connected to (s_i, m) and if (s_k, p) is connected to (s_j, n), then we have:

$$\begin{aligned}
&((s_i, m)\circ(s_j, n))\circ(s_k, p) = \\
&\quad = \{tA^*((s_i, m), x) \mid x \in \text{Prefix}\, r, tA^*((s_i, m), r) = (s_j, n)\}\circ(s_k, p) = \\
&\quad = \{tA^*(tA^*((s_i, m), x), y) \mid x \in \text{Prefix}\, r, tA^*((s_i, m), r) = (s_j, n), \\
&\qquad\qquad y \in \text{Prefix}\, q, tA^*(tA^*((s_i, m), x), q) = (s_k, p)\} = \\
&\quad = \{tA^*((s_i, m), v) \mid v \in \text{Prefix}\, w, tA^*((s_i, m), w) = (s_k, p)\}.
\end{aligned}$$

On the other hand,

$$\begin{aligned}
&(s_i, m)\circ((s_j, n)\circ(s_k, p)) = \\
&\quad = (s_i, m)\circ\{tA^*((s_j, n), x) \mid x \in \text{Prefix}\, r, tA^*((s_j, n), r) = (s_k, p)\} = \\
&\quad = \{tA^*((s_i, m), z) \mid z \in \text{Prefix}\, u, tA^*((s_i, m), u) = (p_k, p) \text{ or} \\
&\quad tA^*((s_i, m), u) = tA^*((s_j, n), x), x \in \text{Prefix}\, r, tA^*((s_j, n), r) = \\
&\quad = (s_k, p) = \{tA^*((s_i, m), v) \mid v \in \text{Prefix}\, w, tA^*((s_i, m), w) = (s_k, p)\}.
\end{aligned}$$

Now, suppose that (s_j, n) and (s_k, p) are connected to (s_i, m), and (s_k, p) is not connected to (s_j, n).

Then

$$\begin{aligned}
&((s_i, m)\circ(s_j, n))\circ(s_k, p) = \\
&\quad = \{tA^*((s_i, m), x) \mid x \in \text{Prefix}\, r, tA^*((s_i, m), r) = (s_j, n)\}\circ(s_k, p) = \\
&\quad = ((s_i, m)\circ(s_j, n)) \cup ((s_i, m)\circ(s_k, p)) \text{ and} \\
&(s_i, m)\circ((s_j, n)\circ(s_k, p)) = \\
&\quad = (s_i, m)\circ\{(s_j, n), (s_k, p)\} = ((s_i, m)\circ(s_j, n)) \cup ((s_i, m)\circ(s_k, p)).
\end{aligned}$$

Let us suppose that (s_j, n) is connected to (s_i, m) and (s_k, p) is not connected to anyone of the other two.

Then

$$\begin{aligned}
&((s_i, m)\circ(s_j, n))\circ(s_k, p) = \\
&\quad = \{tA^*((s_i, m), x) \mid x \in \text{Prefix}\, r, tA^*((s_i, m), r) = (s_j, n)\}\circ(s_k, p).
\end{aligned}$$

The element (s_k, p) is not connected to anyone of $tA^*((s_i, m), x)$, otherwise it would result that (s_k, p) is connected to (s_i, m), which is absurd.

Therefore

$$((s_i, m)\circ(s_j, n))\circ(s_k, p) =$$
$$=\{tA^*((s_i, m), x) \mid x\in \text{Prefix}\, r, tA^*((s_i, m), r)=(s_j, n)\} \cup \{(s_k, p)\}.$$

On the other hand,

$$(s_i, m)\circ((s_j, n)\circ(s_k, p)) = (s_i, m)\circ\{(s_j, n), (s_k, p)\} =$$
$$= ((s_i, m)\circ(s_j, n))\cup((s_i, m)\circ(s_k, p)) =$$
$$= \{tA^*((s_i, m), x) \mid x\in \text{Prefix}\, r, tA^*((s_i, m), r)=(s_j, n)\}\cup\{(s_k, p)\}.$$

If the elements $(s_i, m), (s_j, n)$ and (s_k, p) are not connected, then

$$((s_i, m)\circ(s_j, n))\circ(s_k, p) = \{(s_i, m), (s_j, n)\}\circ(s_k, p) =$$
$$= ((s_i, m)\circ(s_k, p)) \cup ((s_j, n)\circ(s_k, p)) = \{(s_i, m), (s_j, n), (s_k, p)\}$$

and similarly we obtain

$$(s_i, m)\circ((s_j, n)\circ(s_k, p)) = \{(s_i, m), (s_j, n), (s_k, p)\}.$$

Therefore "$\circ$" is associative.

Moreover, $\forall\, (s_i, m) \in A$, we have

$$(s_i, m)\circ A = A = A\circ(s_i, m).$$

Notice that "$\circ$" is not commutative. ■

Using this hypergroup, all the states at which the automaton can possibly be found, at a given moment t, may be effectively determined.

§3. Automata and quasi–order hypergroups

In the following, some basic properties of automata are described, using their corresponding hyperstructures.

From now on, we shall denote an automaton by a triplet (S, M, τ), where S is the set of states, M the alphabet ($M \neq \emptyset$) and $\tau = t^* : S \times M^* \longrightarrow S$ is the extended state transition function, satisfying the two conditions: $\tau(s, \lambda) = s$, $\forall\, s \in S$ and $\tau(s, ab) = \tau(\tau(s, a), b)$, $\forall\, s \in S$, $\forall\, (a, b) \in M^{*2}$.

A subautomaton of the automaton (S, M, τ) is an automaton (S_0, M, τ_0), where $S_0 \subseteq S$; τ_0 is the restriction of τ on $S_0 \times M^*$ and $\forall\, s_0 \in S_0$, $\forall\, a \in M^*$, $\tau(s_0, a) \in S_0$.

If $S_1 \subseteq S$, let us denote:

$$\tau(S_1, M^*) = \{\tau(s_1, a) \mid s_1 \in S_1,\ a \in M^*\}$$

and $\tau(s_1, M^*)$ instead of $\tau(\{s_1\}, M^*)$.

We shall consider only the automata with nonempty state sets.

17. Definition. A nonempty subautomaton (S_0, M, τ) of an automaton (S, M, τ) is called *separated* if $\tau(S - S_0, M^*) \cap S_0 = \emptyset$. An automaton, with no separated proper subautomaton is called *connected.* An automaton (S, M, τ) is called *strongly connected* if $\forall\, (s, t) \in S^2$, $\exists a \in M^*$, such that $\tau(s, a) = t$.

18. Definition. An automaton (S, M, τ) is called *retrievable* if $\forall\, s \in S$, $\forall\, a \in M^*$, $\exists b \in M^*$, such that $\tau(s, ab) = s$.

It holds the following result:

19. Theorem. *An automaton is retrievable if and only if it is a union of its strongly connected subautomata.* ([17]).

With any automaton (S, M, τ) , we can associate a *quasi–order hypergroup* $(S, \circ)$ (that is $\forall\, (s, t) \in S^2$, we have $s \in s^2 = s^3$

and $s \circ t = s^2 \cup t^2$) in the following manner:

$$s \circ t = \tau(s, M^*) \cup \tau(t, M^*).$$

Indeed, $\forall\, (s,t) \in S^2$, $\{s,t\} \subset s \circ t$, so $s \circ t \neq \emptyset$.

Moreover, $s \circ t = s^2 \cup t^2$, since $s^2 = \tau(s, M^*)$. We also have:

$$\begin{aligned} s^3 &= \bigcup_{u \in \tau(s,M^*)} s \circ u = \tau(s, M^*) \cup \bigcup_{u \in \tau(s,M^*)} \tau(u, M^*) = \\ &= s^2 \cup \{\tau(\tau(s,a_1),a_2) \mid (a_1,a_2) \in M^{*2}\} = \\ &= s^2 \cup \{\tau(s, a_1a_2) \mid a_1a_2 \in M^*\} = s^2 \cup \tau(s, M^*) = s^2 \cup s^2 = s^2. \end{aligned}$$

Notice that $\forall\, (s,t) \in S^2$, we have

$$s \circ t = \widetilde{\rho}(s) \cup \widetilde{\rho}(t), \text{ where } \widetilde{\rho} \subset S^2$$

is defined as follows:

$\widetilde{\rho}$ is the transitive closure of ρ, where $\rho \subset S^2$ and
$s \rho t \iff \exists a \in M : \tau(s,a) = t$.

20. Definition. A quasi–order hypergroup $< H, \circ >$ is called an *order hypergroup* if $\forall\, (a,b) \in H^2$, the following implication holds:

$$a^2 = b^2 \Longrightarrow a = b.$$

21. Definition. A commutative hypergroup $< H, \circ >$ is called *inner irreducible* if for every subhypergroups H_1 and H_2 of H, such that $H = H_1 \circ H_2$, we have $H_1 \cap H_2 \neq \emptyset$.

Now, let us see some relationships between some properties of automata and of their corresponding hypergroups.

22. Theorem.

1) *An automaton (S, M, τ) is connected if and only if its state hypergroup $(S, \circ)$ is inner irreducible.*

2) *An automaton (S, M, τ) is strongly connected if and only if its state hypergroup $(S, \circ)$ satisfies the condition* $\forall\, s \in S,\ S = s^2$.

3) *An automaton (S, M, τ) is retrievable if and only if for any inner irreducible subhypergroup $(K, \circ)$ of the state hypergroup $(S, \circ)$, there exists $k \in K$, such that $K = k^2$.*

Proof. 1) "$\Longrightarrow$" Let us consider (S, M, τ) a connected automaton and S_1, S_2 subhypergroups of $(S, \circ)$, such that $S = S_1 \circ S_2$.

We have $\forall\, s_1 \in S_1,\ \forall\, a \in M^*,\ \tau(s_1, a) \in \tau(s_1, M^*) = s_1 \circ s_1 \subset S_1$, so (S_1, M, τ_1) is a subautomaton of (S, M, τ), where $\tau_1 = \tau/_{S_1 \times M^*}$. Since (S, M, τ) is connected, it follows $\tau(S - S_1, M^*) \cap S_1 \neq \emptyset$, whence $\exists (t_1, t_2) \in (S - S_1) \times S_1$, $\exists a \in M^*$, such that $\tau(t_1, a) = t_2 \in \tau(t_1, M^*) = t_1 \circ t_1$. Since $S = S_1 \circ S_2$ it follows that $\exists (u, v) \in S_1 \times S_2$, such that $t_1 \in u \circ v = \tau(u, M^*) \cup \tau(v, M^*)$.

We have $\tau(u, M^*) = u \circ u \subset S_1, \tau(v, M^*) = v \circ v \subset S_2$ and $t_1 \in S - S_1$, so $t_1 \in v \circ v$, hence $t_2 \in t_1 \circ t_1 \subseteq (v \circ v) \circ (v \circ v) = v^3 \circ v = v^2 \subset S_2$. Then $t_2 \in S_1 \cap S_2$, that is $S_1 \cap S_2 \neq \emptyset$, and so it follows that the state hypergroup $(S, \circ)$ is inner irreducible.

"$\Longleftarrow$" Now, let $(S, \circ)$ be an inner irreducible hypergroup and suppose that the automaton (S, M, τ) is disconnected. Then there exists a separated proper subautomaton (S, M, τ_1) of (S, M, τ), that means

$$\tau(S - S_1, M^*) \cap S_1 = \emptyset, \text{ so } \tau(S - S_1, M^*) \subseteq S - S_1,$$

that is $(S - S_1, \circ)$ is a subhypergroup of $(S, \circ)$.

Since $\tau_1(S_1, M^*) \subseteq S_1$, it follows that $(S_1, \circ)$ is also a subhypergroup of $(S, \circ)$.

Moreover, since $\forall\, s \in S$, $\tau(s, \lambda) = s$ it follows that $\tau(S_1, M^*) = S_1$ and $\tau(S - S_1, M^*) = S - S_1$. We have $(S - S_1) \circ S_1 \subseteq S$. On the other hand, if $s \in S - S_1$, then we consider an arbitrary element t of S_1 and, if $s \in S_1$, we consider an arbitrary t in $S - S_1$, We have

$$s \in \tau(s, M^*) \cup \tau(t, M^*) = s \circ t = t \circ s \subset (S - S_1) \circ S_1.$$

Therefore $S = (S - S_1) \circ S_1$, which is a contradiction with the fact that $(S, \circ)$ is inner irreducible.

2) "$\Longrightarrow$" Suppose that the automaton (S, M, τ) is strongly connected. Let $s \in S$. We have $s \circ s \subset S$ and $\forall t \in S$, $\exists a \in M^*$, such that $t = \tau(s, a) \in \tau(s, M^*) = s \circ s$, so $S \subseteq s \circ s$, whence $S = s^2$.

"$\Longleftarrow$" Conversely, for any $s \in S$, we have $S = s^2 = \tau(s, M^*)$ and so $\forall t \in S$, $\exists a \in M^*$ such that $t = \tau(s, a)$, whence (S, M, τ) is a strongly connected automaton.

3) "$\Longrightarrow$" Let (S, M, τ) be a retrievable automaton. It means that (S, M, τ) is a union of its strongly connected subautomata (S_i, M, τ_i), $i \in I$, where $S_i \cap S_j = \emptyset$ for $(i, j) \in I^2$, $i \neq j$, (otherwise, if $S_i \cap S_j \neq \emptyset$, we would have $S_i = S_j$; indeed, if $s \in S_i \cap S_j$ and t is arbitrary in S_i, then $\exists a \in A^*$, such that $t = \tau_i(s, a) = \tau_j(s, a) \in S_j$, so $S_i \subset S_j$ and, similarly, $S_j \subset S_i$).

Moreover, $(T, \circ)$ is a subhypergroup of $(S, \circ)$ if and only if there is $J \subseteq I$, $J \neq \emptyset$, such that $T = \bigcup_{i \in J} S_i$. The subhypergroup $(T, \circ)$ of $(S, \circ)$ is inner irreducible if and only if $\exists j \in I$, such that $T = S_j$.

Indeed, if $T = \bigcup_{k \in J} S_k$ and J is a subset of I, containing at least two elements, then $\forall i \in J$, we have

$$S_i \circ \left(\bigcup_{\substack{k \in J \\ k \neq i}} S_k \right) = T \text{ and } S_i \cap \left(\bigcup_{\substack{k \in J \\ k \neq i}} S_k \right) = \emptyset,$$

contradiction with the fact that $(T, \circ)$ is inner irreducible.

According with 2) we obtain that $\forall i \in I$, $S_i = s_i^2$ for any $s_i \in S_i$, therefore for any inner irreducible subhypergroup $(T, \circ)$ of $(S, \circ)$ we have $T = S_i$ for some $i \in I$ and $S_i = s_i^2$ for any $s_i \in S_i = T$.

"$\Longleftarrow$" Since any inner irreducible subhypergroup $(T, \circ)$ of $(S, \circ)$ can be written as $T = t^2$, for some $t \in T$ it follows, according to 2), that $(T, M, \tau/T)$ is a strongly connected subautomaton of (S, M, τ). On the other hand, we have:

$$S = \bigcup_{t \in S} t^2 = \bigcup_{\substack{(T,\circ) \text{ inner irreducible} \\ \text{subhypergroup of } (S,\circ)}} T$$

so the automaton (S, M, τ) is retrievable. ■

In the following, we shall give necessary and sufficient conditions, such that the state hypergroup $(S, \circ)$ of an automaton (S, M, τ) is a join space.

23. Proposition. *Let (S, r) be a quasi–order set and $(S, \circ_r)$ the quasi–order hypergroup defined as follows:*

$$\forall (s,t) \in S^2, \;\; s \circ_r t = r(s) \cup r(t).$$

Then the following two conditions are equivalent:

1) *the hypergroup $(S, \circ_r)$ is a join space;*

2) *if a and b are arbitrary elements of S such that $\exists x \in S$, for which xra and xrb, then $\exists y \in S$, such that ary and bry.*

Proof. 1) $\Longrightarrow$ 2) Since xra and xrb it follows $\{a, b\} \subset r(x)$, so $x \in a/b \cap b/a$ and since $(S, \circ_r)$ is a join space, we obtain tht $a^2 \cap b^2 \neq \emptyset$, that is $r(a) \cap r(b) \neq \emptyset$, whence $\exists y \in r(a) \cap r(b)$, that means ary and bry.

2) $\Longrightarrow$ 1) Let $(a, b, c, d) \in S^4$, such that $a/b \cap c/d \ni x$, for some $x \in S$. It follows $a \in r(b) \cup r(x)$ and $c \in r(d) \cup r(x)$.

We have the following situations:

1°) $\{a, c\} \subset r(x)$. Then, by 2), it follows $\exists y \in r(a) \cap r(b)$, whence $(r(a) \cup r(d)) \cap (r(b) \cup r(c)) \neq \emptyset$, that is $a \circ_r d \cap b \circ_r c \neq \emptyset$.

2°) $a \in r(b)$ and $c \in r(d)$. Then $a \in r(a) \cap r(b)$ so $a \circ_r d \cap b \circ_r c \neq \emptyset$.

3°) $a \in r(x)$ and $c \in r(d)$. Then $c \in r(c) \cap r(d)$, so $a \circ_r d \cap b \circ_r c \neq \emptyset$.

4°) $a \in r(b)$ and $c \in r(x)$. Then $a \in r(a) \cap r(b)$, so $a \circ_r d \cap b \circ_r c \neq \emptyset$.

Therefore, $(S, \circ_r)$ is a join space. ∎

24. Theorem. *Let (S, M, τ) be an automaton and $(S, \circ)$ the associated state hypergroup. The following conditions are equivalent:*

1) *the hypergroup $(S, \circ)$ is a join space;*
2) *for any $(s,t) \in S^2$, for which $\exists u \in S$, such that $s \circ t \subseteq u^2$, there exists $v \in S$ with the property $v^2 \subseteq s^2 \cap t^2$.*
3) *for any $(s,t) \in S^2$, for which $\exists (a,b) \in M^{*2}$, $\exists u \in S$, such that $\tau(u,a) = s$, $\tau(u,b) = t$, we have that $\exists (c,d) \in M^{*2}$, such that $\tau(s,c) = \tau(t,d)$.*

Proof. 1)$\Longrightarrow$2) Let r be the quasi–order on S, which determines the hyperoperation "$\circ$". Notice that r is the transitive closure of the relation $\rho \subset S^2$ defined as follows:

$$s_1 \rho s_2 \Longleftrightarrow \exists a \in M, \ \tau(s_1, a) = s_2.$$

Let $(s,t) \in S^2$, such that $\exists u \in S : s \circ t \subseteq u^2$, that is $r(s) \cup r(t) \subseteq$ $\subseteq r(u)$, whence we obtain $s \in r(u)$ and $t \in r(u)$. By the previous proposition, it follows that $\exists v \in S$, such that $v \in r(s)$, $v \in r(t)$. Then $r(v) \subseteq r^2(s) \cap r^2(t) \subseteq r(s) \cap r(t)$, that is $v^2 \subseteq s^2 \cap t^2$.

2)$\Longrightarrow$3) Using the above defined quasi–order r, we have

$$\forall s \in S, \ r(s) = \tau(s, M^*).$$

Let $(s,t) \in S^2$ such that $\exists (a,b) \in M^{*2}$, $\exists u \in S : \tau(u,a) = s$ and $\tau(u,b) = t$. Then $s \in r(u)$ and $t \in r(u)$, whence $s \circ t = r(s) \cup$ $\cup r(t) \subseteq r^2(u) \cup r^2(u) = r^2(u) \subseteq u^2$. By 2) it follows that $\exists v \in S$ such that $v^2 \subseteq s^2 \cap t^2$, hence $v \in v^2 \subseteq r(s) \cap r(t)$, that is srv and trv.

By the definition of r, it follows that there exists $(c,d) \in M^{*2}$, such that

$$\tau(s,c) = v \quad \text{and} \quad \tau(t,d) = v,$$

hence $\tau(s,c) = \tau(t,d)$.

3)$\Longrightarrow$1) Considering the relation r defined as follows:

$$srt \Longleftrightarrow \exists a \in M^* : \tau(s,a) = t,$$

we obtain that 3) is exactly the condition 2) of the previous proposition, so 3)$\Longrightarrow$1). ∎

Chapter 7

Cryptography

For ages, cryptography has been used in military and diplomatic communication, in order to make the meaning of transmitted messages incomprehensible to unauthorized users.

As Francis Bacon said, "The art of ciphering, half for relative an art of deciphering, by supposition unprofitable, but as things are, of great use". Lately, W. Diffie and M. Hellman [126] *point out the new directions in Cryptography.*

In this chapter, we have presented some hyperstructures derived from generalized designs and some cryptographic interpretations on hyperstructures. As being a science in a continuous development, ciphering can still be improved, using a relative new theory, that one of Hyperstructure Theory.

§1. Algebraic cryptography and hypergroupoids

The study of sending messages methods, which cannot be read by an unauthorized person, is called *cryptography*.

One of the most famous cryptography code was introduced before 1500 by the Frenchman Blaise de Vigenere. This code was unbreakable for more than three hundred years. The Vigenere square is one of the first algebraic structures of the history, probably the first one; its ancient name was ZIRUPH. This square is isomorphic to the additive group of residues modulo 26.

A Prussian officer broke the Vigenere code in 1860, using a statistical text: the Kasisky text.

In few words, we explain how a message can be sent to a receiver, using an *algebraic cipher system.*

Let us consider a finite set A, called *alphabet*, a subset K of A, called *key-set* and a binary operation on A, called *enciphering*. The message to be sent (called *cleartext*) is written with the elements of the alphabet A. Using the elements of K, we construct a so–called *keyword*, writing the elements of K, one after the other, respecting the lengths of the cleartext words. We transform the cleartext into a *ciphertext*, using the keyword, in the following manner: each element of the cleartext is replaced by the result of enciphering this element with the corresponding element of the keyword. The obtained ciphertext is sent to the receiver.

In order to reconstruct the initial message, $\forall k \in K$, $\forall a \in A$, the equation $k * x = a$ must have a unique solution, that means $\forall k \in K$, $(k * x)_{x \in A}$ must be a permutation of the alphabet A.

Example. Let $(A, *)$ be a finite groupoid and $K \subseteq A$, such that $\forall (k, a) \in K \times A$, the equation $k * x = a$ has a unique solution in A.

Suppose

$$\begin{aligned} A &= \{a, b, c, d, e, f, ..., z, t\} \text{ and} \\ K &= \{u, v, z, t\} \end{aligned}$$

cleartext: hypergroups have applications

keyword: uvztuvztuvz tuvz tuvztuvztuvz

ciphertext:

$(u{*}h)(v{*}y)(z{*}p)(t{*}e)(u{*}r)(v{*}g)(z{*}r)(t{*}o)(u{*}u)(v{*}p)(z{*}s)$
$(t{*}h)(u{*}a)(v{*}v)(z{*}e)$
$(t{*}a)(u{*}p)(v{*}p)(z{*}\ell)(t{*}i)(u{*}c)(v{*}a)(z{*}t)(t{*}i)(u{*}o)(v{*}n)(z{*}s)$

Using the hyperstructures, we can construct some more sophisticated cryptographic systems.

This topic has been investigated by L. Berardi, F. Eugeni and St. Innamorati and more recently, by R. Migliorato and G. Gentile. In the following, we shall present some results of Berardi, Eugeni and Innamorati.

1. In this case, the key-word contains two secrets: the alphabet and the length of ciphering. The enciphering is now a *hyperoperation.*

Example. Let

$$A = \{a, b, c, ..., z\} \text{ and}$$
$$K = \{b, a, d\};$$

the length of ciphering: 2 (corresponding to b), 1 (corresponding to a), 4 (corresponding to d) and let us consider a hyperoperation "$*$" on A, such that $\forall\, k \in K$, $\forall\, (x, y) \in A^2$, we have

$$k * x = k * y \Longrightarrow x = y.$$

Let us consider

$$\begin{array}{lcl} b * h & = & \text{in} \\ a * y & = & \text{t} \\ d * p & = & \text{eres} \\ b * e & = & \text{ti} \\ a * r & = & \text{n} \\ d * g & = & \text{gand} \\ b * r & = & \text{be} \\ a * o & = & \text{a} \\ d * u & = & \text{utif} \\ b * p & = & \text{ul} \end{array}$$

cleartext: hypergroup
keyword: badbadbadb
Therefore, the *ciphertext* is: interestingandbeautiful.

2. Variable–size cipher system. Let $(A, *)$ be a hypergroupoid and H the set of idempotents of $(A, *)$. (h is idempotent if

$h * h = h$.) The alphabet is A. We use two keys: a *main key* (that belongs to $A - H$) and a *special key* (that belongs to H).

The codification consists in ciphering any clearletter m by the secret main key k, and then, in writting the special key h after the cipher $k * m$.

If the *cleartext* is: $m_1 m_2 ... m_t m_{t+1} ... m_s m_{s+1} ...$ the main key is: $k_1 k_2 ... k_s k_1 k_2 ...$ and the special key is: $h_1 h_2 ... h_t h_1 h_2 ...$ then, the *ciphertext* is:

$$[k_1 * m_1] h_1 * h_1 [k_2 * m_2] h_2 * h_2 ... [k_t * m_t] h_t * h_t ...$$

and, since "$*$" is a hyperoperation, we obtain

$$a_1 a_2 ... a_i h_1 b_1 b_2 ... b_j h_2 ... z_1 z_2 ... z_g h_t ...$$

The receiver knows the special key, but he does not know the position of the special key in the ciphertext. Notice that the cipher $k * m$ could contain the special key h, as in the following example:

ciphertext: albdebfanm
special key: bmbm...

We have two possibilities: the ciphers could be:
al; debfan or
albde; fan.

We can avoid this situation, assuming that for every k of the main key, the corresponding row $\{k * x\}_{x \in A}$ is a Sperner family, which does not happen in our case (indeed, we have "al$\subset$albde" and "fan$\subset$debfan"). Remember that a family $\mathcal{R}$ of subsets of A is a Sperner family if

$$\forall (X, Y) \in \mathcal{R}^2, \quad \text{neither } X \subset Y \text{ nor } Y \subset X.$$

3. The next procedure is called "**how to share pieces of messages**". The idea is the following one: the sender transmits the secret message to two receivers using two different algorithms f and g, respectively, such that none of them can read the message without the permission of the other one. None of the receivers knows the algorithms f and g, but they know an algorithm F that computes the secret message m by the two cipher messages $f(m)$ and $g(m)$, that is they know an algorithm F, such that $F(f(m), g(m)) = m$.

§2. Cryptographic interpretation of some hyperstructures

Let us notice that hyperstructures derived from linear spaces can be obtained; these hyperstructures have cryptographic interpretation.

4. Definition. A *geometric space* is a pair $(P, \mathcal{B})$, where P is a finite set of elements, called *points* and $\mathcal{B}$ is a family of subsets of P, called *blocks*.

A *linear space* is a geometrical space for which, through any two distinct points there is a unique block, said *line*.

Let us denote by $L(x, y)$ the line through the different points x and y of P and let us define the following hyperoperation on P:

$$\forall\,(x,y) \in P^2,\ x * y = \begin{cases} \{x\}, & \text{if } x = y \\ L(x,y), & \text{if } x \neq y. \end{cases}$$

The hyperstructure $(P, *)$ is a quasi–hypergroup.

Other examples of hyperstructures, associated with non–projective linear spaces or reducible projective spaces, are presented in [24].

Notice that, from the cryptographic point of view, it is not very useful to consider hyperstructures having a kind of regularity, for reasons we shall present below.

5. Theorem. *A hypergroup $(A, *)$, with $A = \mathbb{Z}_n$, satisfies the following conditions:*

1) $\forall\,(i,j) \in A^2,\ \mathrm{card}(i * j) = i + 1.$
2) $\forall\,(i,h,k) \in A^3,\ h \neq k \Longrightarrow i * h \neq i * k.$

if and only if the hyperstructure "$*$" is defined as follows:

$$\forall\,(i,j) \in A^2,\ i * j = j + \{0, 1, ..., i\}$$

Proof. "$\Longleftarrow$" Immediate.

"$\Longrightarrow$" Set $\forall\,(i,j)\in A^2$, $i*j=\{x_0^{i*j},...,x_i^{i*j}\}$, where $k<h\Longrightarrow x_k^{i*j}<x_h^{i*j}$. Since $(A,*)$ is a hypergroup, the conditions 1) and 2) hold.

The proof consists in the following phases:

I) We shall verify that $\forall\,i\in A,\ 0*i=\{i\}$. We have

$$0*(i*0)=0*\{x_0^{i*0},...,x_i^{i*0}\}=\bigcup_{j=0}^{i}\left(0*x_j^{i*0}\right).$$

But $\mathrm{card}\left(0*x_j^{i*0}\right)=1$ and, by 2), it follows that the elements $0*x_0^{i*0},...,0*x_i^{i*0}$ are different. Therefore, $\forall\,i\in A$, $\mathrm{card}\,0*(i*0)=i+1$.

On the other hand, $(0*i)*0=\{x_0^{0*i}\}*0=x_0^{0*i}*0$, whence $\forall\,i\in A$, $\mathrm{card}((0*i)*0)=x_0^{0*i}+1$. Since $(A,*)$ is a hypergroup, we have $\forall\,i\in A$, $x_0^{0*i}=i$, hence $\forall\,i\in A$, $0*i=\{x_0^{0*i}\}=\{i\}$.

II) Now, we shall check that

$$\forall\,i\in A,\ i*0=\{0,...,i\}.$$

By I), we have $0*0=\{0\}$, so

$$i*(0*0)=i*\{0\}=i*0=\left\{x_0^{i*0},...,x_i^{i*0}\right\}.$$

Therefore, $\forall\,i\in A$, $\mathrm{card}(i*(0*0))=i+1$.

On the other hand, $(i*0)*0=\{x_0^{i*0},...,x_i^{i*0}\}*0=\bigcup_{j=0}^{i}\left(x_j^{i*0}*0\right)$.

Therefore, $\forall\,i\in A$,

$$i+1=\mathrm{card}(i*(0*0))=\mathrm{card}((i*0)*0)=\mathrm{card}\bigcup_{j=0}^{i}\left(x_j^{i*0}*0\right).$$

The set $(i*0)*0$ is a union of non–empty sets. Hence each subset of this union contains at most $1+i$ elements, that is $\forall\,j\in\{0,...,i\}$, $\mathrm{card}\left(x_j^{i*0}*0\right)\le i+1$. By 1), $\left(x_j^{i*0}*0\right)$ has exactly $x_j^{i*0}+1$ elements, so $\forall\,j\in\{0,...,i\}$, $x_j^{i*0}\le i$. The elements $x_0^{i*0},...,x_i^{i*0}$ of $i*0$ are distinct, so, $\forall\,i\in A$, $i*0=\{0,...,i\}$.

III) We prove that

$$\forall (i,j) \in A^2,\ h \in \{0, ..., i\} \text{ we have } h * j \subseteq i * j.$$

By I) and II), $\forall (i,j) \in A^2$, we have

$$i * j = i * (0 * j) = (i * 0) * j = \{0, ..., i\} * j = \bigcup_{h=0}^{i} h * j,$$

whence $\forall h \in \{0, ..., i\},\ h * j \subseteq i * j$.

IV) We shall verify the implication:

$$(\forall h \leq i,\ h + j \in h * j) \Longrightarrow i * j = \{j, j+1, ..., j+i\}.$$

Let $h + j \in h * j$, for all $h \leq i$. Then

$$\{j, j+1, ..., j+i\} \subseteq \bigcup_{h=0}^{i} h * j \subseteq i * j.$$

But $\text{card}(i * j) = i + 1$, so we have

$$i * j = \{j, j+1, ..., j+i\}.$$

V) Notice that, if $\forall (i,j) \in A^2,\ i + j \in i * j$, then the theorem is obtained directly from IV). So, we shall verify that this condition holds.

VI) We prove that $\forall (i,j) \in A^2, j+i \in i*j$. Suppose that $\exists (u_0, v_0) \in A^2$, such that $u_0 + v_0 \notin v_0 * u_0$ and let v be the smallest element of A, such that $\exists u_1 \in A : u_1 + v \notin v * u_1$. By I), it follows $v \neq 0$. Let u be the smallest element of A, such that

$$u + v \notin v * u.$$

By IV), we have

$$\forall (i,j) \in A^2,\ i < v \Longrightarrow j + i \in i * j \Longrightarrow i * j = \{j, j+1, ..., j+i\},$$
$$\forall j \in A,\ j < u \Longrightarrow j + v \in v * j \Longrightarrow v * j = \{j, j+1, ..., j+v\}.$$

Particularly, $(v-1)*u=\{u,u+1,...,u+v-1\}$.

By III), we have $(v-1)*u\subset v*u$.

On the other hand, $\text{card}(v*u)=v+1$ and since $u+v\notin v*u$, the set $v*u-\{u,u+1,...,u+v-1,u+v\}$ has only an element x. Then $v*u=\{u,u+1,...,u+v-1\}\cup\{x\}$.

Possibilities:

1° $x\in\{u+v+1,...,n-1\}$,

2° $x\in\{0,1,...,u-1\}$.

1° If $x\in\{u+v+1,...,n-1\}$, then $v*u=\{x_0^{v*u},...,x_r^{v*u}\}$, where $x_s^{v*u}=u+s$, for $s\in\{0,...,v-1\}$ and $x_v^{v*u}=x$. We have

$$(v*u)*0=\{x_0^{v*u},...,x_v^{v*u}\}*0=\bigcup_{j=0}^{v}\left(x_j^{v*u}*0\right)$$

and using III, we obtain

$$\bigcup_{j=0}^{v}\left(x_j^{v*u}*0\right)=x_v^{v*u}*0.$$

Therefore $(v*u)*0=\{0,...,x_v^{v*u}\}=\{0,...,x\}$, whence $u+v\in(v*u)*0$. Moreover,

$$\begin{aligned}v*(u*0)&=v*\{0,...,u\}=\bigcup_{j=0}^{u}v*j=\left(\bigcup_{j=0}^{u}v*j\right)\cup(v*u)=\\&=\{0,1,...,u+v-1\}\cup\{x_0^{v*u},...,x_v^{v*u}\}=\\&=\{0,1,...,u+v-1\}\cup\{u,u+1,...,u+v-1,x\}=\\&=\{0,1,...,u+v-1,x\},\end{aligned}$$

whence it follows $u+v\notin v*(u*0)=(v*u)*0$, a contradiction.

2° If $x\in\{0,1,...,u-1\}$, then we have $v*u=\{x_0^{v*u},...,x_v^{v*u}\}$, where $x_0^{v*u}=x$ and $x_s^{v*u}=u+s-1$, for $s\in\{1,...,v\}$.

Then, $v*u=\{x_0^{v*u},...,x_v^{v*u}\}=\{x\}\cup\{u,u+1,...,u+v-1\}$, where $v\neq 0$.

If $v>1$, we obtain easily a contradiction. Indeed, let us consider:

$$\begin{aligned}(v-1)*(1*u) &= (v-1)*\{u,u+1\} = \\ &= ((v-1)*u)\cup((v-1)*(u+1)) = \\ &= \{u,u+1,...,u+v-1\}\cup\{u+1,u+2,...,u+v\} = \\ &= \{u,...,u+v\}.\end{aligned}$$

On the other hand,

$$((v-1)*1)*u = \{1,...,v\}*u = \bigcup_{j=1}^{v} j*u = v*u = \{x,u,...,u+v-1\}.$$

Therefore, $\{u,...,u+v\} = \{x,u,...,u+v-1\}$, a contradiction.

By 2), we have $v*(u-1) \neq v*u$, hence $x \neq u-1$, whence $u \notin \{0,1\}$.

We shall prove that $v=1$ implies $u=1$, which is in contradiction with $u \notin \{0,1\}$. First, we prove that if $u > s-1$, then

$$s*(u-1) = \{x,x+1,...,x+s-2,u-1,u\} \text{ for } s \in \{2,...,u\}.$$

For $s=2$, we have

$$\begin{aligned}1*(1*(u-1)) &= 1*\{u-1,u\} = (1*(u-1))\cup(1*u) = \\ &= \{u-1,u\}\cup\{x,u\} = \{x,u-1,u\} \text{ and} \\ (1*1)*(u-1) &= \{1,2\}*(u-1) = (1*(u-1))\cup(2*(u-1)) = \\ &= 2*(u-1),\end{aligned}$$

whence $2*(u-1) = \{x,u-1,u\}$.

Suppose the assertion true for $s-1$ and we shall prove it for s (where $s \leq c$). We have:

$$\begin{aligned}1*((s-1)*(u-1)) &= 1*\{x,x+1,...,x+s-3,u-1,u\} = \\ &= \{x,x+1\}\cup\{x+1,x+2\}\cup...\cup\{x+s-3,x+s-2\}\cup \\ &\cup\{u-1,u\}\cup\{x,u\} = \{x,x+1,...,x+s-2,u-1,u\} \text{ and} \\ (1*(s-1))*(u-1) &= \{s-1,s\}*(u-1) = \\ &= ((s-1)*(u-1))\cup(s*(u-1)) = s*(u-1).\end{aligned}$$

Therefore, $s * (u-1) = \{x, x+1, ..., x+s-2, u-1, u\}$.

Let us consider $u = s$. We have

$$\begin{aligned} 1 * (s * (u-1)) &= 1 * \{x, x+1, ..., x+s-2, u-1, u\} = \\ &= \{x, x+1\} \cup \{x+1, x+2\} \cup ... \cup \{x+s-2, x+s-1\} \cup \\ &\cup \{u-1, u\} \cup \{x, u\} = \{x, x+1, ..., x+s-1, u-1, u\} \text{ and} \\ (1 * s) * (u-1) &= \{x, s\} * (u-1) = (x * (u-1)) \cup (s * (u-1)) = \\ &= s * (u-1) = \{x, x+1, ..., x+s-2, u-1, u\}, \end{aligned}$$

whence we obtain $x + s - 1 = u - 1$ and since $s = u$, it follows $x = 0$. Therefore

$$u * (u-1) = \{0, 1, ..., u-1, u\} = u * 0$$

and by 2) it follows $u - 1 = 0$, that is $u = 1$.

Therefore, the both possibilities for x lead us to contradictions. Then $\forall (i, j) \in A^2$, $j + i \in i * j$, and so, the theorem is completely proved. ■

6. Remark. Notice that for any $k \in \{1, 2, ..., n-1\}$ the elements of the row k (in the composition square) are sets of $k+1$ elements. The advantage of using this hyperstructure is the following one: we use dispositions instead of permutations and there are many more dispositions than permutations.

On the other hand, this hyperstructure is not very interesting from the cryptography point of view: indeed, it is the same as ciphering in such a way as to divide the cleartext in letters and to insert a number of letters equal to the key (because $\forall j \in A$, $j \in i * j$ and $\text{card}(i * j) = i + 1$).

In constructing algebraic cryptosystems, it is very important to remember that: "*Cryptography likes confusion*"

Chapter 8

Codes

In general, Code Theory and more precisely Error–Correcting Code Theory is one branch of applied mathematics, which massively uses algebraic methods and results.

Through a channel, recall that Error–Correcting Code Theory is essential for all types of communications (for instance, telephonic communications, radio communications and so on).

Among the most remarkable codes, we recall Hamming codes, QR-codes, which are important classes of cyclic codes.

We present below a connection between Steiner hypergroups and linear codes. We think that the study of this connection deserves to be studied in depth. For more details on Error–Correcting Codes, see [452], [454] and [457].

G. Tallini established connections between Code Theory and Hyperstructure Theory. We mention some of his results in §1 and in §3.

All the notions mentioned in this chapter are defined and studied in a very interesting book [454] on Combinatorics, Galois geometry and Codes. For a good understanding of the results of this chapter, we suggest the reader to examine this book. Thus, we shall present here some definitions we shall use in the following.

§1. Steiner hypergroupoids and Steiner systems

1. Definition. A hypergroupoid $(H, \cdot)$ is called *n-hypergroupoid of Steiner* if it satisfies the following conditions:

(i) $\forall (x,y) \in H^2,\ x \in xy \ni y.$

(ii) $\forall (x,y) \in H^2,\ \text{card}(x \cdot y) = \begin{cases} 1, & \text{if } x = y \\ n, & \text{if } x \neq y. \end{cases}$

(ii) the associativity law holds for every three elements, not all distinct.

2. Remarks.

1. By (i), it follows that $(H, \cdot)$ is a quasi–hypergroup and by (i) and (ii), we obtain that $\forall x \in H,\ xx = \{x\}$ and $n \geq 2$.
2. By (iii) and 1), it follows:
   $\forall (x,y) \in H^2,\ x(xy) = xy = (xy)y$ and $x(yx) = (xy)x$,
   whence we obtain $\forall (x,y) \in H^2,\ x \neq y,\ \forall (z,u) \in xy,\ z \neq u$,
   we have $xy = zu$ and so, it follows the commutativity.

3. Definition. An *n-system of Steiner* is a pair $(H, \mathcal{R})$, where H is a non-empty set, whose elements will be called *points* and $\mathcal{R}$ is a family of subsets of H, called *lines*, such that the following conditions hold:

(i) any line has exactly n points

(ii) for any two different points there is a unique line which contains them.

Let us see what is a Galois field.

Let $g \in \mathbb{Z}_p[X]$ (where p is a prime), g irreducible, such that g has the degree $h \geq 2$. The field $\mathbb{Z}_p[X]/(g)$ has the order $q = p^h$ and it is called a *Galois field of order q;* we shall denote it by G_q.

Notice that any finite field of order $q = p^h$ is isomorphic with G_q.

4. Example of an n-system of Steiner. Any projective or affine space over a Galois field of order q is an n-system of Steiner, with respect to its lines (where $n = q + 1$ or $n = q$).

5. Theorem. *With any n-system of Steiner, we can associate an n-hypergroupoid of Steiner and conversely.*

Proof. Let $(H, \mathcal{R})$ be an n-system of Steiner. For $\forall\,(x, y) \in H^2$, let

$$x \circ y = \begin{cases} x, & \text{if } x = y \\ \text{the line through } x \text{ and } y, & \text{if } x \neq y. \end{cases}$$

We can easily check that $(H, \circ)$ is an n-hypergroupoid of Steiner and we shall call it the n-hypergroupoid of Steiner, associated with the n-system of Steiner $(H, \mathcal{R})$.

Conversely, if $(H, *)$ is an n-hypergroupoid of Steiner, then we consider the family

$$\{x * y \mid (x, y) \in H^2,\ x \neq y\}, \text{ denoted by } \mathcal{R}.$$

We shall verify that $(H, \mathcal{R})$ is an n-system of Steiner. Indeed, by (ii) of the definition of an n-hypergroupoid of Steiner, it follows that for all $(x, y) \in H^2$, $x \neq y$ we have $\text{card}(x * y) = n$. By (i) of the same definition, $\forall\,(x, y) \in H^2$, $x \neq y$, there exists a line $x * y$, which contains x and y.

Moreover, there exists a unique line, containing x and y.

Indeed, if $(r, s) \in \mathcal{R}^2$ and $\{x, y\} \subset r \cap s$ and if $r = z * t$ (where $(z, t) \in H^2$, $z \neq t$) and $s = u * v$ (where $(u, v) \in H^2$, $u \neq v$), then $z * t = x * y = u * v$.

Hence $r = s$ and therefore, $(H, \mathcal{R})$ is an n-system of Steiner. ∎

6. Definition. A hypergroup $(H, \circ)$ is called a *Steiner hypergroup* if the following conditions hold:

(i) $\forall x \in H,\ x \circ x = x$.

(ii) $\forall\,(x, y) \in H^2$, $x\ ney$, we have $x \in x \circ y \ni y$, $x \circ y \neq H$ and $\text{card}(x \circ y) \geq 3$.

(iii) $\forall (x,y) \in H^2,\ x \neq y,\ \forall z \in x \circ y,\ z \neq x$, we have $x \circ y = z \circ x$.

7. Remarks.

1. Any Steiner hypergroup is a commutative hypergroup (by (iii)).
2. In a Steiner hypergroup $(H, \circ)$, the following condition holds: $\forall (x,y) \in H^2,\ x \neq y,\ \forall \{u,v\} \subset x \circ y,\ u \neq v$, we have $x \circ y = = u \circ v$. Indeed, if $u = x$, the above condition results from (iii) and from the commutativity. If $u \neq x$, then $v \in x \circ y = = u \circ x = v \circ u = u \circ v$.

8. Theorem. *With any finite Steiner hypergroup* $(H, \circ)$, *we can associate a finite irreducible projective space of dimension* ≥ 2 *and conversely.*

Proof. Let $\mathcal{F} = \{x \circ y \mid (x,y) \in H^2,\ x \neq y\}$. Since $\forall (x,y) \in H^2$, $x \neq y,\ \forall \{u,v\} \subset x \circ y,\ u \neq v$ we have $x \circ y = u \circ v$, it follows that $(H, \mathcal{F})$ is a projective space. Moreover, since $\forall (x,y) \in H^2,\ x \neq y$ we have $\text{card}(x \circ y) \geq 3$ and $x \circ y \neq H$, we obtain that $(H, \mathcal{F})$ is a finite irreducible projective space $\mathbb{P}_r$ of dimension $r \geq 2$, that is either $PG(r,q)$, the projective space over the Galois field of order $q = \text{card}(x \circ y) - 1$ or a non-Desarguesian plane of order q (see Theorem 13.1, [454]).

Conversely, let $\mathbb{P}_r$ be a finite irreducible projective space of dimension ≥ 2. We can define on $\mathbb{P}_r$ the following hyperoperation: $\forall (x,y) \in \mathbb{P}_r^2,\ x \neq y,\ x \circ x = x$ and $x \circ y$ is the line through x and y.

Then $(\mathbb{P}_r, \circ)$ is a Steiner hypergroup. ∎

§2. Some basic notions about codes

The theory of codes finds out and corrects the errors, that can be introduced by the transmission of information from a source to a receiver.

Usually, the information is translated in a language with a small number of symbols, which are the elements of $\mathbb{Z}_p$ (where p is a prime natural number).

Often, it is considered $p = 2$. Any element of the message is represented by a finite sequence of symbols, which is called *password.* We shall consider codes with an invariable length, that means codes whose paswords contain the same number of symbols.

This number is called the *code length.*

Let n be the length of a code V over $\mathbb{Z}_p$. We consider a proper subset of $\mathbb{Z}_p^n$ as set of passwords.

The set of passwords must be different from $\mathbb{Z}_p^n$, otherwise it is impossible to correct errors.

Indeed, if instead of a code password we receive another one, different from all the code passwords, then we notice an error. It is important that the set of code paswords contains only one password which is similar with the received password. Thus, any proper subset of $\mathbb{Z}_p^n$ is considered to be a *code of length* n over $\mathbb{Z}_p$.

Let V be such a code. The elements of V are called *passwords.* If $x = (x_1, ..., x_n) \in \mathbb{Z}_p^n$ then the number

$$w(x) = \text{card}\{x_i \mid i \in \{1, 2, ..., n\},\ x_i \neq 0\}$$

is called the *Hamming weight* (or weight) of x.

Let $(x, y) \in \mathbb{Z}_p^n \times \mathbb{Z}_p^n$. The weight of $x - y$ is called the *Hamming distance* (or distance) of x and y and it is denoted by $d(x, y)$.

The following conditions, which characterize the distance notion, are verified: for any x, y, z in $\mathbb{Z}_p^n$:

$$\begin{aligned}
&d(x, y) = 0 \text{ if and only if } x = y \\
&d(x, y) = d(y, x) \\
&d(x, z) \leq d(x, y) + d(y, z).
\end{aligned}$$

The *minimum weight* of V, denoted by w, is the minimum of the nonzero passwords weights of V.

The *minimum distance* of V, denoted by d, is the minimum of distances between two distinct elements of V.

V is called a *linear code* if it is a subspace of the vectorial space $\mathbb{Z}_p^n$.

In a linear code, the minimum distance d and the minimum weight w are equal.

If V is a linear code of length n, dimension r and minimum weight w (which is equal with d), then we say that V is an (n, w, r)–*code.*

If we write the elements of V one below another, then we obtain the so–called *book* of V.

Let V be a linear code of dimension r.

V is uniquely determined by r independent passwords of it. The matrix which has as lines these r passwords is called the *generated matrix of* V.

Let $H = \begin{pmatrix} x_{11} \cdots x_{1n} \\ \ldots\ldots\ldots\ldots \\ x_{r1} \cdots x_{rn} \end{pmatrix}$ be a generated matrix of V and suppose that $\begin{vmatrix} x_{11} \cdots x_{1n} \\ \ldots\ldots\ldots\ldots \\ x_{r1} \cdots x_{rr} \end{vmatrix} \neq 0.$

Let $c_1, c_2, ..., c_r$ be r arbitrary elements in $\mathbb{Z}_p$. Then there is a unique password of V whose first r coordinates are exactly $c_1, c_2, ..., c_r$.

Therefore, the first r coordinates are sufficient to obtain the information.

From H we can obtain the so–called *matrix of information* composed by the first r columns; the others $n-r$ columns form the *check matrix* (or parity check).

The *rate* of the code V is the number r/n.

A reasonable code has a high rate and a high minimum distance.

If d is the minimum distance of V, we can correct any password which has a number h of errors (during the transmission), with $h < d/2$. Indeed, if we receive $\bar{x}$ instead of $x \in V$, such that $d(\bar{x}, x) = h < d/2$, then for any $y \in V$, $y \neq x$, we have $d \leq$ $\leq d(x, y) \leq d(x, \bar{x}) + d(y, \bar{x}) = h + d(y, \bar{x}) < d/2 + d(y, \bar{x})$, whence

$d(y, \bar{x}) > d - d/2 = d/2$. Hence x is the unique password of V, such that $d(x, \bar{x}) < d/2$.

Since the number of errors is $h < d/2$, we can identify x as the right password, obtained from $\bar{x}$.

An open problem is the following one: let n and r be known; we are interested to find the maximum value of d, such that there is an (n, d, r)-code.

§3. Steiner hypergroups and codes

Now, recall some definitions (see [454]).

A *projective plane* is a pair $(\pi, \mathcal{R})$, where π is a set, whose elements are called *points* and $\mathcal{R}$ is a family of subsets of π, called *lines* such that the following conditions hold:

(i) there is a unique line, containing two distinct points;

(ii) any two distinct lines has exactly one common point;

(iii) there exist four points, such that any three points of them are not collinear.

Let π be a projective plane.

A subset K of π is called *arc* if any three points of K are not collinear.

A line is called *tangent* of K if its intersection with K has exactly one point.

An *oval* Ω of π is an arc such that for $\forall\, x \in \Omega$, there is exactly one tangent t_x in x to Ω.

A *hyperoval* is an arc without tangents.

Let $(H, \circ)$ be a Steiner hypergroup with n elements and let $\mathbb{P}_{r,q} = (H, \mathcal{F})$ be the associated projective space, where r is the dimension and $q = \text{card}(x \circ y) - 1$.

Now, we order the points of H and set $n = \nu_r = \sum\limits_{i=0}^{r} q^i$. The characteristic function of each subset X of H determines a vector $\varphi(X)$ of $\mathbb{Z}_2^n$.

Let $\psi : H\times H \longrightarrow \mathbb{Z}_2^n - \{0\} = PG(n-1,2)$ (where $PG(n-1,2)$ is the $(n-1)$ dimensional projective space over $\mathbb{Z}_2$),

$$\psi(x,y) = \varphi(x \circ y).$$

Denote $K = \psi(H\times H - I_H)$, where $I_H = \{(x,x) \mid x \in H\}$. That means K consists of all the points in $PG(n-1,2)$, whose coordinates are the characteristic functions of the lines of $\mathbb{P}_{r,q}$.

Let $\mathcal{L}$ be a family of lines of $\mathbb{P}_{r,q}$. We say that $\mathcal{L}$ is of *even type* if through any point of $\mathbb{P}_{r,q}$, an even number of lines of $\mathcal{L}$ pass. The points of K, which correspond to the lines of $\mathcal{L}$, are linearly dependent.

Conversely, if a subset L' of K is linearly dependent, then the family $\mathcal{L}'$ of lines of $\mathbb{P}_{r,q}$, whose characteristic functions are the coordinates of the points of L', contains a subfamily $\mathcal{L}$ of even type.

9. Theorem. *Let $\mathcal{L} \neq \emptyset$ be an even type family of lines of $\mathbb{P}_{r,q}$. Then* $\operatorname{card}\mathcal{L} \geq q+2$ *and we have* $\operatorname{card}\mathcal{L} = q+2$ *if and only if if q is even and $\mathcal{L}$ is a dual hyperoval on a plane of $\mathbb{P}_{r,q}$, that is a hyperoval in the dual plane.*

Proof. Let ℓ be a line of $\mathcal{L}$. Through any point of ℓ, there passes an odd number of lines of $\mathcal{L}$, different from ℓ. Since $\operatorname{card}\ell = q+1$, it follows $\operatorname{card}\mathcal{L} \geq q+2$.

If $\operatorname{card}\mathcal{L} = q+2$, then through any point of ℓ, there passes exactly one line of $\mathcal{L}$, different from ℓ.

Since ℓ is no special line of $\mathcal{L}$, it follows that the lines of $\mathcal{L}$ are pairwise incident. Therefore, they lie on a plane of $\mathbb{P}_{r,q}$ and form a dual hyperoval. ■

Let P be a point of $\mathbb{P}_{r,q}$ and let $\mathcal{S}_P$ be the set of all lines of $\mathbb{P}_{r,q}$, which pass through P. We shall call $\mathcal{S}_P$ the *star of lines with center P*.

10. Theorem. *There is no star of lines of $\mathbb{P}_{r,q}$ which contains a non–empty set of even type.*

Therefore, the image of a star, under φ, *is a subset* K *of* $PG(n-1,2)$ *which consists of linearly independent points.*

Consequently, if A *denotes the matrix whose columns are the coordinates of the points of* K, *then*

$$\mathcal{V}_{r-1} \leq \operatorname{rank} A \leq \mathcal{V}_r = n \textit{ and } \operatorname{card} K = \mathcal{V}_r \mathcal{V}_{r-1}/\mathcal{V}_1$$

Proof. Let $\mathcal{S}_P$ be a star in $\mathbb{P}_{r,q}$. A unique line of $\mathcal{S}_P$ passes through any point of $\mathbb{P}_{r,q} - \{P\}$. Therefore, $\mathcal{S}_P$ contains no non–empty even type set.

The matrix A has $n = \mathcal{V}_r$ lines and $\operatorname{card} K = \mathcal{V}_r \mathcal{V}_{r-1}/\mathcal{V}_1$ (the number of lines in $\mathbb{P}_{r,q}$) columns. By the previous argument, A has $\mathcal{V}_{r-1} = \operatorname{card} \mathcal{S}_P$ linearly independent columns, so $\mathcal{V}_{r-1} \leq \operatorname{rank} A \leq$ $\leq \mathcal{V}_r = n$. ∎

11. Definition. Let $(s, N) \in \mathbb{Z}^2$, $N > s+1 \geq 2$. A subset K of $PG(r,q)$ is called *N–cap of kind s* if K has N elements, such that any $(s+1)$ elements of K are linearly independent and there are $(s+2)$ elements of K, which are linearly dependent.

From the above two theorems, it follows the following

12. Corollary. *The set* K *is an* N*-cap of kind* s, *which belongs to a space* $PG(t-1,2)$, *where*

$$N = \operatorname{card} K = \mathcal{V}_r \mathcal{V}_{r-1}/\mathcal{V}_1, \; s \geq q \textit{ and}$$
$$\mathcal{V}_{r-1} \leq t = \operatorname{rank} A \leq \mathcal{V}_r = n.$$

We have $s = q$ *if and only if* q *is even and either* $r \geq 3$ *or* ($r = 2$ *and the dual plane* $\mathbb{P}^*_{2,q}$ *contains hyperovals.*

Moreover, if $r \geq 3$, *then* ($q < s \leq 2q$ *if* q *is odd*) *or* ($s = q$ *if* q *is even*).

Let G_q be a Galois field of order q (where $q = p^h$, p prime) and C^k a linear code of dimension k of G_q^n, that means C^k is a vectorial subspace of dimension k, of G_q^n.

The study of linear codes of G_q^n, which correct errors is in connection with the study of n-caps of a Galois space (see [454], 44).

13. Theorem. (Th. 43.2, [454]) *The linear code C^k corrects e errors if and only if* $e = \left[\dfrac{w-1}{2}\right]$, *where w is the minimum weight of C^k and $[x]$ is the integer part of x, that means* $[x] \subseteq x < [x]+1$.

Now, consider the following subspace of $\mathbb{Z}_2^N$: $C^d = \{X \in \mathbb{Z}_2^N \mid AX = 0\}$.

It follows that C^d is a linear (N, w, d)-code with $N = \mathcal{V}_r\mathcal{V}_{r-1}/\mathcal{V}_1$, $w = s+2$ and $d = N-t$ and it corrects $e = [(w-1)/2] = [(s+1)/2]$ errors. Since ($q < s \leq 2q$ if q is odd) or ($s = q$ if q is even) it follows that $q/2 \leq e \leq q$.

Moreover, the following statements can be verified:

1) if $r = 2$, $q \equiv 2(\text{mod}\, 4)$
then $t = (q^2+q+2)/2$ and $d = (q^2+q)/2$.

2) if $r = 2$, $q \equiv 0(\text{mod}\, 4)$
then $t < (q^2+q+2)/2$ and $d > (q^2+q)/2$.

3) if $r = 2$, q odd then $t = n-1 = q^2+q$; $d = 1$.

14. Proposition. *We have* $t \geq \mathcal{V}_{r-1} + q^{r-1} - 1$.

Proof. Let us consider two distinct points P_1 and P_2 of $\mathbb{P}_{r,q}$ and a hyperplane π on P_2 and not on P_1.

The set of all the lines through P_1 and all the lines through P_2 not on π, contains no even type subset.

Therefore, K contains $\mathcal{V}_{r-1}+(\mathcal{V}_{r-1}-\mathcal{V}_{r-2}-1) = \mathcal{V}_{r-1}+q^{r-1}-1$ linearly independent points. ■

Chapter 9

Median algebras, Relation algebras, C-algebras

- *For the first time, median algebras appeared in the late fourties. A.A. Grau [148] characterized Boolean algebras in terms of median operation and complementation, G. Birkhoff and S.A. Kiss [25] discusses the median operation for distributive lattices. The concept of abstract median algebra was introduced by S.P. Avann [12] and later M. Scholander [356], [357], [358] and S.P. Avann [13] performed a detailed study of median algebras. More recently, J. Nieminen [301], E. Evans [139], H.M. Mulder A. Schrijver [297], J.R. Isbell [165], H. Werner [424] worked on this subject.*
- *We shall see that quasi–canonical hypergroups can be characterized as the atomic structures of complete atomic integral relation algebras (§2). Moreover, the Tarski complex-algebra construction gives a full embedding of quasi-canonical hypergroups into relation algebras. Therefore, certain combinatorial properties of quasi-canonical hypergroups transfer to relation algebras. Using this process, results of Monk [295], [296] or McKenzie [263], [453], about relation algebras (or cylindric algebras) turn out to be just interpretations of quasi–canonical hypergroup results.*

- *Let us remember some remarkable C-algebras: the adjancency algebras of association schemes* [441], *S-algebras over finite groups* [31], *and centralizer algebras of homogeneous coherent configurations* [449].

§1. Median algebras and join spaces

In this section, we present a connection between median algebras and join spaces, which was established by H.J. Bandelt and J. Hedlíkovà.

1. Definition. A *ternary algebra* is a set M together with a single ternary operation $(a, b, c) \longrightarrow (abc)$. A ternary algebra M is called *median algebra* if it satisfies the following identities for any $(a, b, c, d, e) \in M^5$:

1) $(aab) = a$;

2) $(abc) = (bac) = (bca)$;

3) $((abc)de) = (a(bde)(cde))$.

2. Theorem. (see [357]) *On every median algebra M, the following hyperoperation*

$$\forall\,(a, b) \in M^2,\ a\circ b = \{x \in M \mid x = (abx)\}$$

satisfies the properties:

(α) $\forall\, a \in M,\ a\circ a = \{a\}$;

(β) *if* $b \in a\circ c$, *then* $a\circ b \subseteq c\circ a$;

(γ) $\forall\,(a, b, c) \in M^3,\ a\circ b \cap b\circ c \cap c\circ a = \{d\}$ (*where* $d = (abc)$).

Conversely, every hyperoperation "$\circ$" which satisfies the properties (α), (β) *and* (γ) *induces a unique ternary operation by which M becomes a median algebra.*

3. Theorem. *Let M be a ternary algebra such that the conditions 1), 2) of Definition 1 and*

4) $\forall (a,b,c) \in M^3,\ ((abc)bc) = (abc)$

are true in M. Set $\forall (a,b) \in M^2$, $a\circ b = \{x \in M \mid x = (abx)\}$. Then $(M,\circ)$ is a join space if and only if M is a median algebra.

Proof. From 1) and 2) it follows that $\forall (a,b) \in M^2$, we have

$$a\circ b \neq \emptyset \neq a/b,\ a\circ b = b\circ a \text{ and } a\circ a = \{a\}.$$

"$\Longleftarrow$" First, suppose that M is a median algebra.

If $x \in (a\circ b)\circ c$, then there exists $y \in M$, such that $x = ((aby)cx) = (a(bcx)(cxy))$, whence the associativity of "$\circ$" follows.

Let us prove now that: if $a/b \cap c/d \neq \emptyset$, then $a\circ d \cap b\circ c \neq \emptyset$.

Let $x \in a/b \cap c/d$. Then $(ad(bdx)) = ((abx)d(bdx)) = (bdx)$. It follows $(bdx) \in a\circ d$, and similarly $(bdx) \in b\circ c$.

Hence, $(M,\circ)$ is a join space.

"$\Longrightarrow$" Conversely, let us assume that $(M,\circ)$ is a join space. If $b \in a\circ c$ and $x \in a\circ b$, then by the associativity of "$\circ$" and by $a\circ a = \{a\}$, for any $a \in M$, we obtain

$$x \in a\circ(a\circ c) = (a\circ a)\circ c = a\circ c.$$

Since "$\circ$" is commutative, the following implication is satisfied:

$$b \in a\circ c \Longrightarrow a\circ b \subseteq c\circ a,$$

that is (β). From 4), we obtain that for $x = (abc)$, we have

$$x \in a\circ b \cap b\circ c \cap c\circ a.$$

On the other hand, if $y \in a\circ b \cap b\circ c \cap c\circ a$, then $b \in x/a \cap y/c$ and $b \in y/a \cap x/c$. Since $(M,\circ)$ is a join space, it follows that there exist $u \in x\circ c \cap a\circ y$ and $v \in y\circ c \cap a\circ x$.

From $x \in a \circ c$ and $u \in x \circ c$, it follows $c \in x/a \cap u/x$, so $x \circ x \cap a \circ u \neq \emptyset$, whence $x \in a \circ u$.

Similarly, we obtain $y \in a \circ v$.

By the associativity of "$\circ$", $u \in a \circ y$ and $x \in a \circ u$ imply $x \in a \circ y$; $v \in a \circ x$ and $y \in a \circ v$ imply $y \in a \circ x$.

Hence $x = (axy) = y$, whence it follows (γ).

By the previous Theorem, we can conclude that M is a median algebra. ■

§2. Relation algebras and quasi–canonical hypergroups

4. Definition. A system $< A, +, \cdot, -, 0, 1, *, ^{-1}, 1' >$ is called a *relation algebra* (RA) if:

$1°$ $< A, +, \cdot, -, 0, 1 >$ is a Boolean algebra;

$2°$ $< A, *, 1' >$ is a semigroup with identity;

$^{-1}$ is a unary operation, which satisfies the following condition:

$3°$ $(x * y) \cdot z = 0 \Longleftrightarrow (x^{-1} * z) \cdot y = 0 \Longleftrightarrow (z * y^{-1}) \cdot x = 0.$

This notion was introduced by Tarski. As examples of relation algebras, we can consider the following system (which is called *proper relation algebra*) $< \mathcal{P}, \cup, \cap, \sim, \emptyset, Y^2, \circ, ^{-1}, I_Y >$ where $\mathcal{P}$ is a family of binary relations on a set Y, such that $\mathcal{P}$ contains $\emptyset$, Y^2 and $I_Y = \{(y, y) \mid y \in Y\}$ and it is closed under $\cup, \cap, \sim$, relation composition $\circ$ and inverse $^{-1}$.

5. Definition. We say that a relation algebra is *representable* if it is isomorphic to a subdirect product of proper relation algebras.

6. Definition. We say that a relation algebra is an *integral* one (IRA) if one of the two following equivalent conditions holds:

1) $x * y = 0 \Longrightarrow x = 0$ or $y = 0$.

2) $1'$ is an atom
(that means there is no element z, such that $0 < z < 1'$.)

Let us consider now $< H, \cdot, ^{-1}, e >$ a quasi–canonical hypergroup and $< \mathcal{P}(H), \cup, \cap, \sim, \emptyset, H >$ the Boolean algebra of all subsets of H.

We shall still denote by "$\cdot$" and "$^{-1}$" the extensions of the quasi–canonical hypergroup operations on subsets.

7. Definition. The following system

$$\mathcal{A}[H] = < \mathcal{P}(H), \cup, \cap, \sim, \emptyset, H, \cdot, ^{-1}, e >$$

is called the *complex algebra* of H.

The following theorem establishes an one–to–one correspondence between quasi–canonical hypergroups and complete atomic IRA's and it is due to St.D. Comer [47].

8. Theorem.

(i) *If H is a quasi–canonical hypergroup, then $\mathcal{A}[H]$ is a complete atomic IRA.*

(ii) *If $\mathcal{A}$ is a complete atomic IRA and $At_{\mathcal{A}}$ is the set of atoms of $\mathcal{A}$, then the system $At(\mathcal{A}) = < At_{\mathcal{A}}, *; ^{-1}, 1' >$ is a quasi–canonical hypergroup.*

(iii) *If H is a quasi–canonical hypergroup and $\mathcal{A}$ is a complete atomic IRA, then*

$$H \simeq At(\mathcal{A}[H]) \text{ and } \mathcal{A} \simeq \mathcal{A}[At(\mathcal{A})].$$

Proof. i) We have to verify only the condition 3° of the definition of a relation algebra. If $(X \cdot Y) \cap Z \neq \emptyset$, then there are $z \in Z$, $x \in X$, $y \in Y$, such that $z \in x \cdot y$. Then $x \in zy^{-1}$, so $(Z \cdot Y^{-1}) \cap X \neq \emptyset$.

Similarly, we prove the other implications, using also the equality $(x^{-1})^{-1} = x$.

ii) Whenever x, y are atoms, notice that $x * y$ and x^{-1} are atoms, too.

The only condition to check is the following:

$$1' \in x * x^{-1} \cap x^{-1} * x, \text{ for all } x \in At_{\mathcal{A}}.$$

We have $(1' * x) \cdot x \neq 0$, so $(x * x^{-1}) \cdot 1' \neq 0$ and since $1'$ is an atom, it follows $1' \in x * x^{-1}$. Similarly, it follows $1 \in x^{-1} * x$.

iii) By the correspondence of x with $\{x\}$, we obtain the first isomorphism and for the second one, we consider the correspondence of $a \in \mathcal{A}$ with the set of atoms $x \leq a$. ■

§3. C-algebras and quasi–canonical hypergroups

The following notion of C-algebra, presented here, is due to Y. Kawada [179] and the connection with quasi–canonical hypergroups is due to St.D. Comer [52].

9. Definition. A *C-algebra* is a pair (A, B) where: A is an algebra and $B = \{x_0, ..., x_d\}$ is a basis for A (as a complex linear space), such that the following conditions are satisfied:

1) for $\forall\,(i,j) \in \{0,1,...,d\}^2$, $x_i \bullet x_j = \sum_k p_{ij}^k x_k$;

2) $\exists e = x_0 \in A$, such that $\forall\,(j,k)$, $p_{0j}^k = \delta_{jk} = p_{j0}^k$;

3) every p_{ij}^k is a real number;

4) there exists a permutation $i \rightsquigarrow i'$ of $\{0,1,...,d\}$, such that $(i')' = i$ and $p_{ij}^k = p_{j'i'}^{k'}$;

5) for $\forall\,(i) \in \{0, 1, ..., d\}$, $\exists k_i$, such that $k_i > 0$,
and $\forall\, j \in \{0, 1, ..., d\}$ we have $p_{ji}^0 = p_{ij}^0 = k_i \delta_{ij'}$;

6) the map $x_i \rightsquigarrow k_i$ induces a linear representation of A.

10. Remark. From 4), it follows that the map $x_i \rightsquigarrow x_{i'}$ extends to an antiautomorphism of A.

A C-algebra is *commutative* if $p_{ij}^k = p_{ji}^k$ for all i, j, k.

11. Lemma. *We have*

1°) $0' = 0$;

2°) $k_0 = 1$;

3°) $k_i = k_{i'}$;

4°) $k_s p_{ij}^s = k_i p_{sj'}^i = k_j p_{i's}^j$.

Proof. We obtain 1°) and 2°) from 2) and 5). 5) also implies $k_i \delta_{ij'} = k_j \delta_{ji'}$ for all i, j; hence $k_i = k_j$ when $j = i'$ and so we obtain 3°). Since $(x_i \bullet x_j) \bullet x_{s'} = x_i \bullet (x_j \bullet x_{s'})$, it follows the first equality of 4°), expressing each of $(x_i \bullet x_j) \bullet x_{s'}$ and $x_i \bullet (x_j \bullet x_{s'})$ as a linear combination of $x_0, ..., x_d$ and comparing the coefficients of x_0.

From the first equality, 4) and 3°) we obtain the second equality, so we have

$$k_i p_{sj'}^i = k_{i'} p_{js'}^{i'} = k_j p_{i's}^j.$$

∎

12. Theorem. *With any C-algebra A with basis B, such that the parameters p_{ij}^k are all non–negative (the Kreim condition), we can associate a quasi–canonical hypergroup $< B, \circ, e >$, where $x_i \circ x_j = \{x_k \mid p_{ij}^k \neq 0\}$ and $x_i^{-1} = x_{i'}$ for all i, j.*

Proof. Since $x_i \bullet x_{i'} = \sum\limits_k p_{ii'}^k x_k = k_i x_0 + \cdots$, it follows $x_0 \in x_i \circ x_{i'}$.

If $x_0 \in x_i \circ x_{j'}$, then $p^0_{ij'} \neq 0$, which implies $j = i$ by 5). Similarly, $x_{i'}$ is the only y such that $x_0 \in y \circ x_i$. Therefore $x_i^{-1} = x_{i'}$ is the unique inverse of x_i.

From 2), we obtain $\forall x \in B$, $e \circ x = x \circ e = \{x\}$. From the previous Lemma, it follows that $x \in y \circ z \Longrightarrow y \in x \circ z^{-1}$ and $z \in y^{-1} \circ x$.

We have to verify only the associativity law.

From 4°) of the previous Lemma, we obtain that $x_u \in (x_i \circ x_j) \circ x_k$ if and only if there exists v, such that $p^v_{ij} p^u_{vk} \neq 0$. Similarly, we have $x_u \in x_i \circ (x_j \circ x_k)$ if and only if there exists v, such that $p^u_{iv} p^v_{jk} \neq 0$. From the equality $\sum\limits_v p^v_{ij} p^u_{vk} = \sum\limits_v p^u_{iv} p^v_{jk}$ (a consequence of 1)) and the Kreim condition, we obtain the associativity law for $< B, \circ, e >$.

Chapter 10

Artificial Intelligence

Weak representations of an interval algebra are the objects of interest in the Artificial Intelligence.

Let us give some words about the Mathematicians who worked on this subject.

Allen [3] defined the calculus of time intervals and Ladkin and Maddux [220] showed the interpretation of the calculus of time intervals, in terms of representations of a particular relation algebra, in the sense of Tarski [178]. They proved that there is, up to an isomorphism, a unique countable representation of this algebra.

Ligozat [241] generalized the calculus of time intervals to a calculus of n-intervals and presented this generalization expressed in terms of relation algebras A_S.

Defining canonical functors between the category of weak representations of $\mathbf{A}_n$ and those of $\mathbf{A}_1$, Ligozat [241] extended the results obtained by Ladkin [219].

Finally, it can be seen that the set of (p, q)-positions can be endowed with natural operations which give rise to a family of quasi–canonical hypergroupoids.

§1. Generalized intervals. Connections with quasi–canonical hypergroups

In this paragraph, the notions of a generalized interval and of a (p,q)-position are presented. These notions have been introduced and studied by G. Ligozat and they have developed the study of "interval calculi" used in Artificial Intelligence for representing temporal knowledge.

It is shown that the set of (p,q)-positions can be endowed with natural operations, which give rise to a family of hypergroupoids or, equivalently, of relations algebras, in Tarski's sense.

Let T be a chain and $(p,q) \in \mathbb{N}^* \times \mathbb{N}^*$.

1. Definition. The element

$$(a_1, a_2, ..., a_p) \in T^p,$$

such that $a_1 < a_2 < \cdots < a_p$, is called a *generalized interval* (or a *p-interval*).

An 1-interval is just a point of T. For any $n \in \mathbb{N}^*$, denote the initial segment of $\mathbb{N}^*$: $\{1, 2, ..., n\}$ by $[n]$; $[n]$ is empty if $n = 0$.

2. Definition. A map $\pi : [p+q] \to \mathbb{N}^*$, which verifies the following conditions:

1) the image of π is an initial segment of $\mathbb{N}^*$;
2) the restrictions of π at $[p]$ and $[p+q]-[p]$ are strictly increasing maps, that is
   $\pi(1) < \pi(2) < \cdots < \pi(p)$ and $\pi(p+1) < \cdots < \pi(p+q)$

is called a *(p,q)-position.*

Let us denote by $\Pi_{p,q}$ the set of (p,q)-positions.

3. Examples.

1° The permutations of $\{1, 2, ..., p+q\}$, that verify 2) are positions, called *general positions.*

2° The position $\mathbf{I}'_{p,p} = (1, ..., p, 1, ..., p)$ is a (p,p) position, called *unit position*. We have $\mathbf{I}'_{p,p}(1)=1, ..., \mathbf{I}'_{p,p}(p)=p$, $\mathbf{I}'_{p,p}(p+1)=1$, $\mathbf{I}'_{p,p}(p+2) = 2, ... \mathbf{I}'_{p,p}(2p) = p$.

3) Let "a" be a p-interval and "b" a q-interval in T. Then the concatenation of a and b is a sequence of $p+q$ elements of T. We shall identify $[p+q]$ with the sequence $(a_1, a_2, ..., a_p, b_1, b_2, ..., b_q)$ and π can be considered a map from the set $\{a_1, ..., a_p, b_1, ..., b_q\}$ into $\mathbb{N}^*$.

We say that (a, b) is a T-realization of π.

We can generalize the definition of a (p, q)-position for an arbitrary finite sequence $(p_1, p_2, ..., p_s)$ of natural numbers, obtaining thus the notion of *$(p_1, p_2, ..., p_s)$-position*.

For $s = 3$, we have the following:

4. Definition. A *(p, r, q)–position* σ is a map $\sigma : [p+r+q] \to \mathbb{N}^*$ such that

1) the image of σ is an initial segment of $\mathbb{N}^*$;
2) the restrictions of σ at the initial, median and terminal subsegments of length p, r and respectively q are strictly increasing maps, that is $\sigma(1) < \cdots < \sigma(p)$; $\sigma(p+1) < \cdots < \sigma(p+r)$, $\sigma(p+r+1) < \cdots < \sigma(p+r+q)$.

Let $\Pi_{p,r,q}$ be the set of all (p, r, q)-positions. We can consider the canonical projections $\mathrm{pr}_{p,r} : \Pi_{p,r,q} \to \Pi_{p,r}$, $\mathrm{pr}_{r,q} : \Pi_{p,r,q} \to \Pi_{r,q}$ and $\mathrm{pr}_{p,q} : \Pi_{p,r,q} \to \Pi_{p,q}$; for instance, if $\sigma \in \Pi_{p,r,q}$, τ is the restriction of σ at $[p+r]$, and $\{t_1, t_2, ..., t_k\}$ is the image of τ, where $t_1 < t_2 < \cdots < t_k$, then $\mathrm{pr}_{p,r}(\sigma)(i) = j$ if and only if $\tau(i) = t_j$.

Operations on $\Pi_{p,q}$:

5. Transposition. If $\pi \in \Pi_{p,q}$, then we can obtain an element $\pi^t \in \Pi_{q,p}$ in the following manner:

$$\pi^t(i) = \begin{cases} \pi(p+i), & \text{if } 1 \leq i \leq q \\ \pi(i-q), & \text{if } q+1 \leq i \leq p+q. \end{cases}$$

We have $(\pi^t)^t = \pi$.

Speaking about generalized intervals a and b, the transposition changes the position of a by that one of b.

6. Symmetry. If $\pi \in \Pi_{p,q}$, such that $\mathrm{Im}\,\pi = \{1, 2, ..., k\}$, then we obtain $\pi^h \in \Pi_{q,p}$, where $\pi^h(i) = (k+1) - \pi(p+q+1-i)$. Speaking about generalized intervals a and b in T, that corresponds to consider the opposite order on T, so we associate at the n-interval $(t_1, t_2, ..., t_n)$, written according the initial order, the n-interval $(t_n, t_{n-1}, ..., t_1)$, written according the opposite order.

Note that the symmetry $s = h \circ t$ is an involution on $\Pi_{p,q}$.

7. Composition.

Remark. Let $\pi_1 \in \Pi_{p,r}$ and $\pi_2 \in \Pi_{r,q}$. Then the set $P = \{\sigma \in \Pi_{p,r,q} \mid (\sigma(1), \sigma(2), ..., \sigma(p)) = (\pi_1(1), ..., \pi_1(p))$ and $(\sigma(p+r+1), ..., \sigma(p+r+q)) = (\pi_2(r+1), ..., \pi_2(r+q)\}$ is not empty.

Definition. Let $\pi_1 \in \Pi_{p,r}$ and $\pi_2 \in \Pi_{r,q}$. Let $\pi_1 \circ \pi_2 = \{\mathrm{pr}_{p,q}(\sigma) \mid \sigma \in P\}$. We say that $\pi_1 \circ \pi_2$ is the *composition* of π_1 and π_2.

According to the preceding remark, P is a finite, nonempty set, so the composition is well–defined. Thus, if (a, c) is a T-realization of π_1 and (c, b) is a T-realization of π_2, then (a, b) is a T-realization of one of its elements of $\pi_1 \circ \pi_2$.

The following properties are easily verified, for any $\pi_1 \in \Pi_{p,r}$, $\pi_2 \in \Pi_{r,q}$ and $\pi_3 \in \Pi_{q,s}$:

8. Proposition.

1) $(\pi_1 \circ \pi_2) \circ \pi_3 = \pi_1 \circ (\pi_2 \circ \pi_3)$;
2) $\pi_1 \circ \mathbf{I}'_{r,r} = \pi_1$ *and* $\mathbf{I}'_{p,p} \circ \pi_1 = \pi_1$;
3) $\mathbf{I}'_{p,p} \in \pi_1 \circ \pi_1^t$ *and* $\mathbf{I}'_{r,r} \in \pi_1^t \circ \pi_1$;
4) $\pi \in \pi_1 \circ \pi_2$ *implies* $\pi_1 \in \pi \circ \pi_2^t$ *and* $\pi_2 \in \pi_1^t \circ \pi$;
5) $(\pi_1 \circ \pi_2)^t = \pi_2^t \circ \pi_1^t$;
6) $(\pi_1 \circ \pi_2)^s = \pi_1^s \circ \pi_2^s$.

Connections with quasi–canonical hypergroups

In the following, we shall use a notion of *simplicial groupoid*, considered by P.J. Higgins [450].

First of all, by a *groupoid* is intended a category in which every morphism (edge) is invertible. Let us see what does it means that a morphism is invertible. Denote by E_{ij} the set of edges from the object i to the object j. The identity elements e_i satisfy the condition:

$$\forall\, a \in E_{ij},\ \ e_i a = a = a e_j.$$

Moreover, in a groupoid, for any $a \in E_{ij}$, there is $a^{-1} \in E_{ji}$, such that $aa^{-1} = e_i$ and $a^{-1}a = e_j$.

Notice that the set of edges from an object i to itself is a group, called the *vertex group at i*.

Now, let I be a set. We denote by $\Delta(I)$ the graph, whose vertex set is I and whose edge set is $I \times I$. Moreover, $\forall\,(i,j) \in I \times I$, there is a unique edge (i,j) from i to j, hence a category structure on $\Delta(I)$ can be uniquely defined, namely by the rule $(i,j)(j,k) = (i,k)$. The identity elements are the edges (i,j) and (j,i) is inverse to (i,j). The groupoid $\Delta(I)$ is called a *simplicial groupoid.*

Let $\Delta(\mathbb{N})$ be the simplicial groupoid on $\mathbb{N}$, that is the groupoid, whose associated graph has $\mathbb{N}$ as vertex set and for any (p,q) in $\mathbb{N}^2$, there is a unique arrow joining p and q.

A subgroupoid of $\Delta(\mathbb{N})$ is characterized by the set S of its arrows, which is an equivalence relation on a subset I of $\mathbb{N}$. Thus, we shall identify S with the corresponding subgroupoid of $\Delta(\mathbb{N})$.

Let S be a subgroupoid of $\Delta(\mathbb{N})$ and:

1) $\Pi_S = \bigcup\limits_{(p.q)\in S^2} \Pi_{p,q}$

2) if $\pi_1 \in \Pi_{p,q},\ \pi_2 \in \Pi_{p',q'}$, then set $\pi_1 \cdot \pi_2 = \begin{cases} \pi_1 \circ \pi_2, & \text{if } q = p' \\ \emptyset, & \text{otherwise.} \end{cases}$

3) $I_S = \{\mathbf{I}'_{p,p} \mid (p,p) \in S\}$.

4) ^{t} is the transposition.

9. Theorem. $(\Pi_S, \cdot, I_S, {}^t)$ *is a quasi–canonical hypergroupoid. It is a quasi–canonical hypergroup if and only if* S *has only one vertex.*

Proof. S is a subgroupoid if the following conditions are satisfied:

1) $(p, q) \in S$ and $(q, r) \in S \Longrightarrow (p, r) \in S$;
2) $(p, q) \in S \Longrightarrow (q, p) \in S$.

It is easily to check for $(\Pi_S, \cdot, I_S, {}^t)$ the conditions of a quasi–canonical hypergroupoid, using the preceding Proposition.

Finally, Π_S is a quasi–canonical hypergroup if and only if S is a group.

10. Remark. Using the standard construction of the associated algebra of complexes, we obtain that for any subgroupoid S of $\Delta(\mathbb{N})$, the complex algebra $\mathbf{A}_S$ of Π_S (see Definition 7, Ch. 9) is a complete, atomic, relation algebra, such that $0 \neq 1$.

If $S = \{n\}$, we write $\mathbf{A}_n$, instead of $\mathbf{A}_{\{n\}}$.

The interest for these algebras is justified by the fact that the objects utilized in Artificial Intelligence are the "weak–representations" of these algebras.

§2. Weak representations of interval algebras

In the following, the notion of *weak representation* of an interval algebra is introduced. G. Ligozat obtained a full classification of the connected weak representations of the algebra $\mathbf{A}_n$ of n-intervals.

First of all, let us recall what a *relation algebra* is.

11. Definition. An algebra $\mathbf{A} = (A, +, 0, \cdot, 1, *, 1', {}^{-1})$, where "+", "·" and "∗" are binary operations on A, "${}^{-1}$" is a unary operation on A and $0, 1, 1'$ are elements of A, is called a *relation algebra* if the following conditions hold:

1) $(A, +, 0, \cdot, 1)$ is a Boolean algebra;

2) $\forall\,(x,y,z) \in A^3$, $(x*y)*z = x*(y*z)$;

3) $\forall\, x \in A$, $1'*x = x = x*1'$;

4) $\forall\,(x,y,z) \in A^3$, we have:
$(x*y)\cdot z = 0 \Longleftrightarrow (x^{-1}*z)\cdot y = 0 \Longleftrightarrow (z*y^{-1})\cdot x = 0.$

12. Example. Let U be a set. Then

$$(\mathcal{P}(U\times U), \cup, \emptyset, \cap, U\times U, \circ, 1_{U\times U}, {}^t)$$

is a relation algebra, where "$\circ$" is the composition, $1_{U\times U}$ is the identity relation and "^{t}" is the transposition.

13. Definition. Let **A** be a relation algebra and U a set. A map $\Phi : A \longrightarrow \mathcal{P}(U\times U)$ is called a *representation* of **A** if:

1) Φ is an one-to-one map;

2) Φ defines a homomorphism of Boolean algebras;

3) $\forall\,(x,y) \in A^2$, $\Phi(x*y) = \Phi(x)\circ\Phi(y)$;

4) $\Phi(1') = 1_{U\times U}$;

5) $\Phi(^{-1}) = {}^t$.

More generally, a *weak representation* is defined by dropping condition 1) and replacing condition 3) by the weaker condition:

3′) $\forall\,(x,y) \in A^2$, $\Phi(x*y) \supseteq \Phi(x)\circ\Phi(y)$.

If **A** is a simple algebra, then we say that a weak representation of **A** into $\mathcal{P}(U\times U)$ is *connected* if $\Phi(1) = U\times U$.

Now, let S be a non-empty subset of $\mathbb{N}$ and Π_S be the disjoint sum of all $\Pi_{p,q}$, where $(p,q) \in S^2$.

In §1, we obtained that $(\Pi_S, \cdot, I_S, {}^t)$ is a quasi–canonical hypergroupoid. Applying to $(\Pi_S, \cdot, I_S, {}^t)$ the standard construction which associates with a quasi–canonical hypergroupoid its complex algebra (see [46]), we obtain the complex algebra $\mathbf{A}_S$.

For $S = \{n\}$, we obtain the relation algebra $\mathbf{A}_n$ of n-intervals.

Now, let Φ be a connected weak representation of $\mathbf{A}_n$ into $\mathcal{P}(U\times U)$, where U is a set.

Recall that the elements of $\Pi_{n,n}$ can be interpreted as maps from the set $\{x_1, ..., x_n, y_1, ..., y_n\}$ into $\mathbb{N}^*$.

For any element π of $\Pi_{n,n}$, which can be considered as an atom of $\mathbf{A}_n$, $\Phi(\pi)$ is a binary relation R_π on U and we have:

1) $(R_\pi)_{\pi\in\Pi_{n,n}}$ is a partition of $U\times U$ and
2) $\forall\,(\pi,\pi')\in\Pi_{n,n}\times\Pi_{n,n}$, we have $R_\pi \circ R_{\pi'} \subseteq R_{\pi*\pi'}$.

For $1\le i,j\le n$, we consider the following elements of $\mathbf{A}_n$:

$a_{i,j}$ which is the sum of all π, such that $\pi(x_i)=\pi(y_i)$;

$b_{i,j}$ which is the sum of all π, such that $\pi(x_i)<\pi(y_i)$.

We obtain the following result:

14. Proposition.

1°) $a_{i,i}\ge \mathbf{I}'_{n,n}$;
2°) $a^t_{i,j}=a_{j,i}$;
3°) $a_{i,j}*a_{j,k}=a_{i,k}$;
4°) $a_{i,j}*b_{j,k}*a_{k,\ell}=a_{i,\ell}$;
5°) $b_{i,j}*b_{j,k}=b_{i,k}$;
6°) $b_{i,j}\cdot b^t_{j,i}=0$;
7°) $b_{i,j}+b^t_{i,j}+a_{i,j}=1$;
8°) *if* $i<j$, *then* $\mathbf{I}'_{n,n}\in b_{i,i}$.

Chapter 11

Probabilities

Using a particular non–standard algebraic hyperstructure, A. Maturo [251] proved that the problems on the coherent assessments of probability and their solutions can be expressed in a very useful and simple form.

Thus, new algorithms to control the coherence can be introduced in this new algebraic context.

In several of their papers, S. Doria and A. Maturo have considered some algebraic structures and hyperstructures of events and contional events. They have studied the properties and the probabilistic meaning of these hyperstructures and they also have considered their associated geometric spaces.

We know that conditional events are used in the Artificial Intelligence to represent partial information and vague data.

In the following, we present some constructions considered by S. Doria and A. Maturo.

Let $E = \{E_1, E_2, ..., E_n\}$ be a finite family of events. Set

$$E_i^{\mathcal{E}_i} = \begin{cases} E_i, & \text{if } \mathcal{E}_i = 1 \\ \overline{E}_i, & \text{if } \mathcal{E}_i = -1 \end{cases}$$

1. Definition. We call *atoms* generated by E, the nonempty intersections $E_1^{\mathcal{E}_1}E_2^{\mathcal{E}_2}...E_n^{\mathcal{E}_n}$, where for $\forall\, i \in \{1, 2, ..., n\}$, $E_i^{\mathcal{E}_i}$ are defined above.

Let $C(E)$ be the set of all atoms generated by E.

Let $\mathcal{E}$ be an algebra of events, so the following two conditions hold:

1. if $A \in \mathcal{E}$, then $\overline{A} \in \mathcal{E}$;
2. if $(A, B) \in \mathcal{E}^2$, then $AB \in \mathcal{E}$.

We define on $\mathcal{E}$ the following hyperoperation

$$\forall\, (A, B) \in \mathcal{E}^2,\;\; A\circ B = C(A, B).$$

Then it follows that:

2. Proposition. $(\mathcal{E}, \circ)$ *is a commutative semihypergroup, called semihypergroup of atoms.*

3. Proposition. *Let E be a family of events, such that $E \cup \{\phi, \Omega\}$ is an algebra of events. Then E is a substructure of $(\mathcal{E}, \circ)$ if and only if the following implication holds:*

$$\phi \in E \Longrightarrow \Omega \in E.$$

Proof. "$\Longleftarrow$" For any $(A, B) \in (E - \{\phi, \Omega\}) \times E$, we have

$$A\circ B \subseteq E - \{\phi, \Omega\}.$$

We can have the following situations:

1. $\phi \notin E$ and $\Omega \notin E$. In this case $E\circ E \subseteq E$, so $(E, \circ)$ is a substructure of $(\mathcal{E}, \circ)$;
2. $\phi \notin E$ and $\Omega \in E$. Then $E\circ E \subseteq E$ and $\Omega\circ\Omega = \{\Omega\}$;
3. $\phi \in E$ and $\Omega \in E$. Since $\Omega\circ\Omega = \{\Omega\}$ and $\phi\circ\phi = \{\phi\}$, it follows $E\circ E \subseteq E$.

"$\Longrightarrow$" If E is a substructure of $\mathcal{E}$ and $\phi \in E$, then $\phi \circ \phi = \{\Omega\}$ and so $\Omega \in E$. ■

Notice that if E is a substructure, then also $E - \{\phi\}$ and $E - \{\phi, \Omega\}$ are substructures.

4. Proposition. *Let $(E, \circ)$ be a subhypergroup of $(\mathcal{E}, \circ)$. Then $\forall (A, B) \in E^2$, we have $(A \subseteq B$ or $A \subseteq \overline{B})$ and $(B \subseteq A$ or $B \subset \overline{A})$.*

Proof. Indeed, there is $Y \in E$, such that $B \in A \circ Y$, whence $B \subseteq A$ or $B \subseteq \overline{A}$.

On the other hand, since $\exists X : A \in B \circ X$ it follows $A \subseteq B$ or $A \subseteq \overline{B}$. ■

5. Corollary. *Let E be a family of events contained in $\mathcal{E}$. Then $(E, \circ)$ is a subhypergroup of $(\mathcal{E}, \circ)$ if and only if $E = \{\Omega\}$ or there exists $A \in \mathcal{E} - \{\phi, \Omega\} : E = \{A, \overline{A}\}$.*

Proof. "$\Longleftarrow$" For any $A \in \mathcal{E} - \{\phi, \Omega\}$, we have that $(\{\Omega\}, \circ)$ and $(\{A, \overline{A}\}, \circ)$ are hypergroups.

"$\Longrightarrow$" Let $(E, \circ)$ be a hypergroup and A, B be two elements of E. Since $\forall A \in E - \{\Omega\}$, it follows $\phi \notin E$. Suppose $\Omega \in E$. Since $\forall A \in E - \{\Omega\}$, we have $\Omega \notin A \circ \Omega$, it follows $E = \{\Omega\}$.

Now, suppose that $E \cap \{\phi, \Omega\} = \phi$.

According to the above proposition, it follows $\forall A, B \in E$, we have $(A \subseteq B$ or $A \subseteq \overline{B})$ and $(B \subseteq A$ or $B \subseteq \overline{A})$. If $A \subseteq B$, then $B \not\subseteq \overline{A}$, otherwise $A = \phi$ and $\overline{A} = \Omega$, a contradiction. Then $A \subseteq B$ implies $B \subseteq A$ and so $A = B$.

Similarly, $A \subseteq \overline{B}$ implies $\overline{B} \subseteq A$ and so $A = \overline{B}$.

Therefore $E = \{A, \overline{A}\}$. ■

Hyperstructures and conditional events

Let $\mathcal{E}$ be an algebra of events.

6. Definition. For any $(A, B) \in \mathcal{E}^2$, we call the *conditional event* A/B the logical entity, which is true if AB is true, false if $\overline{A}B$ is true and it is undetermined if B is false.

Notice that the triplet (X, Y, Z) of events represents a conditional event if and only if X, Y, Z are pairwise incompatible and their union is Ω.

Let CE be the set of triplets (X, Y, Z), which represent conditional events.

Let $U = \{\{A, B\} \subset \mathcal{E} \mid A \subseteq B \text{ or } B \subseteq A\}$.

7. Proposition. *The map $f : CE \to U$, defined as follows:*

$$f(X, Y, Z) = \{X, X \cup Y\}$$

is a bijection.

Proof. Indeed, for any $\{A, B\} \in U$, with $A \subseteq B$, we have that $f^{-1}(\{A, B\}) = \{(A, B - A, \Omega - B)\}$ has cardinality 1. ∎

In this manner, we can represent any conditional event as an element of U.

Now, let $\mathcal{E}$ be an algebra of events. We define on $\mathcal{E}$ the following hyperoperation:

$$\forall (A, B) \in \mathcal{E}^2, \ A \odot B = \{AB, B\}.$$

The hypergroupoid $(\mathcal{E}, \odot)$ is called the *hypergroupoid of conditional events* and it is denoted by CEH.

8. Proposition. *The hypergroupoid $(\mathcal{E}, \odot)$ is a weak–commutative and a regular weak–associative one.*

Proof. For any $(X, Y, Z) \in \mathcal{E}^3$, we have

$$X \odot (Y \odot Z) \supseteq (X \odot Y) \odot Z.$$

Indeed, we have

$$X \odot (Y \odot Z) = $$
$$= \bigcup_{V \in Y \odot Z} X \odot V = (X \odot YZ) \cup (X \odot Z) = \{XYZ, YZ, XZ, Z\}$$

and

$$(X \odot Y) \odot Z = \bigcup_{T \in X \odot Y} T \odot Z = (XY \odot Z) \cup (Y \odot Z) = \{XYZ, YZ, Z\}.$$

Therefore, $(\mathcal{E}, \circ)$ is a weak-associative hypergroupoid.

On the other hand, $\forall (X, Y) \in \mathcal{E}^2$, $X \neq Y$, we have $X \odot Y = \{XY, Y\}$ and $Y \odot X = \{XY, X\}$. Thus, $X \odot Y \neq Y \odot X$ but $X \odot Y \cap Y \odot X \neq \emptyset$. Moreover, $\forall (X, Y, Z) \in \mathcal{E}^3$, we have

(X) $$X \odot (Y \odot Z) = (X \odot Y) \odot Z \cup (Y \odot X) \odot Z \quad ■$$

9. Proposition. *Set $H \subset \mathcal{E}$, $H \neq \emptyset$. Then $(H, \odot)$ is a substructure of $(\mathcal{E}, \odot)$ if and only if*

$$\forall (X, Y) \in H^2, \quad \text{we have } XY \in H.$$

Proof. "$\Longleftarrow$" $\forall (X, Y) \in H^2$, we have $X \odot Y = \{XY, Y\} \subseteq H$, so $(H, \odot)$ is a substructure.

"$\Longrightarrow$" Suppose that $(H, \odot)$ is a substructure of $(\mathcal{E}, \odot)$. Then we have

$$\forall (X, Y) \in H^2, \ XY \in \{XY, Y\} = X \odot Y \subseteq H. \quad ■$$

10. Corollary. *All the conditional events A/B are substructures of $(\mathcal{E}, \odot)$.*

11. Theorem. *A substructure $(H, \odot)$ of $(\mathcal{E}, \odot)$ is a hypergroup if and only if $H = \{X\}$, where $X \in \mathcal{E}$.*

Proof. "$\Longrightarrow$" Let $(H, \odot)$ be a hypergroup. Then $(H, \odot)$ is a quasi–hypergroup, so $\forall (A, B) \in H^2$, $\exists X \in H$, $\exists Y \in H$ such that $B \in X \odot A = \{XA, A\}$ and $A \in Y \odot B = \{YB, B\}$. Since $B \in \{XA, A\}$ it follows $B \subseteq A$ and since $A \in \{YB, B\}$ it follows $A \subseteq B$. Therefore $A = B$ and so H consists of an only one element.

"$\Longleftarrow$" $\forall X \in \mathcal{E}$, we have $X \odot X = \{X\} \subseteq \{X\}$, so $(\{X\}, \odot)$ is a quasi–hypergroup. Moreover, $(X \odot X) \odot X = X \odot (X \odot X) = \{X\}$, so $(\{X\}, \odot)$ is a hypergroup. ■

12. Definition. A weak associative hypergroupoid $(H, \circ)$ is called

(i) *left directed* if $\forall (x, y, z) \in H^3$, $x \circ (y \circ z) \subseteq (x \circ y) \circ z$;

(ii) *right directed* if $\forall (x, y, z) \in H^3$, $x \circ (y \circ z) \supseteq (x \circ y) \circ z$;

(iii) *directed* if it is right and left directed.

The class of left directed (respectively, right directed) weak associative hypergroupoids is denoted by LHD (respectively, by RDH).

Let $(H, \circ)$ be a hypergroupoid.

If $n \in \mathbb{N}^*$ and $(x_1, x_2, ..., x_n) \in H^n$, then we define the set $\mathcal{H}(x_1, x_2, ..., x_n)$ of all hyperproducts generated by $(x_1, x_2, ..., x_n)$ as follows:

$$\mathcal{H}(x_1) = \{x_1\}$$

and for $n > 1$, $\mathcal{H}(x_1, x_2, ..., x_n)$ is the set of all hyperproducts $P = P_1 \circ P_2$, where $P_1 \in \mathcal{H}(x_1, x_2, ..., x_h)$ and $P_2 \in \mathcal{H}(x_{h+1}, ..., x_n)$, where $h \in \{1, 2, ..., n-1\}$.

Let $(H, \circ)$ be a hypergroupoid.

Let $n \in \mathbb{N}^*$ and $(x_1, x_2, ..., x_n) \in H^n$. The *right hypeproduct* $\rho(x_1, x_2, ..., x_n)$ generated by $(x_1, x_2, ..., x_n)$ is defined as follows:

$$\rho(x_1, x_2, ..., x_n) = \begin{cases} \{x_1\}, & \text{if } n = 1 \\ x_1 \circ \rho(x_2, ..., x_n), & \text{if } n > 1. \end{cases}$$

Similarly, the *left hyperproduct* $\lambda(x_1, x_2, ..., x_n)$ generated by $(x_1, x_2, ..., x_n)$ is defined as follows:

$$\lambda(x_1, x_2, ..., x_n) = \begin{cases} \{x_1\}, & \text{if } n = 1 \\ \lambda(x_1, ..., x_{n-1}) \circ x_n, & \text{if } n > 1. \end{cases}$$

13. Theorem. *Let $(H, \circ) \in RDH$. Then, for any $n \in \mathbb{N}^*$, $\mathcal{H}(x_1, x_2, ..., x_n)$ is a finite lattice, with respect to the inclusion.*

Particularly, $(H, \circ)$ is a feebly associative hypergroupoid.

Proof. We prove this by induction on n. By $(*)$ it follows that the theorem is true for $n \leq 3$.

Suppose the statement true for any $n \leq h$, where $h \geq 3$.

Then $\forall k \in \mathbb{N}$, $1 \leq t < h$, we have

$$\begin{aligned} &\lambda(x_1, ..., x_h, x_{h+1}) = \lambda(x_1, ..., x_h) \circ x_{h+1} \subseteq \\ &\subseteq (\lambda(x_1, ..., x_t) \circ \lambda(x_{t+1}, ..., x_h)) \circ x_{h+1}) \subseteq \\ &\subseteq \lambda(x_1, ..., x_t) \circ (\lambda(x_{t+1}, ..., x_h) \circ x_{h+1}) = \\ &= \lambda(x_1, ..., x_t) \circ \lambda(x_{t+1}, ..., x_{h+1}) \subseteq P_1 \circ P_2, \\ &\forall P_1 \in \mathcal{H}(x_1, ..., x_t), \ \forall P_2 \in \mathcal{H}(x_{t+1}, ..., x_{h+1}). \end{aligned}$$

Since $\lambda(x_1, ..., x_h) \circ x_{h+1} \subseteq P_1 \circ x_{h+1}$, $\forall P_1 \in \mathcal{H}(x_1, ..., x_h)$ it follows that $\lambda(x_1, ..., x_{h+1})$ is the minimum of $\mathcal{H}(x_1, ..., x_{h+1})$. Similarly, it follows that $\rho(x_1, ..., x_{h+1})$ is the maximum. ∎

14. Corollary. *$(\mathcal{E}, \odot)$ is a feebly associative hypergroupoid. Moreover, $\forall n \in \mathbb{N}^*$, $\forall (E_1, E_2, ..., E_n) \in \mathcal{E}^n$, $\mathcal{H}(E_1, E_2, ..., E_n)$ is a finite lattice with the minimum $\lambda(E_1, E_2, ..., E_n)$ and the maximum $\rho(E_1, E_2, ..., E_n)$.*

By induction, it follows the following

15. Theorem. $\forall n \in \mathbb{N}^*$, $\forall (E_1, E_2, ..., E_n) \in \mathcal{E}^n$, *we have*

(i) $\lambda(E_1, E_2, ..., E_n) = \left\{ \prod_{s=1}^{n} E_s, \ i \in \{1, 2, ..., n\} \right\}$;

(ii) $\rho(E_1, E_2, ..., E_n) = \left\{ \prod_{p=1}^{n-1} F_p E_n, \text{ where } F_p \in \{E_r, \Omega\} \right\};$

(iii) $\forall P \in \mathcal{H}(E_1, E_2, .., E_n),\ \min P = \prod_{s=1}^{n} E_s$ *and* $\max P = E_n.$

16. Proposition. *Let K_1 and K_2 be two subhypergroupoids of $(\mathcal{E}, \odot)$. Then also $K_1 \odot K_2$ is a subhypergroupoid.*

Proof. We have $K_1K_1 \subseteq K_1$ and $K_2K_2 \subseteq K_2$, whence

$$(K_1 \odot K_2)(K_1 \odot K_2) =$$
$$= (K_1K_2 \cup K_2)(K_1K_2 \cup K_2) \subseteq K_1K_2 \cup K_2 = K_1 \odot K_2.$$

Therefore, $K_1 \odot K_2$ is a subhypergroupoid. ∎

17. Proposition. *For $\forall n \in \mathbb{N}^*$, $\forall (E_1, E_2, ..., E_n) \in \mathcal{E}^n$, every $P \in \mathcal{H}(E_1, E_2, ..., E_n)$ is a subhypergroupoid of $(\mathcal{E}, \odot)$.*

Proof. We prove by induction. For $n = 1$, the statement is clearly true.

Let us suppose the statement true for $n \le k$ and we shall verify it for $n = k+1$. Indeed, $\forall P \in \mathcal{H}(E_1, E_2, .., E_{k+1})$, $\exists s \in \{1, 2, ..., k\}$, $\exists P_1 \in \mathcal{H}(E_1, ..., E_s)$, $\exists P_2 \in \mathcal{H}(E_{s+1}, ..., E_{k+1})$ such that $P = P_1 \odot P_2$. By the above proposition it follows that P is a subhypergroupoid. ∎

Now, let $\mathcal{E}$ be an algebra of events.

18. Definition. For any $(E_1, E_2) \in \mathcal{E}^2$, we define

$$E_1 \,\Box\, E_2 = \{E_1E_2, \overline{E}_1E_2, \overline{E}_2\}.$$

The hypergroupoid $(\mathcal{E}, \Box)$ is called the *hypergroupoid of the atoms of conditional events.*

19. Theorem. $(\mathcal{E}, \square)$ *is a weak associative and a weak commutative hypergroupoid.*

Proof. 1) For any $(E_1, E_2, E_3) \in \mathcal{E}^3$, we have:

$$(E_1 \square E_2) \square E_3 = \bigcup_{F \in E_1 \square E_2} F \square E_3 =$$
$$= E_1E_2 \square E_3 \cup \overline{E}_2 \square E_3 \cup \overline{E}_1E_2 \square E_3 =$$
$$= \{E_1E_2E_3, \overline{E_1E_2}E_3, \overline{E}_3, \overline{E}_2E_3, E_2E_3, \overline{E}_1E_2E_3, \overline{\overline{E}_1E_2}E_3\} \text{ and}$$

$$E_1 \square (E_2 \square E_3) = \bigcup_{K \in E_2 \square E_3} E_1 \square K =$$
$$= E_1 \square E_2E_3 \cup E_1 \square \overline{E}_3 \cup E_1 \square \overline{E}_2E_3 =$$
$$= \{E_1E_2E_3, \overline{E_2E_3}, \overline{E}_1E_2E_3, E_1\overline{E}_3, E_3, \overline{E}_1\overline{E}_3, E_1\overline{E}_2E_3, \overline{\overline{E}_2E_3}, \overline{E}_1\overline{E}_2, E_3\},$$

whence $(E_1 \square E_2) \square E_3 \cap E_1 \square (E_2 \square E_3) \neq \emptyset$

2) For any $(E_1, E_2) \in \mathcal{E}^2$, we have: $E_1 \square E_2 = \{E_1, \overline{E}_1E_2, \overline{E}_2\}$ and $E_2 \square E_1 = \{E_1E_2, \overline{E}_2E_1, \overline{E}_1\}$, whence $E_1 \square E_2 \cap E_2 \square E_1 \neq \emptyset$.■

Bibliography

[1] Alajbegovic, J., Mockor, J., *Valuations on multirings*, Commentarii Mathematici Universitatis Sancti Pauli, vol. 34, no. 2 (1985), ed. Rikkyo Univ/Math. Ikeburo Tokyo.

[2] Al Ali, M.I.M., *Hypergraphs, Hypergroupoids and Hypergroups*, Italian J. Pure and Appl. Math. no. 8, (2000).

[3] Allen, J.F., *Maintaining Knowledge about Temporal Intervals*, Communications of the ACM 26, 11 (1983), 832-843.

[4] Ameri, R., *On categories of hypergroups and hypermodules*, Proc. of VIIth Int. Congress on AHA, in printing in J. of Discrete Math.

[5] Ameri, R., *Topological (transposition) hypergroups*, to appear.

[6] Ameri, R., Zahedi, M.M., *Fuzzy subhypermodules over fuzzy hyperrings*, Proceeding of the Sixth International Congress "Algebraic Hyperstructures and Applications", Prague, Czech Republic, Sept. 1-9, 1996, Publisher: Democritus University of Thrace, Greece, 1–15.

[7] Ameri R., Zahedi M.M., *Hypergroup and Join Space induced by a Fuzzy Subset*, PU.M.A., vol. 8 (1997).

[8] Ameri, R., Zahedi, M.M., *Fuzzy direct sum and fuzzy (weakly) prime and primary subhypermodules*, Discrete Math. Sci. and Cryptography, vol.1, (1998).

[9] Ameri, R., Zahedi, M.M., *Hyperalgebraic Systems*, It. J. Pure and Appl. Math. no. 6 (1999).

[10] Ashrafi, A.R., *About some join spaces and hyperlattices*, Italian Journal of Pure and Applied Mathematics, no. 10, 2001.

[11] Ashrafi, A.R., *Construction of hyperlattices from combinatorial structures*, to appear.

[12] Avann, S.P., *Ternary distributive semilattices*, Bull. Amer. Math. Soc. 54, 1948, 79.

[13] Avann, S.P., *Metric ternary distributive semi–lattices*, Proc. Amer. Math. Soc. 12 (1961), 407–414.

[14] Azzali, E., *Sull'ordinamento della classe delle strutture di un ipergruppo su un insieme finito*, Convegno su Ipergruppi, altre Strutture Multivoche e loro Applicazioni, Editor P. Corsini (Udine 1985).

[15] Bae Jun, Y., Zahedi, M.M., Long Xin, X., Borzoei, R.A., *On hyper BCK-algebras*, Italian Journal of Pure and Applied Mathematics, nr. 8, 2000, 127–136.

[16] Bandelt, H., Hedlíkovà, J., *Median algebras*, Discrete Math. 45, 1983.

[17] Bavel, Z., *The source as a tool in automata*, Inform. Control. 18 (1971), 140–155.

[18] Bednaryk, F., Slapal, J., *On exponentiation of n-ary hyperalgebras*, Discussiones Mathematicae, General Algebras and Applications.

[19] Benado, M., *Sur la théorie de la divisibilité*, Acad. R.P. Române, Bul. Sti. Sect. Mat. Fyz. 6 (1954).

[20] Benado, M., *Bemerkungen zur Theorie der Vielverbande* I, II, III, Math. Nachr. 20, 1959.

[21] Benado, M., *Bemerkungen zur Theorie der Vielverbande* IV, Proc. Camb. Phil. Soc. 56 (1960).

[22] Benado, M., *Teorija multiresotok i jejo znacenie v algebre i geometriji*, Acta Fac. Rerum Nat. Univ. Comenianae, Math. (1961).

[23] Berardi, L., Eugeni, F., Innamorati, S., *Generalized designs, Linear spaces, Hypergroupoids and Algebraic Crypotography*, IV Congress on AHA, Xanthi, 1990.

[24] Berardi, L., Eugeni, F., Innamorati, S., *Remarks on Hypergroupoids and Criptography*, Journal of Combinatorics, Information and System Sciences, vol. 17, no. 3-4, 217–231, 1992.

[25] Birkhoff, G., Kiss, S.A., *A ternary operation in distributive lattices*, Bull. Amer. Math. Soc. 53 (1974), 749–752.

[26] Bonansinga, P., *Sugli ipergruppi quasicanonici*, Atti Soc. Pelor. Sc. Fis. Mat. Nat. vol. 27 (1981).

[27] Bonansinga, P., *Omomorfismi polidromi tra gruppi ciclici finiti*, Riv. di Mat. Univ. Parma, 10 (1984).

[28] Bonansinga, P., Corsini, P., *Sugli omormorfismi di semi–ipergruppi e di ipergruppi*, B.U.M.I., 1-B (1982).

[29] Borzoei, R.A., Hasankhani, A., *HyperK_H-algebras*, Italian Journal of Pure and Applied Mathematics no. 10, 2001, 207–212.

[30] Bose, R.C., Mesner, D.M., *On linear associative algebras corresponding to association schemes of partially balanced designs*, Annals Math. Stat. 36, 1959, 21–38.

[31] Brender, M., *A class of Schur algebras*, Trans. AMS, 248 (1979), 435–444.

[32] Bryant, V.W., Webster, R.J., *Convexity spaces.* I: *The basic properties*, J. Math. Anal. Appl. 37 (1972).

[33] Cameron, P.J., Van Lint, J.H., *Graph Theory, Coding theory and Block Designs*, London Math. Soc. Lecture Notes no. 19, Cambridge Univ. Press, 1975.

[34] Campaigne, H., *A lower limit of the number of hypergroups of a given order*, J. Washington Acad. Sci., 44, 5-7 (1954).

[35] Călugăreanu, G., Leoreanu, V., *Hypergroups associated with lattices*, Italian J. of Pure and Appl. Math., no.9, 2001, 165–173.

[36] Chaopraknoi, S., Kemprasit, Y., *Some linear transformation semigroups which do not admit the structure of a semihyperring with zero,* Proc. of ICAA 2002, Chulalongkorn University, Thailand.

[37] Chvalina, J., *Commutative hypergroups in the sense of Marty and oredered sets*, Summer School on General Algebra and Ordered Sets, Proc. Internat. Conf. Olomouc (Czech Rep.), 1994, 19–30.

[38] Chvalina, J., *Relational product of join spaces determined by quasi–orders*, Sixth Int. Congress on AHA (1996), 15–23, Democritus Univ. of Thrace Press.

[39] Chvalina, J, Chvalinová, L., *Betweenness, automata and commutative hypergroups*, Acta Mathematica e Informatica Universitatis Ostraviensis, 4 (1996), 105–120.

[40] Chvalina, J, Chvalinová, L., *State hypergroups of automata*, Acta Mathematica e Informatica Universitatis Ostroviensis 4(1996), 105–120.

[41] Chvalina, J., Hort, D., *A certain compatibility types of binary relations on hypergroups determined by quasi–orders*, Contributions to General Algebra 11, 113–125.

[42] Chung Sang-Cho, Choi Byung-Mun, *H_v-groups on the set* $\{e, a, b\}$, Italian Journal of Pure and Applied Mathematics no. 10, 2001, 133-140.

[43] Comer, St.D., *Multi–valued loops, geometries and algebraic logic*, Houston Journal of Mathematics, vol. 2, no.3, 1976.

[44] Comer, S.D., *Constructions of color schemes*, Acta Universitatis Carolinae (Math. et Phys.) 34 (1983).

[45] Comer, S.D., *Color schemes forbidding monochromatic triangles*, Congressus Numerantium 39 (1983).

[46] Comer, S.D., *A New Foundation for the Theory of Relations*, Notre Dame Journal of Formal Logic, 24, 2, 1983, 181–187.

[47] Comer, S., *Combinatorial aspects of relations*, Alg. Universalis, 18 (1984), 77–94.

[48] Comer, S.D., *Combinatorial types*, In Algebra Combinatorics and Logic in Computer Science; Colloquia Math. Soc. J. Bolyai, North Holland Publ. (1985).

[49] Comer, S.D., *Multivalued Algebras and Their Graphical Representations*, (Preliminary Draft), Math. and Comp. Sci. Dep. the Citadel Charleston, South Carolina 29409, July 1986.

[50] Comer, S.D., *Weakly regular trees and their color algebras*, Riv. di Mat. Pura ed Appl. no. 3 (1988).

[51] Comer, S.D., *Lattices of conjugacy relations*, Proc. Malcev Conference of Algebra, Novosibirsk, USSR (1989).

[52] Comer, St.D., *Hyperstructures associated with character algebras and color schemes*, Hadronic Press, Palm Harbor, New Frontiers in Hyperstructures, T. Vougiouklis Editor, 1996, 49–66.

[53] Comer, St.D., *Application–oriented hyperstructures*, Hadronic Press, Palm Harbor, New Frontiers Hyperstructures, T. Vougiouklis Editor, 1996, 213–218.

[54] Corsini, P., *Hypergroupes et groupes ordonnés*, Rend. Sem. Mat. Univ. Padova, v. XLVIII (1972).

[55] Corsini, P., *Homomorphismes d'hypergroupes*, Rend. Sem. Mat. Univ. Padova, v. XLVIII (1972).

[56] Corsini, P., *Hypergroupes réguliers et hypermodules*, Ann. Univ. Ferrara, Sc. Mat. Ser. VII, v. XX (1975).

[57] Corsini, P., *Sur les homomorphismes polydromes*, Ann. Univ. Ferrara, Sc. Mat. Ser. VII, v. XXI (1975).

[58] Corsini, P., *Hypergroupes d'associativité de quasigroupes médiaux*, Atti del Convegno su Sistemi Binari e loro Applicazioni, Taormina, 1978.

[59] Corsini, P., *Sur les semi-hypergroupes*, Atti Soc. Pelor. Sc. Mat. Fis. Nat., XXVI, (1980).

[60] Corsini, P., *Contributo alla teoria degli ipergruppi*, Atti Soc. Pelor. Sc. Mat. Fis. Nat., XXVI (1980).

[61] Corsini, P., *Sur les semi–hypergroupes complets et les groupoides*, Atti Soc. Pelor. Sc. Mat. Fis. Nat., XXVI, (1980).

[62] Corsini, P., *Ipergruppi semiregolari e regolari*, Rend. Sem. Mat. Univ. Politecn. Torino, v. 40, 3, (1982).

[63] Corsini, P., *Recenti risultati in teoria degli ipergruppi*, B.U.M.I., 2-A (1983).

[64] Corsini, P., *Feebly canonical and* 1*-hypergroups*, Acta Universitatis Carolinae, Math. et Phys. v. 24, no.2 (1983).

[65] Corsini, P., *(i.p.s.) Ipergruppi di ordine* 7. Atti Sem. Mat. Fis. Univ. Modena, XXXIV, (1985–1986)

[66] Corsini, P., *Sugli ipergruppi canonici finiti con identità parziali scalari*, Rend. del Circ. Mat. di Palermo serie II, T. XXXVI (1987).

[67] Corsini, P., *(i.p.s.) Ipergruppi di ordine* 6, Ann. Sc. de l'Univ. Blaise Pascal, Clermont–Ferrand II, (1987).

[68] Corsini, P., *Themes and problems in Hypergroup Theory*, Proceedings of the Fourth International Congress on Algebraic Hyperstructures and Applications, Xanthi, Greece, World Scientific, (1990).

[69] Corsini, P., *Graphs and join spaces*, J. of Combinatorics, Information and System Sciences, v. 16, n. 4, (1991).

[70] Corsini, P., *Join Spaces, Power Sets, Fuzzy Sets*, Proceedings of the Fifth Congress on Algebraic Hyperstructures and Applications, Iaşi, 1993, M. Ştefănescu Editor, Hadronic Press, Palm Harbor, U.S.A. (1994).

[71] Corsini, P., *Remarks on join spaces associated with graphs*, Riv. Mat. Pura e Appl., no.16, 1995, 115–119.

[72] Corsini, P., *On W-Hypergroups*, Proc. of I.R.B. Int. Workwhops, New Frontiers in Multivalued Hyperstructures, Monteroduni (1995).

[73] Corsini, P., *Mutually associative hypergroupoids*, Sixth Int. Congress on AHA, Democritus Univ. of Thrace Press (1996), 25–33.

[74] Corsini, P., *Hypergraphs and hypergroups*, Algebra Universalis, 36 (1996), 548–555.

[75] Corsini, P., *A remarkable class of F-hypergroups*, Proceedings of "The Eighth Symposium of Mathematics and its Applications", Timişoara, Romania (1999).

[76] Corsini, P., *Rough Sets, Fuzzy Sets and Join Spaces*, Honorary Volume dedicated to Professor Emeritus Ioannis Mittas, Aristotle University of Thessaloniki, 1999–2000.

[77] Corsini, P., *Properties of hyperoperations associated with fuzzy sets and with factor spaces*, Int. Journal of Science and Technology, Kashan University no. 1, 2000.

[78] Corsini, P., *On the Hypergroups Associated with Binary Relations*, J. Multiple - valued Logic, 2000, vol. 5, pp. 407–419.

[79] Corsini, P., *Binary relations and hypergroupoids*, Italian J. Pure and Appl. Math. no.7, (2000).

[80] Corsini, P., *Fuzzy sets, join spaces and factor spaces*, PU.M.A., vol. 11, no. 3, 2000.

[81] Corsini, P., *On Chinese hyperstructures*, Proc. of the VII Int. Congress on AHA, Taormina '99, in printing on J. of Discrete Mathematics.

[82] Corsini, P., *Hyperstructures associated with ordered sets*, Proc. of the Fourth Panhellenic Conference on Algebra and Number Theory, in printing on Bull. of the Greek (Hellenic) Mathematical Society.

[83] Corsini, P. , Borzooei, R.A. , Zahedi, M.M., *Some kinds of positive implicative hyper K-ideals*, Accepted by Journal of Discrete Mathematical Sciences & Cryptography, Delhi, (2001).

[84] Corsini, P., Freni, D., *On the heart of hypergroups*, Mathematica Montisnigri, n.2, (1992).

[85] Corsini, P., Leoreanu, V., *Join spaces associated to Fuzzy Sets*, J. of Combinatorics, Information and System Sciences, 20, no.1 (1995).

[86] Corsini, P., Leoreanu, V., *Le théorème de Lagrange dans les hypergroupes*, Bul. Şt. Timişoara, Ser. Mat.-Fiz., Tom 41 (55), no.2, 1996, 1–22.

[87] Corsini, P., Leoreanu, V., *About the heart of a hypergroup*, Acta Universitatis Carolinae, vol. 37, no. 1 (1996).

[88] Corsini, P., Leoreanu, V., *Hypergroups and binary relations*, Algebra Universalis, 43, 2000, p. 321–330.

[89] Corsini, P., Leoreanu, V., *Fuzzy Sets, Join Spaces associated with Rough Sets*, accepted by Circ. Mat. di Palermo, 2002.

[90] Corsini, P., Mittas, J., *Sur la monotonie des suites hypercompositionnelles*, Rend. del Circ. Mat. di Palermo, ser. II, t. XLIII (1994).

[91] Corsini, P., Mittas, J., *New topics in hypegroup theory*, accepted by Multiple–valued logic and its applications (1995).

[92] Corsini, P., Moghani, G.A., *On finite semi–join spaces and fuzzy sets*, accepted by PU.M.A., 2002.

[93] Corsini, P., Romeo, G., *Hypergroupes complets et t-groupoides*, Convegno su "Sistemi Binari e loro Applicazioni", Taormina (1978).

[94] Corsini, P., Sureau, Y., *Sur les sous–hypergroupes d'un hypergroupe,* Riv. di Mat. Pura e Appl., n. 1, (1987).

[95] Corsini, P., Tofan, I., *On fuzzy hypergroups*, PU.M.A., vol. 8, no.1 (1997).

[96] Corsini, P., Vougiouklis, T., *From groupoids to groups through hypergroups*, Rend. Mat. Ser. VII, vol. 9, Roma (1989).

[97] Corsini, P., Zahedi, M.M., Khorashadi–Zadeh, M.R., *Some results on the category of Lt-fuzzy hypergraphs*, Accepted by PU.M.A., Budapest, (2001).

[98] Croisot, R., *Quelques applications et propriétés des treillis semimodulaires de longueur finie*, Ann. Fac. Sci. Univ. Toulouse, Ser. IV, 16 (1952).

[99] Cupona, G., Madarasz, R.Sz., *Free poly–algebras*, Univ. u Novom Sadu, Zb. Rad. Prirod. Mat. Fak., Ser. Mat. 23, 2 (1993), 245–261.

[100] Dasic, V., *Hyper-near-rings*, Proc. of the Fourth Int. Congress on AHA, Xanthi, Greece, 1990, World Scientific, 75–85.

[101] Davvaz, B., *On Hv-rings and Hv-ideals*, The J. of Fuzzy Math., vol. 6, no. 1, (1998).

[102] Davvaz, B., *Rough sets and Fundamental Rings*, Bulletin of Iranian Mathematical Society, vol.24, no.2, 1998, 49–61.

[103] Davvaz, B., *Interval–valued fuzzy subhypergroups*, Korean J. of Comput. and Appl. Math., vol. 6, no.1 (1999).

[104] Davvaz, B., *Fuzzy Hv-groups*, Fuzzy Sets and Systems, 101, (1999).

[105] Davvaz, B., *Fuzzy hyperideals in semihypergroups*, Italian J. Pure and Appl. Math. no. 8, (2000).

[106] Davvaz, B., *On Fuzzy relations and fuzzy subhypergroups*, PU.M.A. vol. 11, no. 1, (2000).

[107] Davvaz, B., *Rough Polygroups*, Accepted by Italian J. Pure and Appl. Math. (2000).

[108] Davvaz, B., *Strong regularity and fuzzy strong regularity in semihypergroups*, Korean J. of Comput. and Appl. Math. vol. 7, no. 1 (2000).

[109] Davvaz, B., *Interval–valued fuzzy ideals in a hyperring*, Italian J. Pure and Appl. Math. no. 10, 2001.

[110] Darafsheh, M.R., Davvaz, B., *Hv-rings of fractions*, Italian J. Pure and Appl. Math. no. 5, (1999).

[111] Degreef, E., *A few more results on the linear hull in convexity spaces*, A Quartely J. of Pure and Applied Math., vol. 54, no.1 (1980).

[112] De Maria, F., *Gruppi selezioni di ipergruppi*, Atti Sem. Mat. Fis. Univ. Modena, 30 (1981).

[113] De Maria, F., *Ipergruppi* $(\mathbb{Z}_n, q)$, Riv. Mat. Pura ed Appl. no. 5 (1989).

[114] De Salvo, M., *I* K_H*-ipergruppi*, Atti Sem. Mat. Fis. Univ. Modena 31 (1982).

[115] De Salvo, M., *Iperanelli ed ipercorpi*, Ann. Sci. Univ. Clermont II, Sér. Math., Fasc. 22 (1985).

[116] De Salvo, M., *Ipergruppi commutativi di ordine* 5 *e lunghezza costante* 2, Le Matematiche Catania 40, fasc. I–II (1985).

[117] De Salvo, M., *Hypergroups and algebras*, Combinatorics '88, Proc. of the Int. Conference on "Incidence Geometries and Combinatorial Structures 88", 1st volume (1988).

[118] De Salvo, M., *Sull'algebra associata a un ipergruppoide finito*, Rend. Mat., Serie VII, vol. 9, Roma (1989).

[119] De Salvo, M., *Feebly canonical hypergroups*, Journal of Combinatorics, Information and System Sciences, 15, no. 1 (1990).

[120] De Salvo, M., *Hypergroupoids and combintorial structures*, In printing on PU.M.A. (1992).

[121] De Salvo, M., *Hypergroupoids and Combinatorial Structures*, PU.M.A., Ser. A., vol. 3 (1992).

[122] De Salvo, M., *Partial semihypergroups*, Riv. Mat. Pura ed Appl. no. 17 (1995).

[123] De Salvo, M., Lo Faro, G., *Wrapping graphs and partial semihypergroups*, J. of Information & Optimization Sciences, vol. 18, no. 1 (1997).

[124] De Salvo, M., Lo Faro, G., *Hypergroups and binary relations*, in printing on Multivalued Logic and Applications.

[125] De Salvo, M., Lo Faro, G., *A new class of hypergroupoids associated to binary relations*,preprint.

[126] Diffie, W., Hellman, M., *New Directions in Cryptography,* IEEE Transactions on Information Theory, IT 22, 1976, pp. 644–654.

[127] Deza, M.M., Rosenberg, I.G., *General convolutions motivated by designs*, Acta Universitatis Carolinae, Math. et Phys., 27, no. 2 (1986).

[128] Dietzman, A.P., *On the multigroups of complete conjugate sets of a group*, C.R. (Doklady) Acad. Sci. USSR (N.S.) 49 (1946).

[129] Dietzman, A.P., *On multigroups whose elements are subsets of a group* (Russian), Moskov Gos. Ped. Inst. Uc. Zap, 71 (1953) (Mat. Rev. 17 (1950)).

[130] Doria, S., Maturo, A., *A hyperstructure of conditional events for artificial intelligence,* in "Mathematical models for handling partial knowledge in artificial intelligence", Plenum Press, New York, 1995, 201–208.

[131] Doria, S., Maturo, A., *Hyperstructures and geometric spaces associated to a family of events*, Riv. di Mat. Pura e Appl., no. 19, 1996.

[132] Dramalidis, A., *Duals H_v-rings*, Riv. Mat. Pura ed Appl. no. 17 (1995).

[133] Dramalidis, A., Vougiouklis, T., *On a class of fuzzy H_v-structures*, Proc. Int. Congress "C. Carathéodory in his...origins", 2000, Hadronic Press, 2001, 121–130.

[134] Drbohlav, K., *Gruppenartige Multigruppen*, Czech. Math. J., Praha, t. 7, 82 (1957).

[135] Drbohlav, K., Kepka, T., Nemec, P., *Associativity semihypergroups and related problems*, Atti Convegno su Ipergruppi, altre Strutture Multivoche e Applicazioni, Editor: P. Corsini, Udine (1985).

[136] Dresher, M., Ore, O., *Theory of multigroups*, Amer. J. Math., 60 (1938).

[137] Dubois, D., Prade, H., *Rough Fuzzy Sets and Fuzzy Rough Sets,* Int. J. General Systems, 17 (1990).

[138] Eaton, J.E., *Theory of cogroups*, Duke Math. J. 6, 1 (1940).

[139] Evans, E., *Median lattices and convex subalgebras*, Colloq. Math. Soc. János Bolyai.

[140] Falcone, G., *On finite strongly canonical hypergroups*, PU.M.A., vol.11 , no.4, 2000, 571–580.

[141] Falcone, G., *K-hypergroups*, Ital. J. Pure Appl. Math. no. 9 (2001).

[142] Ferri, B., Maturo, A., *Hyperstructures as tool to compare urban projects*, Ratio Math. 12, 1997, 79–89.

[143] Foldes, S., *Lexicographics sums of ordered sets and hypergroupoids*, Proc. of the Fourth Int. Congress on AHA, Xanthi, Greece, 1990, World Scientific, 97–101.

[144] Freni, D., *Semi–ipergruppi e equivalenze regolari*, Atti Sem. Mat. Fis. Univ. Modena, XXIX (1980).

[145] Freni, D., *Une note sur le coeur d'un hypergroupe et sur la cloture transitive β^* di β*, Riv. di Mat. Pura e Appl. no. 8 (1991).

[146] Gionfriddo, M., *Hypergroups associated with multihomomorphisms between generalized graphs,* Convegno su sistemi binari e loro applicazioni, Edited by P. Corsini. Taormina (1978) 161–174.

[147] Golan, J.S., *Semirings of formal series over hypermonoids: some interesting cases*, Kyungpook Math. J. 36, (1996), 107–111.

[148] Grau, A.A., *Ternary Boolean algebra*, Bull. Amer. Math. Soc. 53, 1974, 567–572.

[149] Grätzer, G., *A reprsentation theorem for multi–algebras*, Arch. Math., vol. XIII (1962).

[150] Griffiths,L.W., *On hypergroups, multigroups and product systems*, Amer. J. Math., 60 (1938).

[151] Guţan, C., *Extensions des sémigroupes associées à certains semi-hypergroupes très fins*, Proc. of the Sixth Int. Congress Prague, 1996, Democritus Univ. of Thrace, Greece.

[152] Guţan, M., *On the transitivity of the relation β in semihypergroups*, Rendiconti del Circolo Matematico di Palermo, 45 (1996), 189–200.

[153] Guţan, M., Sureau, Y., *Hypergroupes de type C à petites partitions*, Riv. di Mat. Pura e Appl. no. 13 (1993).

[154] Haddad, L., Sureau, Y., *Les cogroupes et les D-hypergroupes*, Algebra, vol. 118, no. 2 (1988).

[155] Haddad, L., Sureau, Y., *Les cogroupes et la construction de Utumi*, Pacific J. Math., vol. 145, no. 1 (1990).

[156] Haddad, L., Sureau, Y., *Les groupes, les hypergroupes et l'enigme de Murgin*, J. of Pure and Applied Algebra, 87, (1993).

[157] Hansen, D., *An axiomatic characterization of multilattices*, Discrete Math. 33 (1981).

[158] Harrison, D.K., *Double coset and orbit spaces*, Pacific J. Math. 80 (1979).

[159] Hasankhani, A., Zahedi, M.M., *F-hyperrings*, Italian J. Pure and Appl. Math. no.4, (1998).

[160] Hoft, H., Howard, P.E., *Representing multi–algebra by algebras, the axiom of choice and the axiom of dependent choice*, Algebra Universalis, 13 (1981).

[161] Hort, D., *A construction of hypergroups from ordered structures and their morphisms*, Proceedings of Algebraic Hyperstructures and Applications, Taormina, 1999, J. of Discrete Math.

[162] Innamorati, S., Peroni, P., *Hyperclassifications of remote sensing images*, Rivista di Mat. Pura e Applicata, no. 19, (1996).

[163] Ioulidis, S., *Hyperespaces vectoriels exceptionnels*, Bull. Soc. Sci. Lett. Lódz, 40, 1990, no. 1–10, 1991, 131–145.

[164] Ioulidis, S., Mittas, J., *Sur certaines notions préliminaires de l'hyperalgèbre linéaire*, Introduction de l'hypergroupe polysymmétrique, Prokt. Akad. Athenon 58, 1983, Part A, 1984, 361–392.

[165] Isbell, J.R., *Median algebra*, Trans. Amer. Math. Soc. 260, 1980, 319–362.

[166] Jakubik, J., *Sur les axiomes des multistructures*, Czech. Math. J. 6 (1956).

[167] Jakubik, J., *Sur l'isomorphisme des graphes des multistructures*, Acta Fac. Rerum nat. Univ. Comenanae, Math. I (1956).

[168] Jantosciak, J., *Classical Geometries as Hypergroups*, Convegno Su Ipergruppi, Altre Strutture Multivoche e Loro Aplicazioni, Editor P. Corsini, Udine, 15–18 Ottobre 1985, 93–104.

[169] Jantosciak, J., *Reduced hypergroups*, Proc. of the Fourth Int. Congress on AHA (editor T. Vougiouklis), Xanthi, Greece, 1990, World Scientific, 119–122.

[170] Jantosciak, J., *Homomorphisms, equivalences and reductions in hypergroups*, Riv. di Mat. Pura et Appl., no.9 (1991), 23–47.

[171] Jantosciak, J., *A brief survey of the theory of join spaces*, Proc. of 5th Int. Cong. on AHA (editor M. Ştefănescu), Iaşi, Romania, Hadronic Press, Palm Harbor, Florida, 1994, 1–12.

[172] Jantosciak, J., *Hypergroups under contractions into and expansions through idempotents*, Proc. 5th Int. Cong. on AHA (editor M. Ştefănescu), Iaşi, Romania, 1993, Hadronic Press, Palm Harbor, Florida, 1994, 109–122.

[173] Jantosciak, J., *Transposition in hypergroups*, Sixth Int. Congress on AHA, 1996, 77–84, Democritus Univ. of Thrace Press, Greece, 1997, 77–84.

[174] Jantosciak, J., *Transposition hypergroups, noncommutative join spaces*, J. Algebra 187, 1997, 97–119.

[175] Jantosciak, J., Massouros, C.G., *Strong identities and fortification in transposition hypergroups*, Proc. 7th Int. Cong. AHA, Taormina, Italy, 1999, in printing on J. of Discrete Math.

[176] Jasem, M., *On weak isometries in multilattice groups*, Math. Slovaca 40, no.4 (1990).

[177] Jezek, J., *On representable mappings of semigroups into cardinals*, Rivista di Mat. Pura e Appl. no. 6, (1990).

[178] Jonsson, B., Tarski, A., *Boolean Algebras with Operators* II, American J. of Mathematics 74, 1952, 127-162.

[179] Kawada, Y., *Uber den Dualitätssatz der Charaktere nichcommutativer Gruppen*, Proc. Phys. Math. Soc. Japan 24 (1942), 97–109.

[180] Kehagias, Ath., *An example of L-fuzzy join space*, Circ. Mat. Palermo, 2002.

[181] Kehagias, Ath., Konstantinidou, M., *Lattice ordered join space: an Applications-Oriented Example*, accepted for publication in Italian Journal of Pure and Applied Math., 2000.

[182] Kemprasit, Y., *Multiplicative intervalsemigroups on R admitting hyperrring structure*, accepted by Italian J. Pure and Appl. Math. (2000).

[183] Kemprasit, Y., *Hyperring structure of some semigroups of linear transformations*, accepted by Italian J. Pure and Appl. Math. (2001) II.

[184] Kemprasit, Y., Yuwaree, P., *Transformation semigroups admitting hyperring structure*, Italian J. of Pure and Appl. Math. no. 10, (2001).

[185] Kepka, T., *A note on the associativity hypergroups of quasigroups and loops*, Atti Soc. Pelor. Sc. Mat. Fis. Nat., vol. 27, (1981).

[186] Kepka, T., *A note on t-groupoids*, Atti Soc. Pelor. Sc. Mat. Fis. Nat., vol. 27, (1981).

[187] Kepka, T., *Distributive t-groupoids*, Atti Soc. Pelor. Sc. Mat. Fis. Nat., vol. 27, (1981).

[188] Kepka, T., *Associative semi–hypergroups and cardinal–value semigroups*, Proceedings of the Fourth Int. Congress on Algebraic Hyperstructures and Applications, Xanthi, (1990), World Scientific, 123–127.

[189] Kepka, T., *A note on representable mappings*, Rivista di Mat. Pura e Appl. no. 9, (1991).

[190] Kepka, T., Nemec, P., Niemenmaa, M., *On representable pairs*, Ann. Acad. Sci. Fennicae, Ser. A.I., Math. 13 (1988).

[191] Klaukova, O., *Pairs of multilattices defined on the same set*, Math. Slovaca 30 (1980).

[192] Kleene, S., *Representation of events in nerve nets and finite automata,* Automata Studies, Princeton, NJ: Princeton University Press, (1956), 3–41.

[193] Kolibiar, M., *Median groups*, Arch. Math. vol. 25, no. 1-2 (1989).

[194] Kolibiar, M., Marcisová, T., *On a question of J. Hashimoto*, Mat. Casopis Sloven. Akad. Vied. 24 (1974).

[195] Konguetsof, L., *Construction d'hypergroupes, a partir d'ensembles munis d'operations partielles*, C.R. Acad. Sc. Paris, t. 258 (1964).

[196] Konguetsof, L., Vougiouklis, T., *Algebraic structures with multivalued operations*, Gr. Sch. St. Piraens (in Greek) (1982).

[197] Konguetsof, L., Vougiouklis, T., *Constructions d'hyperanneaux à partir d'anneaux*, Acta Univ. Car. Math.Phys. vol. 28 (1) (1987).

[198] Konguetsof, L., Vougiouklis, T., Kessoglides, M., Spartalis, S., *On cyclic hypergroups with period*, Acta Univ. Car. Math. Phys., vol. 28 (1) (1987).

[199] Konstantinidou, M., *Modular hyperlattices*, *Πρακτικα τησ Ακαδημιασ Αθηνων, ετοσ* 1978, *τομ* $53^{οσ}$ (1978).

[200] Konstantinidou, M., *Distributive and complemented hyperlattices*, *Πρακτικα τησ Ακαδημιασ Αθηνων, ετοσ* 1981, *τομ* $56^{οσ}$ (1981).

[201] Konstantinidou, M., *On the hyperlattices–ordered groupoids*, Boll. Un. Mat. Ital. (6), 2-A (1983).

[202] Konstantinidou, M., *Contribution à la théorie de la résiduation dans les hyperstructures ordonnées*, Atti Convegno su Ipergruppi, Altre Strutture Multivoche e loro Applicazioni, Udine (1985).

[203] Konstantinidou, M., *Sur les hypergroupoides résidués généralisés*, Rend. Circ. Mat. Palermo, Ser. II, XXXV (1986).

[204] Konstantinidou, M., *Sur une relation d'équivalence dans des hyperstructures ordonnées*, International Congress "Combinatorics '88", Ravello (1988).

[205] Konstantinidou, M., *Some properties of the semi–modular hyperlattices*, Riv. di Mat. Pura e Appl., no. 3 (1988).

[206] Konstantinidou, M., *On P-hyperlattices and their distributitivy*, Rendiconti del Circolo Matematico di Palermo, Serie II, Tomo XLII, 1993, p. 391–403.

[207] Konstantinidou, M., *Construction d'un hypergroupe par rapport a une topologie sur lui*, Rivista di Mat. Pura e Appl. no. 15, (1994).

[208] Konstantinidou, M., Mittas, J., *An introduction to the theory of hyperlattices*, Mathematica Balcanica, 7 (23) (1977), 187-193.

[209] Konstantinidou, M., Mittas, J., *Contributions à la théorie des treillis avec des structures hypercompositionnelles y attachées*, Riv. Mat. Pura e Appl., no. 14, 1999, 83–119.

[210] Konstantinidou, M., Serafimidis, K., *Sur les filets des hypergroues canoniques strictement réticulés*, Riv. Mat. Univ. Parma (4), 13 (1987).

[211] Konstantinidou, M., Serafimidis, K., *Hyperstructures dérivées d'un treillis particulier*, Rend. Mat. Appl. (7), 13, no. 2, 1993, 257–265.

[212] Konstantinidou, M., Serafimidis, K., *Sur les P-supertreillis*, New Frontiers in Hyperstructures, Hadronic Press, Palm Harbor, FL 34682-1577, U.S.A. (1996).

[213] Koskas, M., *Groupoids, demi–hypergroupes et hypergroupes*, J. Math. Pure et Appl., 49, 1970.

[214] Krasner, M., *Une généralisation de la notion de corps–corpoïde. Un corps–corpoïde remarquable de la théorie des corps valués*, Acad. Sci. Paris 219 (1944), 345–347.

[215] Krasner, M., *Une nouvelle présentation de la théorie des groupes de permutations et ses applications à la théorie de Galois et de produit d'entrelacement (weath product) de groupes*, Math. Balk. 3 (1973).

[216] Krasner, M., *A class of hyperrings and hyperfields*, Internat. J. Math. Sci. 6 (1983).

[217] Krasner, M., Lecomte, P., *Quotients projectifs des demi–hypergroupes et des hypergroupes*, C.R. Acad. Sci. Paris, t. 283 (1976).

[218] Kuntzmann, J., *Operations multiformes, hypergroupes*, C.R. Acad. Sci. Paris, 204 (1937).

[219] Ladkin, P.B., *Time Representation: A Taxonomy of Interval Relations*, Proceedings of AAA I - 86, 360-366.

[220] Ladkin, P.B., Maddux, R.D., *The Algebra of Convex Time Intervals*, Kestrel Institute Technical Report, KES.U. 87.2.

[221] Leoreanu, V., *New results in subhypergroup theory and on the nucleus structures*, Pure Math. and Appl., Budapest, vol. 5, no. 3 (1994).

[222] Leoreanu, V., *On the width of a hypergroup,* Fifth Int. Congress on AHA 93, Hadronic Press, USA, Congress Proceedings, 1994.

[223] Leoreanu, V., *Centralisateur d'un élément dans un hypergroupe,* Circ. Mat. Palermo,Seria II, Tom XLIII, 1994.

[224] Leoreanu, V., *Généralisation de notions et résultats de la théorie des groupes dans les hypergroupes,* Proceedings of IRB International Workshops - New Frontiers in Multivalued Hyperstructures, 1995.

[225] Leoreanu, V., *On the heart of join spaces and of regular hyupergroups*, Riv. Mat. Pura e Appl., no. 17 (1995).

[226] Leoreanu, V., *Sur la limite directe d'hypergroupes*, Ann. Scientifiques de l'Université "Ovidius", Constanţa, no.2, 1995.

[227] Leoreanu, V., *Une note sur les demi–hypergroupes qui sont quasi–hypergroups d'un côté*, Sixth Int. Congress on AHA (1996), 85–87, Democritus Univ. of Thrace Press.

[228] Leoreanu, V., *The heart of some important classes of hypergroups*, PU.M.A., vol. 9 (1998), no. 3-4, 351–360.

[229] Leoreanu, V., *New results on the hypergroups homomorphisms*, J. Int. and Optimization Sciences, vol. 20, 1999, no. 2, 1–12.

[230] Leoreanu, V., *The class of hypergroups in which the heart is the set of identities*, Le Matematiche (Catania), vol. LIV, 1999,m fasc. 1, 49–54.

[231] Leoreanu, V., *Commutative hypergroups associated with arbitrary hypergroups,* Proceedings of the Seventh Congress of AHA, Taormina, 1999, Journal of Discrete Mathematics.

[232] Leoreanu, V., *Direct limits and products of join spaces associated with rough sets*, Honorary Volume dedicated to Professor Emeritus Ioannis Mittas, Aristotle University of Thessaloniki, 1999–2000.

[233] Leoreanu, V., *About the simplifiable cyclic semihypergroups,* Proceedings of the Congress in Algebra, Iaşi, Romania, 1998, Italian Journal of Pure and Applied Mathematics, n.7, 2000.

[234] Leoreanu, V., *Direct limits and products of join spaces associated with rough sets*, Part II, Int. J. Sci. & Technol. Univ. Kashan, vol. 1, no. 1 (2000).

[235] Leoreanu, V., *Direct limit and inverse limit of join spaces associated with fuzzy sets*, PU.M.A., vol. 11, no.3 (2000), 509–516.

[236] Leoreanu, V., *Direct limits and inverse limits of SHR semigroups*, Southeast Asian Bulletin of Math., 2001, 25, 421–426.

[237] Leoreanu, V., Radu, Gh., *Direct limits and inverse limits of hypergroups associated with lattices*, It. J. of Pure and Applied Math., no.11, (2002).

[238] Leoreanu, V., Leoreanu, L., *Hypergroups associated with hypergraphs*, Italian Journal of Pure and Applied Mathematics, 4 (1998), 119–126.

[239] Letizia, A., *Λ-Iperspazi con C.C.A.*, Riv. Mat. Pura e Appl., no.8 (1991).

[240] Ligozat, G., *Intervalles généralisés* I, C.R. Acad. Sci. Paris, t. 310, Série I, p. 225-228, 1990.

[241] Ligozat, G., *Weak representations of Interval Algebras*, AAAI-90, Boston, 1990.

[242] Lihova, J., *Varieties of directed multilattice*, Math. Slovaca 40, no.3 (1990).

[243] Lihova, J., Repaski, K., *Congrunce relations on varieties of directed multilattices*, Math. Slovaca 38, 1988.

[244] Li Hong Xing, Duan Qinzhi, Wang Peizhuang, *Hypergroup* (I), BUSEFAL, 23 (1985).

[245] Machid, H., *Normal form of Hyperoperations and Existence of Sheffer hypeoperations,* Accepted for publication on It. J. of Pure and Appl. Math, 2002.

[246] Marchi, M., *On a new class of geometrical hyperstructures*, Rivista di Mat. Pura e Appl. no. 7, (1990).

[247] Marty, F., *Recherches sur les groupes finis de fonctions algébriques*, Societas Scientiarum Fennica, Commentationes Physicae Mathematicae VII, 12 (1934).

[248] Marty, F., *Sur une généralisation de la notion de groupe*, IV Congrès des Mathématiciens Scandinaves, Stockholm (1934), 45–49.

[249] Marty, F., *Sur le rôle de la notion d'hypergroups dans l'étude des groupes non–abeliens*, C.R. Acad. Sci. 201 (1935).

[250] Marty, F., *Sur les groupes et les hypergroupes attachés à une fraction rationnelle*, Ann. de l'Ecole Normale 53 (1936).

[251] Maturo, A., *On a non–standard algebraic hyperstructure and its application to the coherent probability assessments*, Italian J. of Pure and Appl. Math., no. 7, 2000, 33–50.

[252] Maturo, A., *On some hyperstructures of conditional events*, Proceedings of the Sixth Int. Congress, Prague, Czech Republic, Sept. 1–9, 1996.

[253] Massouros, Ch.G., *Free and cyclic hypermodules*, Ann. Mat. Pura ed Appl., Serie IV, Tomo CL (1988).

[254] Massouros, Ch.G., *Hypergroups and Convexity*, Riv. Mat. Pura ed Appl., no.4 (1989).

[255] Massouros, Ch.G., *Quasicanonical hypergroups*, Proc. of the Fourth Int. Congress on AHA, Xanthi, Greece, 1990, World Scientific, 129–136.

[256] Massouros, Ch.G., *Constructions of hyperfields*, Math. Balk., vol. 5 (1991).

[257] Massouros, G.G., *An automaton during its operation*, Algebraic Hyperstructures and Appl. Proc. 5th Inter. Congress Iaşi, 1993, Hadronic Press, Palm Harbor USA, 1994, 267–276.

[258] Massouros, G.G., *Automata–Languages and hypermoduloids*, Algebraic Hyperstructures and Appl. Proc. 5th Inter. Congress Iaşi, 1993, Hadronic Press, Palm Harbor USA, 1994, 251–256.

[259] Massouros, G.G., *Hypercompositional Structures in the Theory of the Languages and Automata*, Analele Ştiinţifice ale Universităţii "Al.I. Cuza", Iaşi, Tomul III, Informatică, 1994, 65–73.

[260] Massouros, G.G., *The subhypergroups of the fortified join hypergroup*, Ital. J. Pure Appl. Math. no. 2 (1997).

[261] Massouros, G.G., Mittas, J., *Languages – Automata and Hypercompositional Structures*, Fourth Int. Congress on AHA (1990), World Scientific, 137–147.

[262] McAlister, D.B., *On multilattice groups*, II, Proc. Camb. Phil. Soc. 62 (1966).

[263] McKenzie, R., *Representations of integral relation algebras*, Michigan Math. J. 17, 1970, 279–287.

[264] McMullen, J.R., *An algebraic theory of hypergroups*, Bull. Austral. Math. Soc., vol. 20 (1979).

[265] McMullen, J.R., Price, J.F., *Reversible hypergroups*, Rend. Sem. Mat. Fis. Milano 47 (1977).

[266] Migliorato, R., *Semi–ipergruppi e ipergruppi n-completti*, Ann. Sci. Univ. Clermont II, Sér. Math., fasc. 23 (1986).

[267] Migliorato, R., *Sugli 1-Ipergruppi*, Riv. di Mat. Pura e Appl. Univ. Udine, no. 5 (1989).

[268] Migliorato, R., *Finite hypergroupoids and combinatoric spaces*, Proc. of the Fourth Int. Congress on AHA, Xanthi, Greece, 1990, World Scientific, 149–159

[269] Migliorato, R., *Some topics on the feebly associative hypergroupoids*, Proc. of the Vth AHA Congress Iaşi, 1993, Hadronic Press.

[270] Migliorato, R., *Fundamental relation on nonassociative hypergroupoids*, Italian J. of Pure and Appl. Math. no. 6, 1999.

[271] Migliorato, R., Gentile, G., *Feebly associative hypergroupoids*, Atti di "Giornate di Geometrie Combinatorie", Perugia, 1993, 259–269.

[272] Migliorato, R., Gentile, G., *Hypergroupoids and cryptosystems*, in printing on J. of Discrete Math. and Cryptography, 2002.

[273] Migliorato, R., Gentile, G., *Hyperstructures and Measurable Security of Criptosystems*, preprint.

[274] Migliorato, R., Spartalis, S., *Feebly associative P–hypergroupoids*, Rend. Circ. Mat. Palermo 2, 49, 2000, no.1, 27–40.

[275] Mittas, J., *Hypergroupes et hyperanneaux polysymmétriques*, C.R. Acad. Sc. Paris, t. 271, sér. A (1970).

[276] Mittas, J., *Contributions à la théorie des hypergroupes, hyperanneaux et les hypercorps hypervalués*, C.R. Acad. Sc. Paris, t. 272, sér. A (1971).

[277] Mittas, J., *Hypegroupes canoniques valués et hypervalués*, Math. Balk, t. 1, Beograd (1971).

[278] Mittas, J., *Hypergroupes canoniques*, Math. Balk, t. 2, Beograd (1972).

[279] Mittas, J., *Contributions à la théorie des structures ordonnées et des structures valués*, Πρακτικα τησ Ακαδημιασ Αθηνων, ετοσ 1973, τομ 48σ, Αθηναι 1973, 319–331.

[280] Mittas, J., *Certains hypercorps et hyperanneaux définis à partir de corps et anneaux ordonnés*, Bull. Math. de la Soc. Math. de la R.S. de Roumanie, t. 15 (63), no.3 (1971), Bucharest (1973).

[281] Mittas, J., *Espaces vectoriels sur un hypercorps. Introduction des hyperspaces affines et cuelidiens*, Math. Balk., t. 5, Beograd (1975).

[282] Mittas, J., *Hypergroupes fortement canoniques et supérieurement canoniques,* Proc. II of the Int. Symposium on Applications of Mathematics in System Theory, Braşov, Romania (1978).

[283] Mittas, J., *Hypergroupes canoniques valués et hypervalués. Hypergroupes fortement et supérieure canoniques*, Bull. of Greek Math. Soc., vol. 23, Athens (1982).

[284] Mittas, J., *Sur les hyperanneaux et les hypercorps*, Math. Balk., to. 3, Beograd (1983).

[285] Mittas, J., *Hypergroupes polysymmétriques canoniques*, Atti del convegno su Ipergruppi, Altre Strutture Multivoche e loro Applicazioni, Udine (1985).

[286] Mittas, J., *Sur les structures hypercompositionnelles*, Proc. of the Fourth Int. Cong. on AHA, Xanthi, Greece, 1990, World Scientific, 9–31.

[287] Mittas, J., *Introduction a l'analyse hypercompositionelle*, Rivista di Mat. Pura e Appl. no. 13, (1993).

[288] Mittas, J., Ioulidis, S., *Sur les hypergroupes polysymétriques commutatifs*, Rend. Ist. Mat. Univ. Trieste, V, XVIII (1986).

[289] Mittas, J., Konstantinidou, M., *Introduction à l'hyperalgèbre de Boole*, Math. Balk., t. 6, Beograd (1976).

[290] Mittas, J., Konstantinidou, M., *Sur une nouvelle généralisation de la notion de treillis. Les supertreillis et certaines de leurs propriétés générales*, Ann. Sci. Univ. Clermont Ferrand (1988).

[291] Mittas, J., Konstantinidou, M., *Contributions à la théorie des treillis en liaison avec des structures hypercompositionnelles y attachées*, Riv. Mat. Pura e Appl. no. 12 (1993).

[292] Mittas, J., Massouros, Ch., *Hypergroupes defined from a linear space*, Bull. of Greek Math. Soc., Athens (1989).

[293] Mockor, J., *A realization of d-groups*, Czech. Math. J.27 (102), (1977).

[294] Mockor, J., *Semivaluations and d-groups*, Czech. Math. J. 32 (107), (1982).

[295] Monk, J.D., *Model-theoretic methods and results in the theory of cylindric algebras*, in "Theory of Models", Proceed. 1963 Internat. Symp. at Berkeley, North–Holland Publ. Co., 1965, 238–250.

[296] Monk, J.D., *On representable relation algebras*, Michigan Math. J., 11, 1964, 207–210.

[297] Mulder, H.M., Schrijver, A., *Median graphs and Helly hypergraphs*, Discrete Math. 25, 1979, 41–50.

[298] Nakano,T., *Rings and partly ordered systems,* Math. Zeitschrift, 99 (1967), 355–376.

[299] Nakassis, A., *Recent results in hyperring and hypeerfield theory,* Int. J. Math. and Math.Sci., 1988.

[300] Niemenen, J., *Join space graphs*, J. of Geometry, 33(1988), 99–103.

[301] Nieminnen, J., *The ideal structure of simple ternary algebras*, Colloq. Math. 40, 1978, 23–29.

[302] Nistor, S.C., Tofan, I., *On Heyting hyperalgebras*, New Frontiers in Hyperstructures, Hadronic Press, Palm Harbor, FL 34682-1577, U.S.A. (19942), 223–229.

[303] Nordo, G., *An algorithm on number isomorophism classes of hypergroups of order* 3, Italian J. of Pure and Appl. Math. no. 2 (1997).

[304] Olson D.M., Ward V.K., *A note on multiplicative hyperrings,* Italian J. of Pure and Appl. Math., n.1 (1997), 77–84.

[305] Ore, D., *Structure and group theory,* Duke Math. J., 3 (1937).

[306] Orsatti, A., *Equivalenze regolari a destra dentro un ipergruppo, Sotto-ipergruppi F-reversibili*, Rend. Sem. Mat. Univ. Padova (1963).

[307] Pasquali Coluzzi, D., Eugeni, F., *H_v-groups without incidence property*, Italian J. of Pure and Appl. Math. no. 4, 1998, 45–60.

[308] Pawlak, Z., *Rough Sets. Theoretical Aspects of Reasoning about Data*, Kluwer Academic Publisher 1991.

[309] Pelea, C., *On the fundamental relation of a multialgebra*, Italian J. of Pure and Appl. Math. no. 10, 2001, 141–146.

[310] Pickett, H.E., *Homomorphisms and subalgebras of multialgebras*, Pacific j. Math., vol. 21, no. 2 (1967).

[311] Prenowitz, W., *Projective Geometries as Multigroups*, Amer. J. Math. 65 (1943), 235–256.

[312] Prenowitz, W., *Descriptive Geometries as Multigroups*, Trans. Amer. Math. Soc. 59 (1946), 333–380.

[313] Prenowitz, W., *Partially ordered fields and geometries*, Amer. Math. Monthly 53 (1946).

[314] Prenowitz, W., *Spherical Geometries and Multigroups*, Canad. J. Math. 2(1950), 100-119.

[315] Prenowitz, W., *A contemporary approach to classical geometry*, Amer. Math. Monthly 68 (1961).

[316] Prenowitz, W., Jantosciak, J., *Geometries and Join Spaces,* J. reine und angewandte Math. 257, 1972, 100–128.

[317] Procesi Ciampi, R., Rota, R., *Vectorial hyperspaces of type (n, s)*, Riv. di Mat. Ser. VII, vol. 9 (1989).

[318] Procesi Ciampi, R., Rota, R., *Stone's representation theorem for Boolean hyperalgebras*, Proc. of the Fourth Int. Congress on AHA, Xanthi, Greece, 1990, World Scientific, 161–165.

[319] Rahnamai, Barghi, A., *The prime ideal theorem and semiprime ideals in meet–hyperlattices*, Italian J. of Pure and Appl. Math. no. 5, 1999.

[320] Rahnamai, Barghi, A., *The prime ideal theorem for distributive hyperlattices*, Italian J. of Pure and Appl. Math. no. 10, 2001.

[321] Rahnamai, Barghi, A., *On commutative quasi–hyperrings*, Italian J. of Pure and Appl. Math. no. 10, 2001.

[322] Romeo, G. – *Limite diretto di semi–ipergruppi i ipergruppi di associatività*, Riv. Mat. Univ. Parma, 1982.

[323] Rosenberg, I.G., *An algebraic approach to hyperalgebras*, Proc. 26th Int. Symp. Multiple–Valued Logic, Santiago de Compostela, IEEE (1996).

[324] Rosenberg, I.G., *Wall monoids,* New Frontiers in Hyperstructures (Molise, 1995), 159–166, Ser. New Front. Adv. Math. Ins. Ric. Base, Hadronic Press, Palm Harbor FL, 1996.

[325] Rosenberg, I.G., *Hypergroups induced by paths of a directed graph,* Italian J. of Pure and Appl. Math., no. 4, 1998, 133–142.

[326] Rosenberg, I.G., *Hypergroups and join spaces determined by relations,* Italian Journal of Pure and Applied Mathematics, no. 4(1998), 93–101.

[327] Rosenberg, I.G., *Mutiple–valued hyperstructures,* Proc. 28th Int. Symp. Multiple–Valued Logic, Fukuoka, (1998).

[328] Rosenfeld, A., *Fuzzy groups,* J. Math. Anal. Appl., 35, 1971, 512–517.

[329] Ross, K., *Hypergroups and centers of measure algebras,* Symposia Mathematica, 22, Academic Press (1977).

[330] Rossi, F., *Sule congruenze in multigruppi associati a spazi proiettivi,* Rend. Ist. Mat. Trieste (1972).

[331] Rossi, F., *Sui multigruppi risolubili,* Rend. Ist. Mat. Trieste (1972).

[332] Roth, R., *Character and conjugacy class hypergroups of a finite group,* Ann. di Mat. Pura e Appl., 105 (1975).

[333] Roth, R., *On derived canonical hypergroups,* Riv. Mat. Pura e Appl., no. 3 (1988).

[334] Sade, A., *Entropie demosienne de multigroupoides et de quasigroupes,* Ann. Soc. Sci. Bruxelles, Ser. I, 73, 1959.

[335] Sade, A., *Multigroupide sur un système demosien de groupoides,* Colloq. Math. 7 (1959/60).

[336] Santilli, R.M., Vougiouklis, T., *Isotopies, genotopies, Hyperstructures and their Applications,* New Frontiers in Hyperstructures, Molise, 1995, 1–48, Hadronic Press, Palm Harbor, FL 34682-1577, U.S.A. (1996).

[337] Scafati–Tallini, M., *A ipermoduli e spazi ipervettoriali,* Riv. Mat. Pura e Appl. no.3, 1988, 39–48.

[338] Scafati–Tallini, M., *Hypervector spaces,* Proceedings of the Fourth Int. Congress on AHA, Xanthi, Greece, 1990, World Scientific, 167–174.

[339] Scafati–Tallini, M., *Matroidal hypervector spaces,* J. of Geometry, vol. 42, 1991, 131–140.

[340] Scafati–Tallini, M., *Spazi ipervettoriali fortemente distributivi a sinistra,* Rendiconti di Matematica, Serie VII, vol. 11, Roma (1991), 633–648.

[341] Scafati–Tallini, M., *La categoria degli spazi ipervettoriali,* Quaderno no. 114 Seminario di Geometrie Combinatorie, Dip. Mat. Univ. Roma "La Sapienza" (Ottobre 1993), 1–17.

[342] Scafati–Tallini, M., *Spazi ipervettoriali deboli e norme in tali spazi,* Quaderno nr. 111 Seminario di Geometrie Combinatorie, Dip. Mat. Univ. Roma "La Sapienza" (Novembre 1993), 1–14.

[343] Scafati–Tallini, M., *La categoria degli spazi ipervettoriai,* Riv. di Mat. Pura e Appl. no. 15, 1994, 97–109.

[344] Scafati–Tallini, M., *Weak hypervector spaces and norms in such spaces,* Atti Convegno AHA, Iaşi, Romania (1993), Hadronic Press, Palm Harbor, Florida (USA) (1994), 199–206.

[345] Scafati–Tallini, M., *Strutture algebriche multivoche e strutture geometriche,* Quaderno n. 127 Seminario di Geometrie Combinatorie "G. Tallini" (Novembre 1996), 1–23.

[346] Scafati–Tallini, M., *Sottospazi, spazi quozienti ed omomorfismi tra spazi ipervettoriali,* Riv. di Mat. Pura e Appl., Univ. di Udine, nr. 18 (1996), 71–84.

[347] Scafati–Tallini, M., *Characterization of remarkable hypervector spaces,* to appear.

[348] D. Schweigert, D., *Congruence relations of multialgebras,* Discrete Math. 33 (1985).

[349] Scorzioni, R., *An algebraic characterization of trees and cycles,* Proccedings of the Fourth Int. Congress on AHA, Xanthi, Greece, 1990, World Scientific, 175–185.

[350] Serafimidis, K., *Sur les hypergroupoides réticulés,* Boll. Un. Mat. Ital. (6), 5-B (1986).

[351] Serafimidis, K., *Sur les hypergroupoides résidués,* Boll. Un. Mat. Ital. (6), 5-A (1986).

[352] Serafimidis, K., *Sur les L-hyperidéaux des hypergroupes canoniques strictement réticués,* Rend. Circ. Mat. Palermo, Ser.II., XXXV (1986).

[353] Serafimidis, K., *Sur les hypergroupes canoniques ordonnés et strictement ordonnés*, Rend. Mat. e delle sue Appl., Ser. VII, vol. 6, no. 1-2 (1986).

[354] Serafimidis, K., *Sur les hypergroupes canoniques strictement réticulés*, Riv. Mat. Pura ed Appl., no.2 (1987).

[355] Serafimidis, K., Kehagias, Ath., Konstantinidou, M., *The L–fuzzy Corsini join hyperoperation*, accepted by Italian J. of Pure and Appl. Math. 2002.

[356] Sholander, M., *Medians and betweenness*, Proc. Amer. Math. Soc. 5, 1954, 801–807.

[357] Sholander, M., *Medians, lattices, and trees*, Proc. Amer. Math. Soc. 5, 1954, 808–812.

[358] Sholander, M., *Trees, lattices, order, and betweenness*, Proc. Amer. Math. Soc. 3, 1952, 369–381.

[359] Skordev, D., *On multivalued homomorphisms in the theory of computability*, Int. Congress of Logic Methodology and Philosophy of Science (Salzburg, Austria (1983), vol.1).

[360] Skordev, D., *On multivalued homomorphisms computation book*, Lecture Notes in Computer Sciences (1985).

[361] Slapal, J., *On exponentiation of universal algebras*, Algebra Universalis 44 (2000).

[362] Spartalis, S., *A class of hyper–rings*, Riv. di Mat. Pura ed Appl., no. 4, (1989).

[363] Spartalis, S., *(H,R)-Hyperring*, Proccedings of the Fourth International Congress on AHA, Xanthi, Greece, 1990, World Scientific, 187–195.

[364] Spartalis, S., *Homomorphisms on (H,R)-H_v-rings*, Proc. of Sixth Congress of AHA, 1996, 133–138, Democritus Univ. of Thrace Press.

[365] Spartalis, S., *On hyperstructures associated with binary relations,* to appear in the Proceedings on 8th AHA Congress, Samothraki, 2002.

[366] Spartalis, S., *Hypergroupoids obtained from groupoids with binary relations*, in printing.

[367] Spartalis, S. Dramalides, A, Vougiouklis, T, *On H_v-group Rings*, Algebras, Groups and Geometries, 15 (1998), 47–54.

[368] Spartalis, S., Vougiouklis, T., *The fundamental relations on H_v-rings*, Riv. di Mat. Pura ed Appl., no. 14 (1994), 7–20.

[369] Ştefănescu, M., *On hypernearrings*, An. Şt. Univ. Al.I. Cuza, Iaşi (1992).

[370] Stefanescu, M., *Constructions of hypergroups,* New Frontiers in Hyperstructures, Hadronic Press, Palm Harbor, Fl. 34682–15777, U.S.A., T. Vougiouklis Editor, 1996, 67–83.

[371] Stratigopoulos, D., *Hyperanneaux non commutatifs, hyperanneaux, hypercorps, hypermodules, hyperspaces vectoriels et leurs propriétés élémentaires*, C.R. Acad. Sci., Sér. A, 269 (1969).

[372] Stratigopoulos, D., *Hyperanneaux non commutatifs: le radical d'un hyperanneaux, somme sous-directe des hyperanneaux, hyperanneaux artiniens et théorie des éléments idempotents*, C.R. Acad. Sci., Sér. A, 269 (1969).

[373] Stratigopoulos, D., *Hyperanneaux non commutatifs: hyper-anneaux artiniens, centralisateur d'un hypermodule et théorème de densité*, C.R. Acad. Sér. A, 269 (1969).

[374] Stratigopoulos, D., *Certaines classes d'hypercorps et d'hyperanneaux*, Atti Convegno su Ipergruppi, Altre Strutture Multivoche e loro Applicazioni, Udine, 1985.

[375] Stratigopoulos, D., *Sur le hypercorps et les hyperanneaux*, Proceedings of the Fourth International Congress on AHA, Xanthi, Greece, 1990, World Scientific, 33–53.

[376] Stratigopoulos, D., *Sur le hypercorps et les hyperanneaux.* Proc. of the Fourth Int. Congress on AHA, Xanthi, Greece, 1990, World Scientific, 33–53.

[377] Stratigopoulos, D., *Starlike hypergroups of type* (m, n), Proc. of Sixth Congress of AHA, 1996, 139–149, Democritus Univ. of Thrace Press.

[378] Stratigopoulos, D., Massouros, Ch. G., *On a class of fields*, Math. Balk, t. 12.

[379] Sureau, Y., *Une notion de conjugaison dans les hypergroupes et hypergroupes homomorphes à groupes*, C.R. Acad. Sci., Paris, t. 284, 1977.

[380] Sureau, Y., *Extension d'un groupe par un hypergroupe*, Convegno su Sistemi Binari e loro Applicazioni, Taormina, 1978.

[381] Sureau, Y., *Hypergroupes opérant transitivement sur un ensemble*, Ann. Sci. Univ. Clermont II, Sér. Math., Fasc. 18 (1979).

[382] Sureau, Y., *Structures d'hypergroupes induites sur des familles de parties d'un hypergroupe et théorèmes d'isomorphisme s'y repportant*, Riv. di Mat. Univ. Parma, 1985.

[383] Sureau, Y., *Hypergroupes de type C*, Rend. Circ. Mat. Palermo, vol. 40 (1991).

[384] Tallini, G., *Introduzione alla teoria dei codici*, Appunti redatti da O. Ferri, Sem. Ist. Mat. Univ. l'Aquila (1979), 1–11.

[385] Tallini, G., *Ipergruppoidi di Steiner e geometrie combinatorie,* Convegno "Sistemi binari e loro applicazioni", Taormina, 1978.

[386] Tallini, G., *Geometric hypergroups and line spaces*, Acta Universatis Carolinai, Math. Phys., 25, 1 (1984), 69–73.

[387] Tallini, G., *On Steiner hypergroups and Linear codes*, Convegno Ipergruppi, Altre Strutture multivoche e loro applicazioni, Udine, 1985, 87–91.

[388] Tallini, G., *General multivalued algebraic structures and geometric spaces*, Proc. of the Fourth Int. Congress on AHA, Xanthi, Greece, 1990, World Scientific, 197–202.

[389] Tallini, G., *Dimensione negli ipergruppi*, Quad. Sem. Geom. Comb. Dip. Mat. "G. Castelnuovo" Univ. Roma "La Sapienza", 107 (1993), 1–20.

[390] Tallini, G., *Dimensione negli ipergruppi*, in *Scritti in onore di Giovanni Melzi*, Ed. Vita e Pensiero (1994), 367–390.

[391] Tallini, G., *Dimensions in multivalued algebraic structures*, Italian J. of Pure and Appl. Math., 1, 51–64.

[392] Tallini, G., *Dimensions in multivalued algebraic structures*, Ital. J. Pure and Appl. Math. no. 1 (1997).

[393] Tallini, G., Ceccherini, P.V., *Codes, caps and linear spaces*, In P.J. Cameron et al. *Finite Geometries and Designs*, Proc. of the 2nd Isle of Thorns Conference, Cambridge University Press (1981), 72–80 (London Math. Soc., LNS 49).

[394] Tofan, I., Volf, A.C., *On some connections between hyperstructures and fuzzy sets*, Italian J. of Pure and App. Math., no. 7, 2000, 63–68.

[395] Tomkova, M., *Graph isomorphisms of modular multilattices*, Math. Slovaca, 30 (1980).

[396] Utumi, Y., *On hypergroups of group right cosets*, Osaka Math. J. 1, 1949, 73–80.

[397] Varlet, J.C., *Remarks on Distributive Lattices,* Bull. de l'Acad. Polonnaise des Sciences, Série des Sciences Math., Astr. et Phys., vol. XXIII, no. 11, 1975.

[398] Vikhrov, AG., *Theory of extension of ultragoups*, Ucenye Zapiski Moskov Gos. Univ., 100, Mathematik (1946).

[399] Vougiouklis, T., *Cyclicity in a special class of hypergroups*, Acta Univ. Car. Math. Ph., vol. 22 (1) (1981).

[400] Vougiouklis, T., *Generalization of P-hypergroups*, Rend. Circ. Mat. Palermo, Ser. II, t. 36 (1987).

[401] Vougiouklis, T., *Representations of hypergroups by hypermatrices*, Riv. Mat. Pura e Appl., no. 2 (1987).

[402] Vougiouklis, T., *Groups in hypergroups*, Anals Discrete Math., vol. 37 (1988).

[403] Vougiouklis, T., *Fundamental relations and hypergroup representations*, U.C.N.W. Maths Preprint 89, 14 (1989).

[404] Vougiouklis, T., *On representations of algebraic multivalued structures*, Riv. Mat. Pura e Appl., no. 7 (1990).

[405] Vougiouklis, T., *The fundamental relation in hyperrings. The general hyperfield,* Proc. of the Fourth Int. Congress on AHA, Xanthi, Greece, 1990, World Scientific, 203–211.

[406] Vougiouklis, T., *The very thin hypergroups and the S-construction*, Combinatorics '88, Ann. of Discrete Math. Incidence Geometries and Combinatorial Structures, vol. 2 (1991), 471–477.

[407] Vougiouklis, T., *H_v-groups defined on the same set,* Combinatorics (Acireale, 1992), Discrete Math. 155, 1996, no. 1-3, 259–265.

[408] Vougiouklis, T., *Representations of hypergroups by generalized permutations*,Algebra Universalis, 29 (1992), 172-183.

[409] Vougiouklis, T., *Hyperstructures and their representations,* Monographs, Hadronic Press, 1994.

[410] Vougiouklis, T., *H_v-vector spaces*, 5th. Int Congress on A.H.A., Iaşi, Romania (1993), Hadronic Press (1994), 181–190.

[411] Vougiouklis, T., *Some remarks on hyperstructures*, Contemporany Mathematics , Amer. Math. Society, 184, (1995),427–431.

[412] Vougiouklis, T., *On H_v-fields,* Sixth Int. Congress on AHA (1996), 151–159, Democritus Univ. of Thrace Press.

[413] Vougiouklis, T., *Constructions of H_v-structures with desired fundamental structures,* Proc. Int. Workshop: New Frontiers in Hyperstructures and Related Algebras, Hadronic Press, 1996, 177–188.

[414] Vougiouklis, T., *Convolutions on WASS hyperstructures,* Discrete Math., Elsevier, V.174, (1997), 347–355.

[415] Vougiouklis, T., *Enlarging H_v-structures,* Algebras and combinatorics (Hong Kong, 1997), 455–463, Springer, Singapore, 1999.

[416] Vougiouklis, T., *Fundamental relations in hyperstructures,* Bulletin Greek Math. Society, 42, (1999), 113–118.

[417] Vougiouklis, T., *An H_v-rings and H_v-representations,* Discrete Mathematics, 208-209 (1999), 615–620.

[418] Vougiouklis, T., *On hyperstructures obtained by attaching elements,* Proc. Int. Congress "C. Carathéodory in his...origins", 2000, Hadronic Press, 2001, 197–206

[419] Vougiouklis, T., Dramalidis, A., *H_v-modulus with external P-hyperoperations*, 5th Int. Congress on A.H.A., Iaşi, Romania, (1993), Hadronic Press (1994), 191–197.

[420] Vougiouklis, T., Konguetsof, L., *P-hypergroups*, Acta Univ. Car. Math. Ph., vol. 28 (1) (1987).

[421] Vougiouklis, T., Spartalis, S., *P-cyclic hypergroups with three characteristic elements*, Annals Discrete Math., vol. 37 (1988).

[422] Vougiouklis, T., Spartalis, S., Kessoglides, M., *Weak hyperstructures on small sets*, Hyperstructures and their applications in cryptography, geometry and uncertainty treatment (Chieti, 1994), Ratio Math., no.12, 1997, 90–96.

[423] Wall, H.S., *Hypergroups*, Amer. J. Math. (1937), 77–98.

[424] Werner, H., *A duality for weakly asociative lattices* , Coll. Math. Soc. Janos Bolyai 28, Finite Algebra and Multiple–Valued Logic, Szeged (Hungary), 1979, North Holland, Amsterdam, 1981, 781–808.

[425] Yatras, C.N., *M-Polysymmetrical hypergroups*, Riv. Mat. Pura e Appl., no. 11 (1992).

[426] Yao Bingxue, Li Hongxing, *Weak HX-rings on a ring*, Italian J. of Pure and Appl. Math. no. 10, 2001.

[427] Yavari, M., *Hypervector Spaces, Generalized Permutations and Join Spaces,* to appear in Italian Journal of Pure and Applied Mathematics.

[428] Yavari, M., Zadeh, A.H., Ashrafi, A.R., *Hypergraphs and join spaces*, accepted for publication in Italian Journal of Pure and Applied Math.

[429] Zadeh, L., *Fuzzy Sets*, Information and Control 8, (1965), 338–353.

[430] Zahedi, M.M., Ameri, R., *Hypergroup and join space induced by a fuzzy subset*, PU.M.A. no. 8, 1998.

[431] Zahedi, M.M., Ameri, R., *On the prime, primary and maximal subhypermodules*, Italian Journal of Pure and Applied Mathematics, no.5 (1998).

[432] Zahedi, M.M., Bolurian, M., Hasankhani, A., *On Polygroups and Fuzzy Subpolygroups*, The J. of Fuzzy Mathematics, vol. 3, no. 1 (1995).

[433] Zappa, G., *Suggli ipergruppi di corispondenza ad indici limitati sopra una curva algebrica*, Ann. Math. Pura e Appl., 20 (1941).

[434] Zappa, G., *Aspetti algebrici della teoria dei codici correttori di errori*, Convegno "Sistemi binari e loro applicazioni", Taormina, 1978.

[435] Zhang Zhen Liang, *The properties of HX-groups,* Italian Journal of Pure and Applied Mathematics, vol.2 (1997).

Books on Hyperstructures and Applications

[436] Corsini, P., *(i.p.s.) Hypergroups of order 8*, Aviani Editore, 1989, 1-106.

[437] Corsini, P., *Prolegomena of Hypergroup Theory*, Second Edition, Aviani Edittore, 1993.

[438] Mockor, J., *Groups of Divisibility*, Mathematics and its Applications, D. Reidel Publishing Company, Dordrecht/Bonston/-Lancaster III, 1983.

[439] Prenowitz, W. and Jantosciak, J., *Join Geometries,* Springer–Verlag UTM, 1979.

[440] Vougiouklis, T., *Hyperstructures and their Representations,* Hadronic Press, Inc., 1994.

Books on other topics

[441] Bannai, E., Ito, T., *Algebraic Combinatorics* I: *Association Schemes*, Benjamin Cummings Publishing Co., 1984.

[442] Berge, C., *Graphes et Hypergraphes*, Dunod, Paris, 1970.

[443] Berge, C., *Hypergraphes, combinatoire des ensembles finis*, Gauthiers–Villars, Bordas, Paris, 1987.

[444] Biggs, N.L., *Algebraic Graph Theory*, Cambridge Univ. Press, 1974.

[445] Birkhoff, B., *Lattice Theory*, 3rd. ed., AMS Coll. Publ. Providence, 1967.

[446] Eilenberg, S., *Automata, Languages and machines*, Academic Press, New York and London, 1974.

[447] Grätzer, G., *Universal Algebra*, Second Edition, Springer–Verlag, New York, Inc., 1979.

[448] Hall, M.Jr., *Combinatorial theory*, J. Wiley & Sons Inc. (1967).

[449] Higman, D.G., *Combinatorial considerations about permutation groups*, Lecture Notes, Math. Institute, Oxford, 1972.

[450] Higgins, P.J., *Notes on categories and groupoids*, Van Nostrand, Londres, 1971.

[451] Hong Xing Li, Vincent C. Yen, *Fuzzy sets and fuzzy decision–making*, CRC Press Inc., 1999.

[452] Mac Williams, F.J., Sloane, N.J.A., *The theory of error–correcting codes*, North Holland, Amsterdam, 1978.

[453] McKenzie, R., *The representation of relation algebras*, PhD Thesis University of Colorado, 1966.

[454] Scafati, M., Tallini, G., *Geometria di Galois e teoria dei codici*, 1995, Centro d'informazioni e stampa universitaria di Enzo Calamartini s.a.s., Roma.

[455] Shafer, G., *A Mathematica Theory of Evidence*, Princeton Univ. Press, 1976.

[456] Sikorski, R, *Boolean Algebras*, Springer-Verlag, Berlin, 2nd Edition, 1964.

[457] Sloane, N.J.A., *A short course on error–correcting codes*, Springer–Verlag, 1975.

[458] Van Lint, J.H., *Introduction to Coding Theory*, Springer–Verlag, Berlin, 1982.